Making India Disaster Resilient

Vishwa Raj Sharma · Chandrakanta
Editors

Making India Disaster Resilient

Challenges and Future Perspectives

Springer

Editors
Vishwa Raj Sharma
Shaheed Bhagat Singh College
University of Delhi
New Delhi, Delhi, India

Dr. Chandrakanta
Assistant Professor
Department of Geography
North Eastern Hill University (NEHU)
Shillong, Meghalaya, India

ISBN 978-3-031-50112-8 ISBN 978-3-031-50113-5 (eBook)
https://doi.org/10.1007/978-3-031-50113-5

This Springer imprint is published by the registered company Springer Nature Switzerland AG
The registered company address is: Gewerbestrasse 11, 6330 Cham, Switzerland

Foreword

It is a matter of immense pleasure that my colleagues Prof. (Dr.) Vishwa Raj Sharma (Faculty, Shaheed Bhagat Singh College, University of Delhi) and Dr. Chandrakanta (Faculty, Shaheed Bhagat Singh College, University of Delhi), planning to publish an edited volume entitled *Making India Disaster Resilient: Challenges and Future Perspectives* from Springer, Switzerland. I would like to congratulate both colleagues for taking such excellent initiative.

Geography studies the characteristics of space and place. So, geographers identify and analyze the spatial patterns of human–environment interaction that shape our lives and society. UN Sustainable Development Goals in 2015 set the stage for the next 15 years of global development. All of the different goals apply to cities, and goal number 11 is about sustainable cities and communities especially "make cities inclusive, safe, resilient and sustainable". Geographers ask relevant questions in order to understand human responses. The discipline of geography promotes transdisciplinary research incorporating co-design processes, facilitating engagement with and involvement of different stakeholders together with developing scientific networks under the science–policy interface (SPI).

The importance of disaster studies becomes of paramount significance in present context. This volume emphasizes on the various disasters our country is facing and how we can cope up these hazards and disasters. Now there is emphasis on disaster risk reduction at various international forums to make world disaster resilient. The Sendai Framework for Disaster Risk Reduction 2015–2030 was launched in March 2015, supported by the United Nations Office for Disaster Risk Reduction. Recently, the UN has a passed resolution to implement Sustainable Development Goals on September 25, 2015, adopted by 193 countries of the UN General Assembly.

This book deals with various sections with case studies associated with floods, climate change and land use, fire/smoke and earthquake hazards in Indian cities. It also deals with gender perspective related to disasters and human dimensions associated with its impact, vulnerability and governance. On behalf of the IGU, I welcome the book.

September 2020

Late Prof. R. B. Singh
Former member of IUGG-IGU Joint National Committee, Indian National Science Academy (INSA) Springer Series Editor (former)—Advances in Geographical and Environmental Sciences and SDGs
Ex-Secretary General-IGU and International Council of Science (ICSU)
Scientific Committee Member-Urban Health and Wellbeing
Ex. Head-Department of Geography
Delhi School of Economics
University of Delhi
New Delhi, Delhi, India
rbsgeo@hotmail.com

Preface

Disasters in different forms and intensities portray the omnipresence from global to regional and local levels. With the continuous increase in population and its constant interference with natural systems has increased the intensity and frequency of disasters. It showcases instances of all categories of disasters including floods, fire, cyclones, droughts, or even biological disasters like the COVID-19 pandemic. The casualties that occurred due to COVID-19 across the globe are a sufficient indicator to explain the furious and adverse aspects of the disasters. Adversities besides inflicting causalities also impact the lives of human beings on the economic, socio-cultural, environmental and psychological fronts.

Despite indistinguishable characteristics of disasters, developing countries are more adversely affected by them. Referring to South Asia, particularly India is a multi-disaster-prone country. Despite remarkable progress on the economic and human development front, India has to face adversity to disasters at a very frequent rate. Owing to its geoclimatic diversity and socio-cultural setting, a majority of states and union territories (UTs) of the nation fall under multiple disaster-prone zones. A comprehensive understanding of multiple disasters in India is required to develop a disaster resilient society. With the adherable consideration of its significance, this book includes chapters on various types of disasters that happen in the country every year.

The chapters of this book offer in-depth coverage of themes like flood disaster, climate change, fire hazards, geohazards like earthquakes, gender and disaster, and institutional framework and initiatives for its management. Under the theme of flood disaster urban flooding, flood plain mapping and zoning with the use of remote sensing and GIS have been comprehended. Climate change and related aspects are included in chapters on global warming, cyclone hazards, drought, etc. Fire is one of the devastating forms of disaster in India and particularly in the urban sphere and forest area; therefore, chapters on different incidences of fire in India have been painstakingly put. Sections under the theme geohazards include disasters like earthquakes, landslides, flash floods, etc. There is a positive correlation between disaster and adverse impact on specific gender particularly in the case of developing countries like India so a specific section of this book covers chapters on the impact or relationship of disaster and gender.

Finally, a section is devoted to the programs and policies that cover chapters on various aspects like governance issues, regional cooperation, policies and initiatives of the government.

Overall, this book advances the scientific understanding of various forms of disasters in India with the inclusion of modern-day technology to deal with and manage them effectively to make India disaster resilient.

This practical, comprehensive and state-of-the-art book is a significant resource for students, researchers, teachers, scientists, policymakers and practitioners who are interested in disaster-related trainings. Further, this book is going to be very helpful in understanding the use and application of remote sensing and GIS in disaster mitigation planning and management and to develop a comprehensive model based on a three-tier model to deal with disasters for a resilient India and the world.

New Delhi, India — Vishwa Raj Sharma
Shillong, India — Chandrakanta

Contents

Editors and Contributors

About the Editors

Dr. Vishwa Raj Sharma is Professor in the Department of Geography, Shaheed Bhagat Singh College, University of Delhi, New Delhi, with over 20 years of teaching and research experience. His field of specialization is urban environmental issues. He has published several research papers in reputed national and international journals. He has also participated and presented research papers in many conferences in India and abroad like International Geographical Union (IGU) regional conferences held in Kyoto, Japan, in 2013, Krakow, Poland, in 2014, IGU Moscow in 2015 and IGU Beijing in 2016. Presently, he is co-investigator of international research project "Emerging Mega Regions and Socio-Economic Developments in Contemporary India: A Study of Delhi Mega Region", sponsored by Hiroshima University, Japan. He is also the corresponding member of IGU and Urban Commission on Urban Challenges in a Complex World. Currently, he is working on major research project "Sustainable Tourism Management in Agra: Carrying Capacity Assessment and Modeling" sponsored by ICSSR.

Dr. Chandrakanta is presently working as a faculty in Department of Geography, NEHU, Shillong, and former faculty, Shaheed Bhagat Singh College, University of Delhi, New Delhi. She received her M.Phil. and Ph.D. in population studies from Jawaharlal Nehru University (JNU), New Delhi. In addition, she has done PG course on "International Migration, Ethnicity and Gender" from Linköping University, Sweden and PG Diploma in "Disaster management" and "Urban Planning and Development". She received Satpaul Mittal Award for her research work (for M.Phil.) from Indian Association of Parliamentarians on Population and Development (IAPPD), Honorary mention of the responsibility Paper Development Award by the Oikos Young Scholars Development Academy 2012, Nairobi-Kenya for her doctoral thesis and UNFPA and UNDP grant to present her research in Paris and Nairobi. She has presented her research at various national and international conferences. Her current research interests and writing are concerned with the disaster management, urban and migration issues. Recently, she has authored edited book entitled *Making Cities Resilient* published by Springer. She is trained in disaster risk reduction issues from NIDM, India. She has been the member of the organizing committee of various national and international conferences, conclave and workshops. She has been the coordinator of "School of Happiness" and also coordinated various events organized by Centre for Disaster Management Studies (CDMS) at Shaheed Bhagat Singh College. She is a life member of National Association of Geographers of India (NAGI) and Association of Geographical Studies (AGS).

Contributors

Rajesh Kumar Abhay Department of Geography, Dyal Singh College, University of Delhi, New Delhi, India

R. Aravinth Institute of Environment Education and Research, Bharati Vidyapeeth University, Pune, India

Kavita Arora Department of Geography, University of Delhi, New Delhi, India

Neha Arora Department of Geography, Shaheed Bhagat Singh College, University of Delhi, New Delhi, India

A. Athul Department of Disaster Management, Pondicherry University, Pondicherry, India

Kamal Bisht Shaheed Bhagat Singh College, University of Delhi, New Delhi, India

Chandrakanta Department of Geography, Shaheed Bhagat Singh College, University of Delhi, New Delhi, India

Suman Das Department of Geography, Kirori Mal College, University of Delhi, New Delhi, India

Upma Gautam University School of Law and Legal Studies, Guru Gobind Singh Indraprastha University, New Delhi, India

Sushma Gulria National Institute of Disaster Management (NIDM), New Delhi, India

Anjana Mathur Jagmohan Department of Geography, Dyal Singh College, University of Delhi, New Delhi, India

Shahid Jamal Department of Geography, University of Delhi, New Delhi, India

Varinder S. Kanwar Department of Civil Engineering, Chitkara University, Himachal Pradesh, India

Kshetrimayum Krishnadas Department of Mathematics, Shaheed Bhagat Singh College, University of Delhi, New Delhi, India

Prabuddh Kumar Mishra Department of Geography, Shivaji College, University of Delhi, New Delhi, India

Varun Narayan Mishra Amity Institute of Geoinformatics and Remote Sensing (AIGIRS), Amity University, Noida, Uttar Pradesh, India

Vishal Mishra Department of Civil Engineering, Indian Institute of Technology, Roorkee, Uttarakhand, India

Jag Mohan Department of Geography, Aditi Mahavidyalaya, University of Delhi, New Delhi, India

B. Nagarajan National Centre for Geodesy, Indian Institute of Technology, Kanpur, India

Bindhy Wasini Pandey Department of Geography, Delhi School of Economics, University of Delhi, New Delhi, India

C. Prakasam Department of Geography, School of Earth sciences, Assam University, Diphu Campus (A Central University), Diphu, India;
Department of Civil Engineering, Chitkara University, Himachal Pradesh, India

Abhay Shankar Prasad Department of Geography, Dyal Singh College, University of Delhi, New Delhi, India

Priyanka Department of Geography, Delhi School of Economics, University of Delhi, New Delhi, India

Rashmi Department of Geography, Delhi School of Economics, University of Delhi, New Delhi, India

Ashis Kumar Saha Department of Geography, Delhi School of Economics, University of Delhi, New Delhi, India

Vineka Sanoria School of Education, Christ University, Bangalore, India

Shubham Kumar Sanu Indraprastha College for Women, University of Delhi, New Delhi, India

R. Saravanan Department of Civil Engineering, Chitkara University, Himachal Pradesh, India;
Ecofirst Services Ltd, Tata Consulting Engineers Limited, Bangalore, India

Anjan Sen Department of Geography, Delhi School of Economics, University of Delhi, New Delhi, India

Soma Sengupta Department of Commerce, Kamala Nehru College, University of Delhi, New Delhi, India

Mamta Sharma Aditi Mahavidyalaya, University of Delhi, New Delhi, India

Vishwa Raj Sharma Department of Geography, University of Delhi, New Delhi, India

Shashwati Singh Computational Biology Division, CSIR-Central Institute of Medicinal and Aromatic Plants, Lucknow, Uttar Pradesh, India

Swarnima Singh Department of Geography, Deen Dayaal Upadhyay University, Gorakhpur, Utter Pradesh, India

Varnav Somwal Jindal Global Law School, O. P. Jindal Global University, Sonipat, Haryana, India

Deeksha Bajpai Tewari Department of Geography, Dyal Singh College, University of Delhi, New Delhi, India

Anuj Tiwari Geomatics Section, Department of Civil Engineering), Indian Institute of Technology, Roorkee, Uttrakhand, India

Kanwar S. Varinder Department of Civil Engineering, Chitkara University, Baddi, India

Chapter 1
An Introduction to Disaster Management: Evidences from India

Vishwa Raj Sharma and Chandrakanta

Making India Disaster Resilient: Challenges and Future Perspectives

For the past two years, people around the globe have been continuously under the wrath of the pandemic. This global outrage of the novel coronavirus turned out to be a challenging "Biohazard". The unexpected outbreak of the virus bowed out to be one of the worst disasters that have affected the worst so far. Disasters do not arise of proclamation. Prediction of a possible mishap cannot always be detected. For instance, the recent Ukraine–Russia crisis that began in late February 2022 resulted in war. This unpredicted consequence resulted in a huge number of refugees including other vigilant loss of life, infrastructure, property, flora-fauna, and it was even attributed to micro-climate changes in the region due to nuclear attacks.

Resilience when discussed in disaster perspective is the ability of a system, community or society exposed to hazards to resist, absorb, accommodate to and recover from the effects of a hazard in a timely and efficient manner, including through the preservation and restoration of its essential basic structures and functions.

Disasters are dynamic, and the essence of this dynamism comes with changes. These include recorded changes in its nature and its impact over time and space. According to Arora (2022) Disasters overlap or it can be said that a place can be prone to multiple hazards at a single point of time. To further establish the concept, the best-referred example turns out to be the tropical cyclone on the eastern coast of India during the first wave of the corona virus. Numerous factors are responsible for multiple hazards at one point. Other than geological instability, factors such

V. R. Sharma (✉)
Department of Geography, University of Delhi, New Delhi, India
e-mail: vrsharma2002@gmail.com

Chandrakanta
Department of Geography, Shaheed Bhagat Singh College, University of Delhi, New Delhi, India
e-mail: chandrakantajnu@gmail.com

V. R. Sharma and Chandrakanta (eds.), *Making India Disaster Resilient*,
https://doi.org/10.1007/978-3-031-50113-5_1

as deforestation, construction-related activities, and excessive plucking of natural resources result in landslides and erosion on mountain slopes. Hence, it is important to conceive of disasters in totality to create a resilient country. Being vary of one disaster in isolation will not be fruitful unless there is an established amalgamation of the totality of the disasters. Numerous efforts for long years have been set along international, national, and local lines. These planted plans and goals have been seriously addressed. The respective bodies have been going ahead an extra mile to tackle the challenges posed by disasters. Upon the recent COP-Glasgow meeting's conclusion, all the carved-out SDGs that have already begun are in progress. However, it is questionable if the SDGs can be achieved, in the present situation around the globe.

Disaster resilience refers to a person's, a community's, an organization's, or a state's capacity to respond to and recover from dangers, shocks, or pressures without jeopardizing long-term development prospects.

According to the Hyogo Framework for Action (UNISDR, 2005), the degree to which people, communities, and public and private organizations can organize themselves to learn from previous disasters and lower their risks of experiencing similar ones in the future, at an international, regional, national, and local level, is what defines disaster resilience.

Disasters do not differentiate and strike both developed and developing countries altogether. In the last two decades, disasters have claimed more than three million lives and affected about one billion people. The frequency of natural disasters in Asia and the Pacific region is comparatively of greater influence than in the rest of the world. South Asia is a homogenous region characterized by a developing economy, agrarian society mostly, and monsoon type climate where dependency is more on rainfall which is already seasonal and erratic. Changing climate has been continuously adding serious challenges to the region as these disasters hinder many developmental projects as huge amounts have to be diverted for relief and recovery (Singh & Singh, 2013).

Amongst the South Asian countries, India has witnessed remarkable progress in human and economic development since independence. However, the path of this economic development and growth has rather been challenging. The concept of disaster management is a buzzword in the present context arising out of increasing disastrous events. India is one amongst many countries that have witnessed natural and human-induced disasters quite frequently over the years. India is the fifth most vulnerable country in terms of climate change (*Global climate risk index report 2020*). Urban areas are a distinctive reflection of development and opportunities in a nation. Therefore, on one hand, they offer the pleasure of vibrant opportunities and on the other hand pose serious challenges.

According to NDMA Annual Report 2020–21 *"India, due to its unique geo-climatic and socio- economic conditions, is vulnerable, in varying degrees, to floods, droughts, cyclones, tsunamis, earthquakes, urban flooding, landslides, avalanches, and forest fire. Out of 36 States and Union Territories (UTs) in the country, 27 are disaster-prone. 58.6% of the landmass is prone to earthquakes of moderate to very high intensity; 12% of the land is prone to flood and river erosion; out of 7,516 km of coastline, 5,700 km is prone to cyclones and tsunamis; 68% of the cultivable land is vulnerable to drought, hilly areas are at risk from landslides and avalanches, and 15% of the landmass is prone to landslides. A total of 5,161 Urban Local Bodies (ULBs) are prone to urban flooding. Fire incidents, industrial accidents, and other manmade disasters involving chemical, biological and radioactive materials are additional hazards, which have underscored the need for strengthening mitigation, preparedness and response measures".* Hence, it is established that disaster intensity and vulnerability in India are further compounded by increasing vulnerabilities related to changing population dynamics and socio-economic conditions of the masses. Deteriorating environmental conditions, changing climate, geologically induced hazards, and other biohazards and pandemics have all resulted in worse situations. Undoubtedly, all of these instances add to a circumstance where disasters seriously threaten India's economy, population, and sustainability.

In the prevalence of this background, it is worth mentioning that since 2005 India has been following the legal capacities of its constitution by laying in the Disaster Management Act, 2005. The Indian Prime Minister recently announced ten-point agenda as one of the initiatives for a resilient India. This agenda aims to reduce the impact of disasters by 2030. India's model to manage disasters is hierarchical and works at the national, state, district, and local levels with inclusive participation of various relevant ministries, government departments' administrative bodies, and local communities (Fig. 1.1).

Education is an instrument for building the knowledge and necessary skill set to prepare and cope with disasters. It also is a crucial tool in helping learners and the community return to a normal life. The government of India has been constantly thriving to put measures in the educational reach to minimize the effects of disasters. The recent announcement of compulsory disaster management papers (as per the India-New Education Policy) at undergraduate and postgraduate levels is a progressive sign. The University Grant Commission (UGC) of India has been actively participating throughout and has now sent disclaimers to every university and college (dated February 24, 2022) to implement disaster risk reduction (DRR) courses in their curriculums. A national action plan on climate change has been adopted, and 8 missions by far have been announced. By working on these missions, various challenges can be sorted out in the field of carbon emission, water resources, sustainable agriculture, and the Himalayan ecosystem.

The vulnerability maps of India demonstrate vital information including large population base including children and elderly numbers, social-economic characteristics/diversity of the population, large sector (agriculture) which is climate-sensitive, poor urban and rural infrastructure, attitude and behaviour of the masses to perceive the risk. The intensity and frequency of various disasters have increased recently and hence require a lot of funds to cope with the situation. So, in the given situation, it has become a hell of a challenge to make the country disaster resilient.

Resilience being the principal focus of the book gives it the liberty to cover issues in all phases of the disaster management cycle. Resilience is a complex and dynamic system-based concept used differently in a variety of disciplines and is also a simple concept referring to the ability of a system to return to a previous or improved set of dynamics followed by shock (Sharma & Chandrakanta 2019).

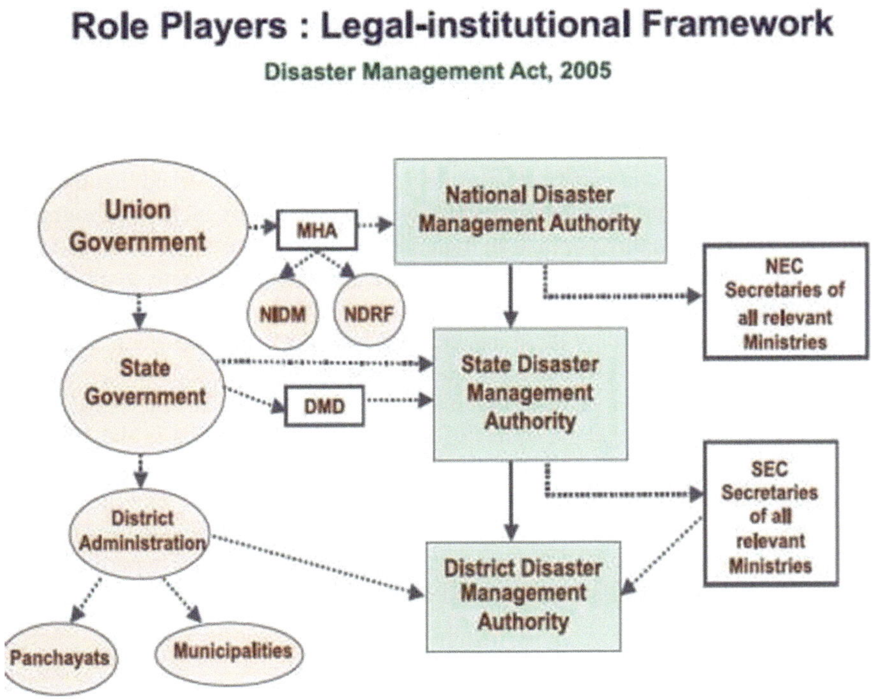

Fig. 1.1 Institutional framework for disaster management in India. *Source* www.ndma.gov.in

The national team of authors who have contributed their chapters in this book combines their years of experience in research and in the field for micro-level survey/case studies to offer vital lessons for practitioners, academics, and students alike. The book aims to highlight the challenges of different types of disasters including identification of vulnerability, risk mapping and risk reduction, disaster preparedness, resilient infrastructure, the communities, and problems during implementation due to lack of local-level capacities. This book also focuses on some of the cities of India with special reference to the National Capital Region of Delhi.

The book has an introductory chapter followed by six prominent theme-based parts which include 17 chapters. This State of articulate compilation is unique as it has nominated research papers that are based on field/primary surveys at the micro-level covering a wide range of geohazards, climatic-weather hazards, biohazards, and hydrological hazards, including various man-induced disasters.

Cities in India have rich economic, demographic, and socio-cultural heritage and have great tourism potential. Hence, literature on urban management and resilience has also been included. The contributions embrace the intellectual research work of the research scholars, professors, academicians, planners, and scientists from various universities and institutions in India. The papers are thoroughly reviewed and presented in their original forms. The editors have taken sincere care in compiling the papers most justifiably and rationally. The book is an unpretentious effort to highlight and justify its theme and objective.

Part 1 focuses on **floods** and includes studies on the recent Kerala floods (2018), floods in Begusarai-Bihar, Chennai (Tamil Nadu) urban flooding, and floods in Varanasi (UP). Chapter 1 comprises the vulnerability assessment of recent floods in the state of Kerala. The state became vulnerable due to its location along the sea coast and the changing climatic dynamics. In the next chapter, the risk of flooding amalgamated with measures to limit the effects of the flood on various livelihood activities in the Begusarai district of Bihar is discussed. This state remained backward in agricultural productivity due to frequent floods in the region. In recent years, urban flooding has turned out to be a very recurrent phenomenon disturbing the daily life of Indian cities leading to catastrophic loss of human life and property. Chennai has also witnessed flooding in recent years. What can be done to prevent such events? What should the steps of administration involve? Urban encroachment and unauthorized construction activities are responsible for such kinds of floods in Chennai city has been covered in this chapter. In the next paper, flood plain mapping and zonation have been depicted in the Varanasi (Banaras) city using geospatial techniques where the role of remote sensing and Geographical Information System (GIS) technique has been emphasized.

Climate change and related hazards are the second part of the book which focuses on environment-friendly fuels to contend with global warming, cyclone hazards towards the Arabian Sea, and agricultural drought in Himachal Pradesh. In a chapter on contending global warming by popularizing environment-friendly fuel, compressed natural gas (CNG) the role of polluted gases in the deterioration of the environment has been highlighted which ultimately leads to global warming and climate change. In the subsequent chapter, increasing vulnerability of the Arabian Sea to cyclone and storms have been assessed in it. The next chapter is associated with agricultural drought vulnerability assessment and management in Mandi District, Himachal Pradesh. Remote sensing and GIS techniques have been used for the assessment of horticulture drought susceptibility.

The third section is on **fire** hazards where the urban fire in the residential colony of Delhi, forest fire, and fire in commercial areas of Delhi including a fire in the institutional building has been discussed. Assessment of urban Fire hazard: A case study of Malviya Nagar-Delhi is a very unique paper on the serious fire incident that happened in south Delhi in 2018. Due to heavy congestion and a very dense population, fire brigades couldn't reach the location of the fire. This fire was later controlled only when water was released by an aircraft. Forest Fire severity mapping using geospatial techniques in Bandipore reserve forest, India, is yet another important paper. Forest fire maps have been prepared based on the burn severity index. Fire hotspot regions have been identified through geospatial techniques. Another significant research paper is on the vulnerability and resilience of fire hazards in Anaj Mandi (Grain market) of old Delhi. Also pronounced as the "Disastrous Devil of December", This incident of Anaj Mandi, occurred on the 8th of December 2019, which caused the lives of 45 people and wedged economic, social, and psychological impacts on thousands of people. The author of this paper has developed a three-tier model at the individual, community, and government levels to combat the fire hazards in the cities. Assessment of Fire Disaster Risk Reduction in Higher Educational Institutions in Delhi is another significant investigation and paper advocated for the safety and security of educational institutions in India. The authors emphasized the awareness issues and action plans which can be taken at the time of emergencies that may arise in the institutions.

The fourth section of the book addresses **geohazards** such as earthquakes in the Yamuna region of Delhi, multi-hazards including the earthquake in the congested area of Delhi, and land subsidence in Himachal Pradesh. Authors have mentioned that regions which are close to the bed of River Yamuna are more vulnerable to earthquakes and are largely illegal and unauthorized constructions are a cause of worry at the time of disasters like earthquakes. Mapping fire, earthquake, and biohazards in Delhi is another micro-level study. Delhi lies in Seismic Zone IV, which makes the area sensitive to disasters, i.e. earthquakes. Illegal and unauthorized construction has made cities highly vulnerable to fire, earthquake, and biohazards. The next chapter is based on a study of Shimla Town which is highly prone to disasters such as earthquakes, landslides, flash floods, etc.

The fifth section of the book arouses the theme of **gender**. Gendered spaces, climate change, and resilience in a squatter slum of the global south are another interesting case study from India. Constantly increasing temperatures lead to global warming and cities of the global south are badly affected by these disasters. Flash floods, droughts, and high temperatures first affect the women and children of the household who are the worst affected due to malnutrition and weak immunity. Women, who are constrained both by the space and vulnerable economic situation and still take care of the household goods and the kids, find themselves, in severely affected conditions at the time of the hazards and disasters. Gender issues are also very significant in the study of hazards and disasters.

Section sixth encompasses human aspects and impact, vulnerability, and **governance issues**. The regulatory framework for regional cooperation on Disaster Risk Reduction (DRR) in South Asia is the last but important chapter with significant studies emphasizing regional cooperation regarding disasters in the region. All the South Asian Association for Regional Cooperation (SAARC) member countries are prone to various disasters whether natural or manmade. All South Asian countries are densely populated, and the loss of life and property is very high during disasters. Sustainable synchronization, alliance, and mainstreaming of a comprehensive legal and institutional framework are required which ensures a rapid response mechanism dealing with disasters.

This book has been produced for teachers, researchers, policymakers, and practitioners. There is a constant gap between theory and practice/implementation, between academic research and active professionals in the field. This fissure needs to be filled, valuable lessons should be learned from the past events and experiences, and people of all other losses can be avoided. This book opens a discourse between theory and practice.

Values of Case Study Method

A variety of quantitative and qualitative research techniques, from desk-based literature reviews to in-depth and thorough case studies, are included in the case study approach. Here are my opinions on the benefits and difficulties of using case studies in social science research, which are based on experience. A case study is a detailed and in-depth analysis of one specific person, group, event, or time frame. They combine a variety of qualitative and quantitative research techniques to look at the underlying ideas behind an event in a real-world setting. The ability to conduct a comprehensive analysis of the incident is the case study method's most important advantage. A case study gives the possibility for a researcher to apply a variety of tools on one issue, in contrast to solitary research approaches that provide more of a snapshot, such as surveys. Due to the ample time and room provided, it is possible to develop a thorough grasp of the subject, providing a solid foundation from which to examine the elements impacting the case study in more depth. In contrast to the singular perspective, you obtain from a survey or interview, case studies collect a variety of viewpoints. Because the motive of one particular person is less clear, this increases the chance of comprehending the topic at hand and lowers the likelihood of prejudice.

Through summaries of prior studies, case study research enables the examination and comprehension of difficult subjects. It may be regarded as a reliable research technique, especially when a comprehensive, in-depth inquiry is needed. Although the case study method has been used in many social science studies, it comes into its own when discussing issues related to sociology, education, and community-based issues like poverty, unemployment, drug abuse, and illiteracy (Gulsecen & Kubat, 2006), amongst others. Researchers' growing concerns about the shortcomings of quantitative approaches in offering comprehensive and in-depth explanations of the social and behavioural issues under investigation led to the acknowledgment of case studies as a research method. A researcher can go beyond the quantitative statistical findings and comprehend the behavioural conditions from the actor's point of view by using case study approaches. Through thorough observation, reconstruction, and analysis of the instances under inquiry, case studies assist explain both the process and consequence of a phenomena by using both quantitative and qualitative data (Tellis, 1997).

There are several types of case studies. Yin (1984) lists three types of case studies: exploratory, descriptive, and explanatory. First, exploratory case studies are designed to investigate any phenomena in the data that piques the researcher's curiosity. As an illustration, a researcher doing an exploratory case study on a person's reading process may pose basic inquiries such, "Does a student use any strategies when he reads a text?" and "If so, how often?". These broad inquiries are aimed to pave the way for a deeper investigation of the phenomena seen. Prior fieldwork and small-scale data collecting may be done in this case study as well before the research questions and hypotheses are put out. This preliminary effort aids in creating the study's structure. A pilot study is essential in choosing the protocol that will be followed and is regarded as an example of an exploratory case study (Yin, 1984; McDonough and McDonough, 1997).

Advantages of Case Study Method

When doing a case study, extensive information about each specific business is gathered, often utilizing a variety of instruments and methodologies. When using the case study technique, resources from both online and offline sources are possible. One of the biggest benefits of the case study is having access to sophisticated knowledge. Case studies provide a variety of verified facts gathered through direct or indirect observations of the particular entity that may be further incorporated into concise information and then concluded with derivatives. The observed data is simply based on the input–output approach, which makes it simple to infer findings about other identical things.

The whole study's data was gathered and observed in real time. Since the same might depend on data-driven findings such as good or negative consequences, the majority of it is also based on researchers' views. For instance, in order to educate people about the benefits and drawbacks of solar energy, researchers must uncover the evidence for and against it using examples from everyday life.

The earliest case studies were performed in the medical sciences, when selected people were to be observed experimentally. Every participant, every spectator, even the parties for whom these results were created, had something to gain from these investigations. As a result, the case study approach effectively compels people to choose a side in the debate and then support that choice with evidence.

Every piece of information, including handwritten notes, internet evaluations, historical data, and real-time experimental activities, is crucial for conducting case studies. The majority of case study methodologies use both indirect and direct sources of information, such interviews and direct observation. With this approach, case studies from a records database and questionnaires may be employed. A typical case study based on daily observations includes writing notes on the entity in diaries and notebooks.

Case studies may be an extremely economical approach to do research in a remote place. To use the available case study approach, researchers are not need to transfer from one location or institution to another. It is possible to do and wrap up research over the phone, email, or messaging services. The entire conversation may occur using a conference programme like Zoom or Skype. Therefore, even if it must be managed remotely, the case study approach is effective for exploratory and formative research.

Disadvantages of the Case Study Method

Given the dynamic structure of the consumer market, the case study method, which was first developed for the medical sciences, has now been applied to every industry. These case studies are crucial for examining critical outcomes, such as the benefits or drawbacks of certain services and products that have already been introduced or are about to do so. However, a case study approach isn't always as fruitful or effective. Before researchers begin a case study, there may be drawbacks that need to be taken into account.

The case study approach is mostly supported by factual evidence. The values of the data are defined by the researcher or data collector. When recorded records are inaccurate or real-time information is unavailable, it can be challenging to distinguish between what is factual and what is not. There will be a mistake in the final report if the data is based solely on subjective judgements. Data redundancy is another drawback of case studies. The process of conducting case studies takes a long time from beginning to completion. It gets more complex as more researchers get involved in it. A tremendous quantity of data has been accumulated. In this kind of study approach, the results are not merely subject to the researchers' control. By providing erroneous or deficient information in response to questions, participants might potentially affect the outcome. This implies that consumption will increase during the course of the study. The simplest way to handle this issue is to rigorously define each researcher's responsibilities and divide the case study procedure into manageable parts.

The case study approach might be ineffective and expensive if there are insufficient data, participants, or researcher engagement. The effectiveness of case study methodologies depends on the involvement of the parties concerned. The quality of the information a researcher receives from various sources is influenced by their skills. This study approach can be rendered ineffective and ineffective by untrained or unskilled researchers. The remaining harm will be caused by insufficient data if researchers are unsure how to handle this procedure.

It is preferable to concentrate on a particular entity and develop conclusions based on the feedback and characteristics rather than covering a whole group or community. For the case study approach to produce an adequate amount of data for analysis, a modest sample size is necessary. The case study approach becomes particularly ineffective if there are multiple demographics associated with the business or if there are distinct demands that must be assessed. This is comparable to the mathematical integration and differentiation techniques. The results of analysing a single unit are identical to those of researching the complete system. The process of gathering data for a large system or group can be fraught with mistakes, ambiguity, and inaccuracy.

The tedious task of going door to door or visiting the participants in various areas might get more intense as the amount of data increases. To properly gather data using the case study technique, researchers need to have a high degree of language proficiency. Every step of the data collecting process requires direct interaction between the researchers and the participants. The concepts and topics of this method rely greatly on the amount of labour each researcher is willing to invest into the actual world, from evaluating files or entries to conducting in-person interviews.

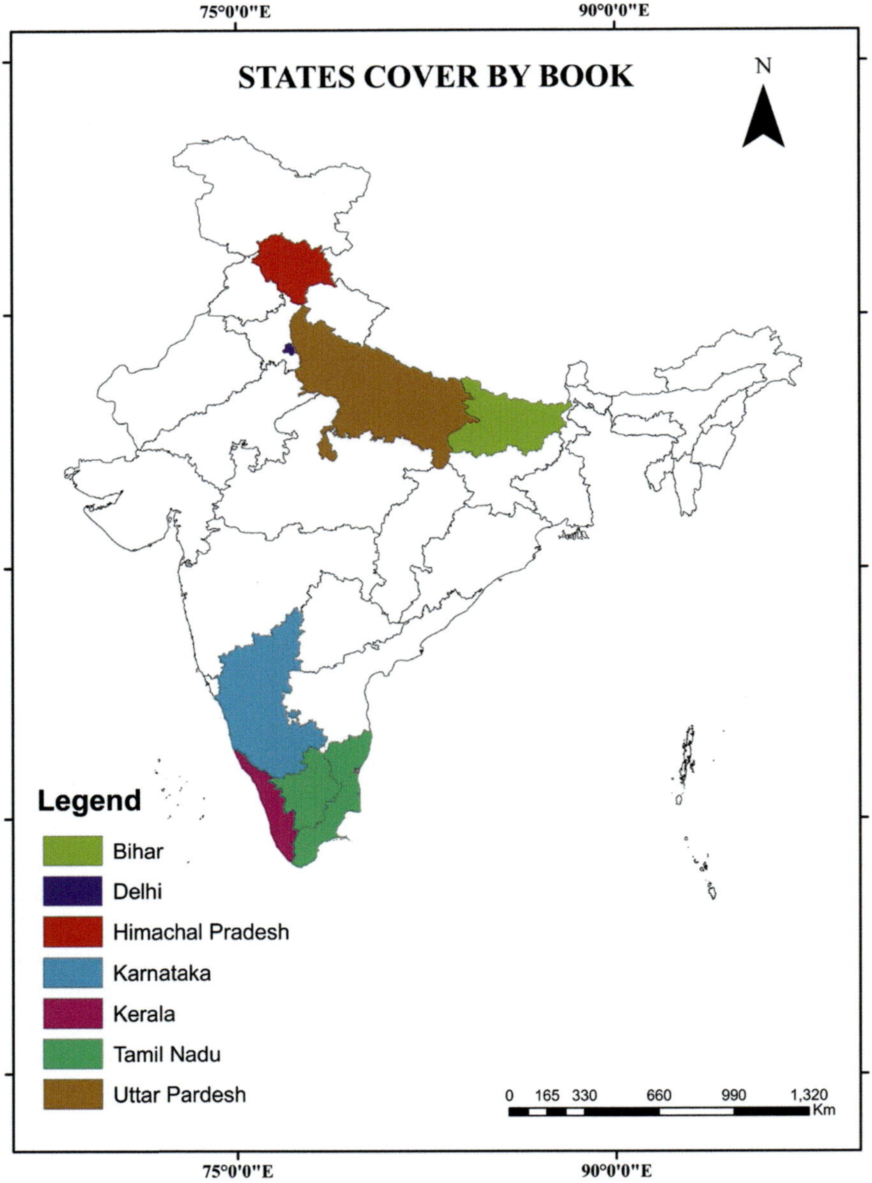

Map 1.1 States covered by book. *Source* Composed by Kamal Bisht

Despite these drawbacks, case study methods are a great technique to assess the efficacy of this research methodology. Finding true and correct information about whether research is short or long may be achieved by overcoming the drawbacks and improving the researcher's abilities (Map 1.1).

References

Arora, N. (2022) [lecture on 29/03/2022 during online training Organized by National institute of disaster management (NIDM)-Delhi].

Singh, S., & Singh, J. (2013). *Disaster Management (First)* (p. 2020). Pravalika Publications.

NDMA annual report 2020-21. Retrieved March 30, 2022 from https://ndma.gov.in/sites/default/files/PDF/Reports/NDMA-Annual-Report_20-21.pdf

Sharma, V. R., & Chandrakanta. (2019). On Making Cities Resilient, Springer, ISBN-10, 3319949314.

Part I
Flood

Chapter 2
Urban Flooding as an Emerging Challenge: Evidences from Chennai City

Shahid Jamal and Anjan Sen

Abstract Urban encroachment has increased with very high pace in the late twentieth century, which has resulted in the agglomeration of people in a very small area. This in turn leads to negative impact on aquatic as well as terrestrial ecosystem. Adverse effect like floods, water logging, water scarcity, water-borne diseases and other negative impacts destroys human life and socio-economic prosperity to a great extent. Such problems occur because of inadequate control of urban area and haphazard growth which shows bad impact on the urban drainage system, solid water management, hygiene, potable water and river flooding to urban flooding. Researchers have been and are underway in the area of natural and human-made disasters to study the exclusive effect of such unforeseen events. Urban flooding is becoming a recurrent phenomenon in various Indian cities like Delhi, Mumbai and Chennai. With the key objective to identify the causes of urban floods in Chennai and its socio-economic impact on the entire city, the local perception about the urban floods has been analysed as a reference for further adaptation. What can be done to prevent such events and what administration has to do? Primary survey as well as secondary data sets has been referred for collecting various data related to flood event. Reports from Indian Meteorological Department citing very strong El-Nino and various other factors led to very heavy rainfall has also been accessed. The study is based on descriptive and analytical research, whereby both qualitative and quantitative approaches have been used to analyse the data. There is no doubt about that state government of Tamil Nadu has failed to control the water drainage system in Chennai. Based on the international experience in terms of urban floods and comprehend indigenous knowledge about the urban floods, and its impact, a household survey was conducted in December 2016 at Chennai in Tamil Nadu. Arc GIS, several statistical techniques have been used to get the desired result. After the discussion, it was evaluated that plastics, poor drainage system, blockage in the runoff, concretisation of the city and settlements on both sides of river channel were

S. Jamal (✉)
Department of Geography, University of Delhi, New Delhi, India
e-mail: shahidjamalkmc@gmail.com

A. Sen
Department of Geography, Delhi School of Economics, University of Delhi, New Delhi, India
e-mail: asen@geography.du.ac.in

V. R. Sharma and Chandrakanta (eds.), *Making India Disaster Resilient*,
https://doi.org/10.1007/978-3-031-50113-5_2

some of the factors obstructing the outflow of water from the city to the sea, resulting in urban floods.

Keywords Carrying capacity · Terrestrial life · El-Nino · Concretisation · Urban floods · Wetlands and marshlands

Introduction

Urban encroachment and growth are one of the most important factors that has accelerated the intensity of natural hazards including urban floods (Parker, 1978). Risk is considered as the probability and natural feature of the hazards (Chatterjee et al., 2009). How the risks are involved in the disaster and its connection with the vulnerability and its impact on the people life (Biotechin Asia, 2015)? During the floods, physical phenomenon contains time span of floods and the depth of geographical location (Blaikie et al., 1994). Hydrological process has changed by urbanisation process and in this regard most of the rainwater does not percolate to the ground due to concretisation of the city surface (Sangster et al., 1990). Surface-runoff has increased by volume and artificial improvement in the drainage network system results in increased and quick runoff followed by intense downpour (Zhong et al., 2014). Intensification in the floods occurrence has also been documented globally (Hollis, 1975).

Climate change is an important potential factor in analysing the further tendencies of floods in the urban locality. However, the state climate model is not competent enough to produce reliable and correct predictions (Chatterjee et al., 2009). Acceleration in the floods and storm lines are also caused by global warming (Gilbert & Horner, 1985). During November and December, the shocking floods submerged the road, smashes the city and different parts of Tamil Nadu and cause huge loss to the life and property. The natural and artificial drainage system of the city are becoming ineffective due to various reasons like decrease in the porthole due to bridge construction, poor designing and accumulation of the sand on the river mouths (Chandan et al., 2014). Illegal dumping of waste materials, inadequate connection of sewers with the micro drainage system, illegal urban encroachment and construction convert the city into floods areas (Tripathi et al., 2014). In order to tackle such situations the successive state governments have emphasised on the process of desilting the drains and cleaning out the river beds (Rajan, 2016).

Study Area

Chennai is almost flat like a pancake with the mean elevation of 6 m above the sea level (Thakkar, 2016). It is one of the important cosmopolitan city of Tamil Nadu, which plays a vital role in the social, intellectual and cultural development of South

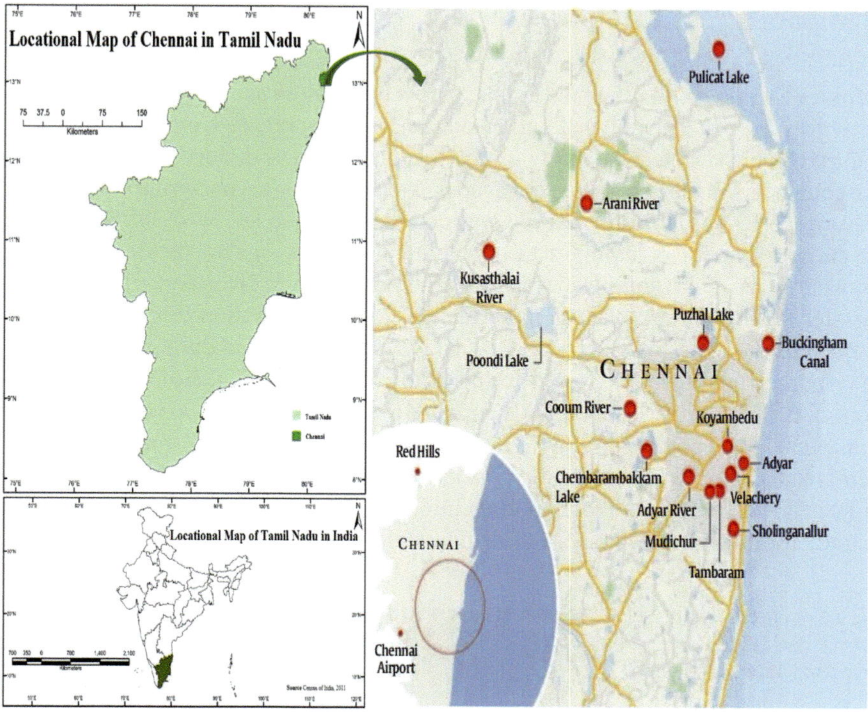

Fig. 2.1 (**a**) Locational map of Tamil Nadu. (*Adopted* from Census of India, 2011). (**b**) Locational map of Chennai. (*Adopted* from NRSC, 2015)

India (Sinha, 2015). Around seventeenth century, East India Company established it's headquarter at St. George that was an important trading hub and the main cause for the majestic transformation in Chennai (District Census Handbook-Chennai, 2011). It lies between 13° 06″ to 13° 44″ N latitude and 80° 23″ to 80° 14″ E longitude (Department of Economics and Statistics, 2016) (Fig. 2.1). The city is urbanising at a very rapid pace over the past few decades (Kohli, 2016). In 1901, the city has the total population of 5 lakh which has increased to more than 45 lakh in 2011 and spread over an area of 178 sq. km (Kotteswaran, 2015). Today, it has 100% urban population while Tamil Nadu has 49% urban population (Census of India, 2011). Due to rapid urbanisation, there is constant fall in the infiltration process of the city hydrological cycle like gradual loss of the many important local water bodies which is a major concern (The Hindu, 2015).

Objective

Identify the causes of urban floods and its socio-economic impact on the entire city.

Methodology and Database

Primary Survey With reference to this, purposive random survey was conducted in December 2016 at Chennai in Tamil Nadu. Twenty respondents from different strata of the city were selected and interviewed. Respondents have divided into three group like youngsters in the first group, old aged and female in the second group, and rescue team in the last group. Pilot survey was conducted in the last week of September for the greater understanding of the flood-affected areas and floods victims where army played a significant role in rescue operations (Fig. 2.3). Interview with National Disaster Response Force was an important part of data collection (Fig. 2.2).

Secondary Data Several secondary data sources were used during the analysis. Few sources that have been used in this study are Indian Meteorological Department, District Census Handbook-Chennai, Greater Chennai Corporation, Disaster Management Support and National Remote Sensing Centre. Apart from the sources mentioned above, important literary works like books, magazines, articles, journals, eminent newspapers and other important information from online websites have been denoted for the in-depth analysis of deluge-affected areas.

Fig. 2.2 Interview with national disaster response force team member (Author, 2016)

Fig. 2.3 Army personnel in rescue operation (*Adopted* from Senthil Kumar, 2015)

Data Collection Method Methodology is a path to discover the answer to several questions in a systematic way (Bailey, 1978). Analytical and descriptive methods have been used for inclusive understanding of the flood-affected victims. Primary survey and secondary data have been used for collection of data from the flood-affected areas. Both qualitative and quantitative approaches have used for the analysis of the collected data.

Analysis and Discussion

Factors Responsible for Heavy Rainfall

The north-east monsoon that brings rainfall in the peninsular India during the month of October to December, brought continuous torrential rainfall over the past 12 days in 2015 (Rajan, 2016). Chennai has received the century heaviest downpour on 1st December, and according to the state meteorological department, the city has received rainfall of 1200 mm within 24 hours (Thakkar, 2016). Although the amount of rain received during 2nd and 3rd December has dropped to 300 mm and on 4th December, it further declines to 160 mm (Regional Meteorological Centre, 2016). Low pressure trough was one of the main reasons behind the heavy downpour which continues over the Bay of Bengal, but after few days, it has shifted 280 km away from the Tamil Nadu coast (Fig. 2.4, Jones, 2016). During the north-east monsoon, inverted trough of low has developed which is the main feature of south-west monsoon (The Indian Express, 2016). In the normal trough of low, there is decrease in the atmospheric pressure from south to north while in the inverted trough of low atmospheric pressure decreases from north to south (Indian Meteorological Department, 2015). The inverted trough of low helps in the swift flow of north-easterly winds towards the coast and brought heavy rainfall in the coastal areas (Rajasekaran, 2015). A trough of either normal low or extended low is an extended form than the atmospheric low pressure which is more localised with the greater intensity (Tripathi et al., 2014). Tropical cyclone often developed when there is adequate trough of low pressure (Jones, 2016).

Primary survey suggests that about 75% households replied that they didn't saw such kinds of floods in their whole life. The level of water that they had experienced during the floods was very dangerous, and they pray to the God that as soon as possible they get rid of the deadly floods. About 20% households replied that such kinds of floods had once occurred about a century ago. They had just read about such kinds of floods in some books, magazines or in internet, and this time they have experienced it and this is their worst experience that they had ever face in their whole life. They hope that they will never ever want to experience such kinds of a devastating floods in the near future. About 5% replied that this floods was caused by the development of low pressure in the Bay of Bengal which was further supported by the different factors like movement of winds with abundant moisture content over the Bay of Bengal.

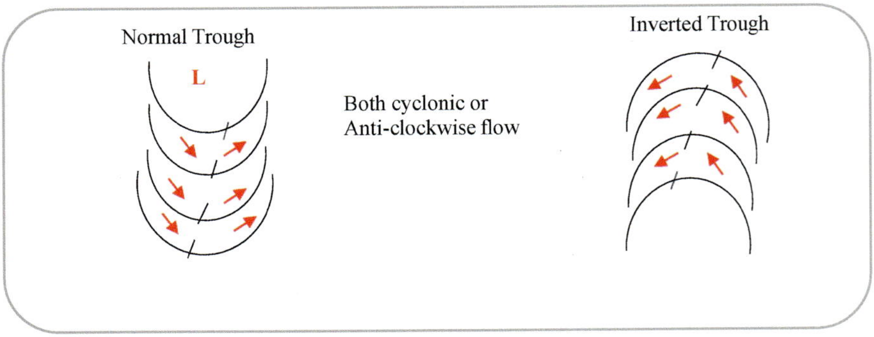

Fig. 2.4 The monsoon story 2016 (*Adopted* from North Florida Weather Blog by Jones)

Conditions of Chennai Metropolitan Area (CMA)

Chennai Metropolitan Area (CMA) is spread over 1190 sq. km (Census of India, 2011). It has three rivers, i.e. Cooum, Kosasthalaiyar and Adyar that flow through the whole CMA (Fig. 2.2). The origin of Kosasthalaiyar River is Kaveripakkam which lies in the northern Arcot town of the city (Narasimhan et al., 2016). The river generally flows in the eastward direction which bifurcates into Kosasthalaiyar as the main branch and Cooum River in Kesavaram anicut (Statistical Handbook, 2016). Then, the main branch enters into the Poondi reservoir. Cooum River after bifurcating from the Kosasthalaiyar River flows towards Kanchipuram district of the state and finally drains into the sea (Chennai Municipal Development Authority, 2016). Adyar River originates from Kavanur and Pillapakkam tanks in Kanchipuram and flows through the CMA and then drain into the sea (Narasimhan, 2016). Buckingham Canal was constructed about 200 years back as salt water navigation canal in the city (Narasimhan, 2015). The total length of the canal is 421 km which is spread from Kakinada in Andhra Pradesh to Puducherry in Tamil Nadu (Table 2.1, Narasimhan, 2016). It connects all the three rivers of CMA, i.e. Kosasthalaiyar in the north, Cooum in the middle and Adyar River in the south (Coastal India Development Council, 2016).

Primary survey suggests that about 80% of the respondents replied that the present conditions of the city drainage are not in a good position. Adyar, Cooum and Kosasthalaiyar are the three main rivers of the city which acts as the natural drainage as well as water source but in the recent times these rivers are drying up rapidly. The catchment areas are shrinking at a very rapid pace due to conversion of the water bodies (natural environment) into various built up land like residential areas, commercial complexes and other economic units (human-made environment). About 15% households replied that the government officials, Tollywood celebrities and businessman are the major defaulters among the encroacher to these river basins. As the cost of the land, flats, apartments and farm house in these areas are very expensive and such an expensive areas are far beyond the reach of mass middle and lower class locals. Within all these mess, a ray of hope came up when about 5%

Table 2.1 Macro drainage of Chennai Metropolitan Area (CMA)

Macro drainage	Drainage area (km^2)	Total length (km)	Length within CMA	Width (m)	Bankfull discharge (m^3/s)
Adyar river	720	42	24	10–200	2038
Cooum river	502	72	24	40–120	991
Kosasthalaiyar river	3741	136	16	150–200	3540
Buckingham canal	NA	421	48	30	43

Source www.cmdachennai.gov.in/SMPS/SMPS.htmal and Narasimhan (2016)

of the respondents replied that the government has taken several initiatives such as removing all kinds of human encroachment from either side of the rivers and also working on new projects to revive and rejuvenate these rivers which is considered as water supply life-line of the city.

In Chennai, 96% households have closed drainage connectivity for the passage of wastewater followed by 3% households have no drainage connectivity and 1% have open drainage connectivity (Fig. 2.5, District Census Handbook-Chennai, 2011). These 3% households do not have any drainage connectivity and harm the city most because wastewater of these households mixed with river, lakes and canal water which pollute all the local water bodies of the city. It was also identified that floods situation in Tamil Nadu worsened when Andhra Pradesh government released a huge amount of water from their different reservoirs without reviewing the deluge condition in northern districts of the state. Water released by the Andhra Pradesh immediately mixed with various drainage system and waterbodies of Kancheepuram, Chennai and Tiruvallur. Most of the drainages for these three places are through Chennai via. Poondi, Cholavaram and Red Hills that provide catchment basin to Chennai which flooded all the low-lying places of Chennai. It results in the widespread destruction of almost all the infrastructure of the city like rail, road, airport, bridge, electricity supply, mobile connectivity, internet connectivity, water supply and many more.

Economic Loss Due to Natural Disasters

During 2015, the entire loss bear by Indian economy from all natural disaster was ₹63,700 cores out of which floods alone cause 75% of it (Krishnamurthy & Desouza, 2015). In this regard, Chennai floods caused loss of more than ₹14,950 crores during November–December which is much higher than any other states of the country (The Hindu, 2015). Because there are several new places which are experiencing the consequences of floods such as Udhampur, Poonch, Lakhimpur, Sonitpur, Gonda, Bahraich, Azamgarh, Sheikhpur, Muzaffarpur, etc. (Lakshmana, 2015). Geographical change, human encroachment on the either side of the water bodies, unplanned

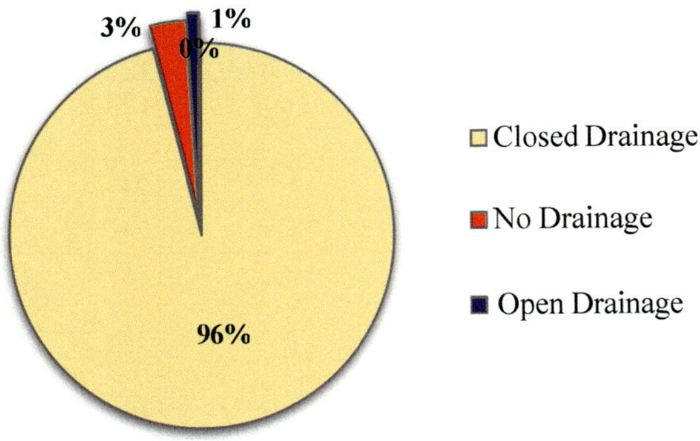

Fig. 2.5 Percentage of households type of drainage connectivity (*Source* District Census Handbook- Chennai, 2011)

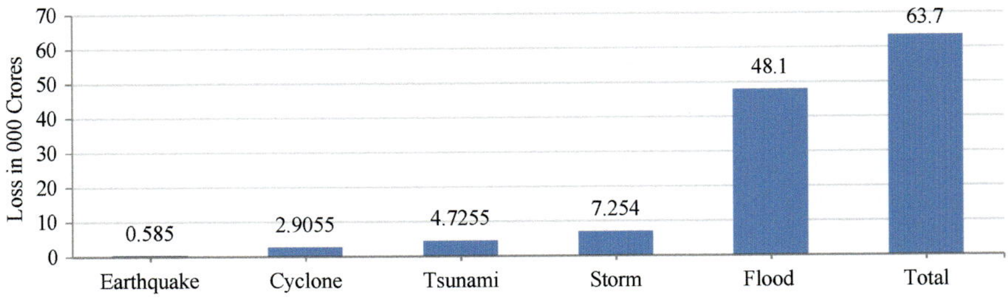

Fig. 2.6 India's average annual loss by natural disaster (*Source* Global Assessment Report of UN Office for Disaster Risk Reduction, 2015)

development of huge built-up are one of the main reasons behind emergence of new floods areas (Lavanya, 2015). In the context of India's average annual natural disaster, floods caused the maximum damage followed by tsunami, storm surge, cyclone and earthquake affects lowest in the monetary terms (Fig. 2.6).

Socio-Economic Loss

The heavy downpour took the life of more than thousands of people and about millions were missing, around millions families suffered, more than lakhs of households were damaged. Besides human beings, crops and livestock of the Chennai people were also suffered a lot which has been shown in the following (Table 2.2).

Primary survey suggests that about 25% households replied that their family members were killed in this heavy downpour and they were the main bread earner for their family. After their death, who will take care of their entire family? About 15% households reported missing of their family members and even today, i.e. after one year, they didn't get evidence from the local search agencies that their missing family members are alive or dead. About 25% reported partial or total damage of the

Table 2.2 Impact of urban floods in Chennai

People	Families	Household	Crops and livestock	Financial loss
Over 1000 people were killed. Around 1.8 million were missing	About 30 lakh families suffered partial or total damage of their dwellings	More than 1lakh homes damaged Each household faced losses of ₹2–20 lakh	About 3. 28 lakh hectares of crops lost Over 98,000 livestock died	Financial loss of more than ₹20 thousand crore

Source ASSOCHAM, Loyola Rain Relief Service and Tamil Nadu Government, 2015

dwellings and most of them are not in a position to re-build the damaged structures of their dwellings. About 20% households had suffered due to the loss of their agricultural and cash crops. The government crop insurance schemes like National Agriculture Insurance Scheme, which had promised to compensate their crop loss from the natural disasters are not giving the compensation amount at the time of the crisis. About 15% households reported that their livestock were died because the flood water entered to their livestock's house and they were tied-up with poles and unable to move to the safer places. These livestock were the main source of their income, and most of their livestock were not insurance.

Role of Government

Chennai was hit by two floods in 2015, one hit on its usual time and the state government thought that the floods is over. But, after four days another devastation flood hit the city which caused a huge loss of life and property (Tamil Nadu Migration Survey, 2015). However, Indian Meteorological Department (IMD) already forecasted that there are possibility of heavy deluge which may occur in the whole month of November and December (Deccan Chronicle, 2015). In the light of such continuous warnings, adequate steps were not taken by the state government like resettlement and rehabilitation of the people to the safer place (Williams, 2015). The process of flooding starts when water was released from the Chembarambakkam reservoir because it was overflowing and water level in the reservoir has already crossed the danger level mark, the biggest reservoir on Chennai border (Amalorpavanathan et al., 2016).

Primary survey suggests that about 30% households replied that the rescue operation starts after 2 days of the floods, and in that case helicopter facilities was started at the end; it should start in the beginning of the disaster in order to minimise the damage occurred caused by the floods. About 40% households replied that not a single politician was there to help the flood-affected people. Politicians house are in upper areas of the city, while the house of the locals is in the lower areas which are very prone to the floods. About 15% households replied that these floods cause a huge loss of life and property due to the ignorance from the state government because

IMD had already warned the local authorities to take adequate steps about the occurrence of such a deadly disaster. About 5% households replied that overflowing of Chembarambakkam reservoir was the main cause of the city floods because in the surroundings of the reservoir several built-up areas has developed. Earlier reservoir's surroundings were dominated by the wetlands and marshlands which act as the check and balance of the reservoir. About 10% households reported that the city infrastructures are not adequate to cope up such kinds of disaster. There is urgent need to upgrade the city infrastructure in accordance to the population of the city requirement.

Suggestions and Recommendations

There is an urgent need to revive the traditional lakes and tanks in the way to tackle such kinds of floods. Latest technologies are to be used in the survey, protection and conservation of wetlands and marshlands. Efficient and effective delivery of medical services are to be provided to the flood's victims of water-borne diseases like diarrhoea, cholera, dysentery, etc. Stop unplanned tourism activities and banned illegal mining activities. Allow local community to govern and surveil the good physical health of the existing canals, wetlands and lakes to stop any kind of further encroachments. Some lands must be kept in common for use by all called as Pocomoke in Tamil language. Involvement of geographers/geologist and experts in policy and planning.

Conclusion

Either be the Chennai or any other city, there is a need to change the urban governance approach with the scientific temperament to tackle such kind of new unforeseen incidences. Repetitive assessment of the various city infrastructures including sewage system, drainage system, maintenance of buildings, potable water supply, transportation services and so on which are often poor or if done, then many loopholes found in the future assessment process. Municipalities often suffer from under staff and inadequate funds, need latest technologies, skills and suitable funding for every kind of city governance in the way to make the city livable. It is unfortunate that the municipalities and political leaders blame the unpredicted rain for the loss of precious life and property in Chennai. Unpredicted rains often hit the city; it is a ferment call to the city that the growth has touched beyond its carrying capacity.

Acknowledgements The authors are thankful to the University Grants Commission (UGC) for providing financial assistance during the study and writing the paper.

References

Amalorpavanathan, J., Ramakumar, M., & Sivasubramanian, S. (2016). Preparedness in disaster situations lessons from Chennai floods 2015. *Economic & Political Weekly, L1*(8), 30–34.
Bailey, D. (1978). *Methods of social research.* New York, USA.
Biotechin Asia (2015). https://biotechin.asia/2015/12/05/health-concerns-dos-and-donts after-chennai-floods/.
Blaikie, P., Cannon, T., Davis, I., & Wisner, B. (1994). *At risk: Natural hazards, people vulnerability, and disasters.* London.
Census of India. (2011). Administrative Set-Up in Tamil Nadu.
Chandan, C., Bharath, H., & Ramachandra, V. (2014). Qualifying urbanisation using geospatial and spatial metrics-a case study of madras. In *Conference on Conservation and Sustainable Management of Wetland Ecosystems in Western Ghats.*
Chatterjee, C., Patro, S., Mohanty, S., Singh, R., & Raghuvanshi, N. (2009). Floods inundation modelling using MIKEFLOOD and remote sensing data. *Journal of Indian Society Remote Sensing, 37*(1), 107–118.
Chennai Municipal Development Authority. (2016). Chennai Metropolitan Area Profile. Retrieved from http://www.cmdachennai.gov.in/
Coastal India Development Council. (2016). Buckingham canal tourism project from Kakinada to Pondicherry, The Society to Develop the Core Strength of the Coastal Districts of India, pp. 1–8.
Deccan Chronicle. (2015). Retrieved from https://www.deccanchronicle.com/151203/nation-currentaffairs/article/when-chennai-was-logged-out-and-how
Department of Economics and Statistics. (2016). Statistical Handbook, Government of Tamil Nadu.
District Census Handbook. (2011). Brief History of Chennai, Government of Tamil Nadu.
Gilbert, S., & Horner, R. (1985). *The Thames Barrier.* London: Thomas Telford.
Hollis, G. E. (1975). The effect of urbanisation on floods of different recurrence interval. *Water Resources Research, 11*, 431–434.
Indian Meteorological Department. (2015). Heavy Rainfall over Southeast India during November & Early December, Ministry of Earth Sciences, Government of India.
Jones, C. (2016). North Florida Weather Blog.
Kohli. A. (2016). How automotive supply chains prepare for chennai-like disasters. Pure Research Newsletter.
Kotteswaran, C. (2015). *Tamil Nadu: Chennai floods cause a loss of ₹ 50k crore.* Deccan Chronicle.
Krishnamurthy, R., & Desouza, C. (2015). City Profile-Chennai, India. *Cities, 42*(A), 118–129.
Lakshmana, K. V. (2015). *Chennai rains have sunk business worth Thousands of Crores, Hindustan Times.*
Lavanya, K. (2015). Urban floods management–a case study of Chennai city. *Architecture Research,* 2(6), 115–121.
Narasimhan, T. E. (2015). *Automakers in Chennai face logistics nightmare. business standard.* Retrieved on 27 April 2016.
Narasimhan, B. (2016). *Indian institute of technology, Madras, interdisciplinary centre for water research indian institute of science, Bangalore.* pp. 1–5.
Narasimhan. B., Bhallamudi. Murty, S., Mondal, A., Ghosh, S., & Majumda, P. (2016). *Chennai floods 2015: A rapid assessment. interdisciplinary centre for water research Indian Institute of Science, Bangalore.*
Parker, D. J. (1978). Floods in Cities: Increasing exposure and rising impact potential. *Built Environment, 21*, 2/3.
Rajan Vishnu, C. (2016). A case study on impact of Chennai floods: Supply chain perspective. *Researchgate IIIE Industrial Engineering Journal, 10*(8). Retrieved from https://www.researchgate.net/publication/309194370
Rajasekaran, I. (2015). How the city braved it out. *Frontline, 32*(23).

Regional Meteorological Centre Chennai. (2016). Government of India, earth system science organization, Indian Meteorological Department, College Road, Chennai. Retrieved from http://imd chennai.gov.in/

Sangster, E., Tunstall, S., Doizy, A. (1990). *Flooding in the maidenhead areas in February 1990 and its effects*. Report for the National Rivers Authority, Enfield.

Sinha, K. (2015). *Chennai Floods: Insurance Companies likely to Face Claims Worth ₹ 1000 crore*. The Economic Times.

Tamil Nadu Migration Survey. (2015). *Centre for development studies, Thiruvananthapuram, Loyola institute of social sciences training and research, Chennai, Centre for Diaspora Studies, MS University, Tirunelveli*, pp. 76–80.

Thakkar, K. (2016). Chennai floods: carmakers like ford and BMW struggle to start work after screening halt. The Economic Times.

The Hindu. (2015). Residents Stock Up for the Rainy Day.

The Indian Express. (2016). *Chennai floods: insurance cost Get ₹ 4.8k crore Claims*.

Tripathi, R., Sengupta, S. K., Patra, A., Chang, H., & Jung, W. H. (2014). Climate change urban development and community perception of and extreme floods. *Applied Geography, 46*, 137–146.

Williams, M. (2015). Flooding in Chennai causes plant closures and supply disruptions. Automotive, Retrieved 2017.

Zhong, G., Liu, S., Han, C., & Huang, W. (2014). Urban floods mapping for Jiaxing City based on hydrodynamic modelling and gis analysis. *Journal of Coastal Research, 0749–0208*, 168–175.

Chapter 3
Understanding Flood Risk and Livelihood Resilience in Begusarai

Rashmi

Abstract Flooding is the deluging of a normally dry area caused by a high flow or discharge of water into an established stream, such as a river, stream or drainage channel, or a water pond near the node where there is occurrence of precipitation and overflow from the banks result in diffusion of water in the plains. The past study of flood trends shows that humans have gradually moved focus from flood control to adaption in order to limit the effects of flooding on various livelihood activities. This is in part due to the realization that flood control measures have constraints because of the fluctuating climatic changes, policy implementation challenges and limitations of flood prediction methods. Also some factors are often non-controllable, like the fact that major tributaries originate from neighbouring countries and as a result water flow is not always in control of the home state or nation. The study is based around the sustainable livelihood framework which assesses residents' access to livelihood assets across five categories with an aim of better understanding of the ways in which people develop and sustain their livelihoods.

Keywords Floods · Community · Agriculture · Livelihood · Resilience · Life · Asset pentagon

Introduction

Almost every year, natural disasters around the world cause deaths, injuries to humans and animals, and the destruction and loss of property, financial and economic resources. The occurrence of floods causes disruptions in human life (Rufat et al., 2015). It disrupts cultivation in the plains, interrupts field irrigation systems, causes damage to properties leading to displacement of families as well as livestock which is an important source of income for rural areas, and hinders transportation systems disrupting the transfer of goods and services. This increased frequency of disasters such as floods throughout the world is an indicator of unsustainable development. In

Rashmi (✉)
Department of Geography, Delhi School of Economics, University of Delhi, New Delhi, India
e-mail: rashmithakur1990@gmail.com

V. R. Sharma and Chandrakanta (eds.), *Making India Disaster Resilient*,
https://doi.org/10.1007/978-3-031-50113-5_3

general, river floods affect most areas of the country, although the Kano, Nyatike, Budalang and Lower Tana plains show severe flooding (GoK, 2009).

The past study of flood trends shows that humans have gradually moved focus from flood control to adaption in order to limit the effects of flooding on various livelihood activities. This is in part due to the realization that flood control measures have constraints because of the fluctuating climatic changes, policy implementation challenges and limitations of flood prediction methods. Also, some factors are often non controllable, like the fact that major tributaries originate from neighbouring countries and as a result water flow is not always in control of the home state or nation (Mashebe et al., 2016).

Developing areas in a huge nation like India also face challenges in adjusting to higher frequency of floods due to low levels of access to natural capital, physical capital and social capital which limits response capacity of government institutions. This ultimately results in higher vulnerability of households to flood shocks, as most of them have high dependence on natural resource-based livelihood options.

Resilience is termed as the community's ability to adapt by changing or resisting when exposed to a hazard to achieve an acceptable level of functioning, structure and organization (Jelena et al., 2016). Flood resilience in specific talks about the measures and changes adapted by community to achieve an acceptable level of functioning during and after the event of flood occurrence (Viz et al., 2003). To break this down further, communities and urban systems should have the capability to resist, recover and introspect from each flood occurrence. There are five key elements of flood risk management that are important from a flood resilience point of view—Reflect, Relief, Resist, Response and Recovery (Batica et al., 2016).

The study is based on the sustainable livelihood framework which assesses residents' access to livelihood assets across five categories. The livelihoods framework takes into account the assets and skills (both social and material) and the methods which are typically used by communities as well as individuals in order to survive. Sustainability in this context implies that these communities or individuals can challenge and overcome flashes of stress and/or crisis, and that they are able to sustain or even improve present and forward-looking skills and assets without misusing their supply of natural resources (Islam et al., 2015).

Objective

The main objective of this study is to understand the region's exposure to floods, understand the livelihood approach and impact of the floods on key livelihood pillars in Begusarai.

Study Area

Begusarai district was one of the subdivisions within the Munger district until the 1971 census and was given the status of a separate district on 2 October 1972. It was previously the northwest subdivision of the Munger district located across the Ganges. Begusarai is the head of the city and the district headquarters. It was an inn (sarai) from which the city probably derives its name. The neighbourhood is in the central Bihar region, bordered to the north by the Samastipur district, to the south of Munger and Lakhisarai Districts, to the east again by Munger and Khagaria Districts district and western districts of Samastipur and Patna. Begusarai is one of the districts among a total of 38 districts in the state of Bihar. Begusarai district enters under Munger Division. It is also known for its industrial recognition; the main industrial units are Indian oil. Refinery, Railings, Power Plant, Railings and hundreds of small and industrial units. Begusarai district is spread across an area of 1918 km^2. The district which is a part of Munger division now is situated on the northern bank of river Ganga. The district is located in the central part of the North Bihar plain between latitudes 25.15 and 25.45N and longitudes 85.45E & 86.36E. It has been divided into two sub-micro regions, namely (a) Ganga-Burhi Gandak Flood Plain of Begusarai and (b) Burhi Gandak Kareha flood plain of Begusarai on the basis of geographical factors like drainage, soil, relief, geology, climate and natural vegetation. The Burhi Gandak river forms the dividing line between the two sub-micro regions.

(1) Ganga-Burhi Gandak Flood Plain of Begusarai
(2) Burhi-Gandak-Kareha Flood Plain of Begusarai.

The most important rivers traversing the district are the Ganga, the Burhi Gandak, the Baya, the Balan, the Bainty and the Chandrabhaga.

Data Source

This paper is primarily based on secondary datasets like the District Census Handbook of Begusarai, Economic Survey of Bihar 2018, District Fact book of Begusarai, Disaster Management, Government of Bihar.

Methodology

Access to the five types of capital was calculated using a set of representative indicators for each of these capital categories such as overall literacy rate in percentage and working population in percentage for human capital, per capita income in INR for financial capital, access to National Highway (NH), State Highway (SH) in kilometre for Physical capital, etc. Similar principles applied by Motsholapheko (2011) in the development of the rural livelihood vulnerability index were used with changes done to focus on assessing capability of households to adapt.

The representative indicator values from different base ranges were transformed to a set of unit free indices with values between zero (lowest) and one (highest). These indices were calculated using equation (3.1) adapted from the standardization of all indicators comprising the human development index (HDI) developed by UNDP (1990):

$$C_i = \frac{C_i - C_{\min}}{C_{\max} - C_{\min}} \quad (3.1)$$

where C_i is the value of the reference indicator measured in any unit, $C_{\min}$ and $C_{\max}$ are its pre-determined minimum and maximum values. The allocation of minimum and maximum values was done as follows:

a. Values measured in percentage have been assigned a minimum value of 0 and a maximum value of 100
b. Number of healthcare facilities per village have been assigned a minimum value of 0 and a maximum value of 2 in healthcare units
c. Access to National Highway (NH), State Highway (SH) & district road network the values assigned from 250 to 1000 kms
d. Livestock wealth per district range from a minimum value of 0 to 1200 in thousand livestock units
e. Small savings range from 50 to 500 in INR crores
f. Average per capita income per month range from 7500 to 15,000 INR

All the representative indicators for each type of capital were averaged separately to arrive at the different capital scores for these three districts using the following equation (3.2):

$$\text{Capital}_a^i = \sum_{i=1}^{n} \frac{\text{Index}_{ai}}{n} \quad (3.2)$$

where Capitala is the type of capital (human, social, etc.) and Indexai is the ith representative indicator for these capital types calculated using equation (3.1) and n is the total number of indicators used for estimating Capitala.

Results and Discussion

The Begusarai district is located in the central part of North Bihar. In general, the plain has a tilt from south to southeast slope. This factor governs the flow of streams in this region. Geomorphology suggests that this region is a part of the Gandak-Kosi interfluve. The southern part of district generally has elevated landmasses except for the low-lying areas prone to flood majorly from in comparison to adjacent districts of Khagaria and Samastipur. The Flood Hazard Zones of Begusarai Districts can be seen in Fig 3.1. Thus, it turns out to be a comparatively safer zone between the other flood-affected areas. The district is drained by a number of major rivers like the Ganga, Bagmati, Balan, Burhi Gandak and a host of small streams, dhars, nalas with local origination sources and act as rain water preservation sources. The main flood affected C.D blocks of the districts are Teghra, Matihani, Bachwara, Barauni, Shamho Akha, Balia, Sahebpur Kamal. The blocks near to the river have more chances of floods in comparison to other blocks. Begusarai comes under the list of extremely flood-prone districts among other. Out of these 28 districts, 15 districts are considered extremely flood prone. Table 3.1 represents the list of extremely flood-prone districts of Bihar (Source: Bihar Flood Reports 2016, Bihar State Disaster Management Authority, Government of Bihar).

Table 3.2 shows the flood damages from 1998 to 2012 with special reference to human lives lost, damage to crops, damages to houses and total damages. In the category of loss to human lives, 2007 is the year where 54 people lost their lives which leads to Rs. 16,710 lakhs worth of crops damage, 29,000 damage to houses. Based on the above data we can say 2007 was major in the history of Bihar which has extremely impacted and affected the life and livelihood of the locals.

Livelihood Activitities in the Flood Affected Region

The concept of livelihood has a wider connotation among the researchers, policymakers and development practitioners in the context of survival. The livelihood strategies and activities of the rural economy are often complex and diverse (Ashley et al., 1999). Rural people usually engage in a combination of different strategies when the availability of resources is inadequate. As the majority of the people in Saran plain lives in rural areas, their lives and livelihoods are directly or indirectly dependent on land and water. Frequent and devastating floods endanger the lives and livelihoods of the native population more when the area is a rural area and when the majority of the people is engaged in agricultural sector (Madhuri et al., 2014).

Livelihood activities are what an individual or the household undertakes to make a living (Francis, 1999). Therefore, the main livelihood activity a household is involved in is the activity that the household depends on for its major source of earning a living. Figure 3.2 depicts the Livelihood activities of the population of Begusarai. This helped to establish whether households had changed their main source of livelihoods over the periods of time or not. In the region, 42% people are engaged in cultivation, 27% are engaged in agriculture activities, that means 70% people are directly engaged in agriculture activities and if flood comes their livelihood gets affected and their family suffers a lots because of natural phenomenon. In the region the percentage of labourer engaged in household industry is 4%, allied industry 0.2% and in other category the percentage is 26%. Asset Capital of Begusarai is represented with the help of Fig. 3.3.

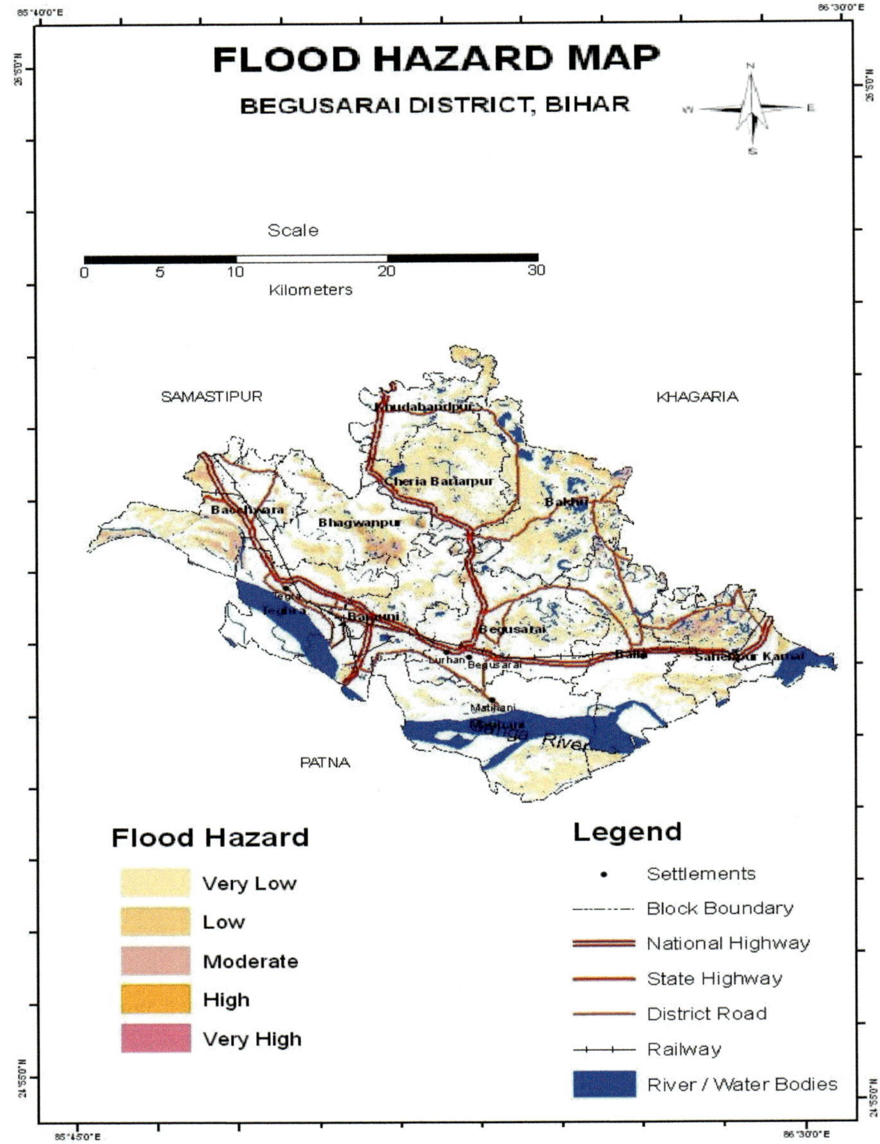

Fig. 3.1 Food hazard map of Begusarai district. *Source* Disaster Management, Government of Bihar

These indicators were transformed to indices between zero (lowest) and one (highest) using equations (3.1) and (3.2) and compared for the district in the study area. Hypothetically, the district with highest access to all the representative indicators under a particular category of capital will have a composite index of one and the lowest will have a composite index of zero. Determinants and indicators of access

Table 3.1 Extremely flood-prone districts

Sl No.	Extremely flood-prone districts
1	Muzaffarpur
2	Vaishali
3	East Champaran
4	Sitamarhi
5	Sheohar
6	Darbhanga
7	Madhubani
8	Samastipur
9	Saharsa
10	Supaul
11	Madhepura
12	Khagaria
13	Begu sarai
14	Bhagalpur
15	Katihar

Table 3.2 Flood damages in respect of Begusarai district

Year	Loss of human lives	Crop damage (INR lakh)	Damage to houses	Total damages (INR lakh)
1998	5	2338.71	102	6880.12
1999	13	240.00	15	671.96
2000	–	560.84	–	1412.29
2001	–	0.58	165	1.35
2002	10	1031.90	1006	2947.28
2003	9	346.48	309	7220.78.163
2004	10	2516	27,257	
2005	No flood			
2006	No flood			
2007	54	16,710.07	29,391	52,151.57
2008	No flood			
2009	No flood			
2010				
2011	11	888.92	193	1022.28
2012				

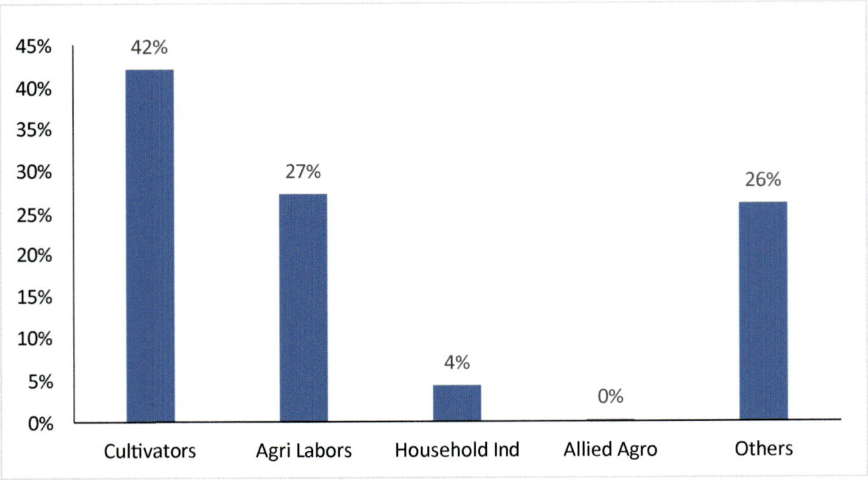

Fig. 3.2 Livelihood activities in Begusarai. *Source* District Census Handbook Begusarai

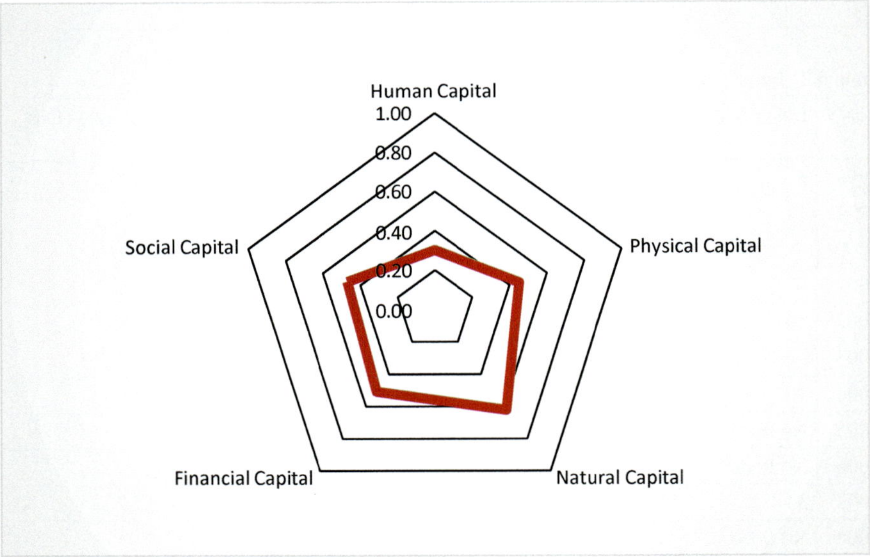

Fig. 3.3 Asset capital of Begusarai. *Source* Author's analysis

to capital in Begusarai are represented by Table 3.3. Access to natural capital and financial capital is moderately high 0.62. and 0.50 with reference to other capitals in the districts. Access to human capital is 0.30. Access to physical capital is 0.45. Access to social capital is 0.47. Impact of Flood on Livelihood Resources & Options and Physical and Financial Capital can be accessed using Figs. 3.4, 3.5, respectively.

Table 3.3 Determinants and indicators of access to capital in Begusarai

Capital	Reference indicators	Unit	Begusarai
Human capital	Overall literacy rate	Percent	63.87
	Working population	Percent	31.58
	# Healthcare facilities per village (Average)[a]	Unit	0.435
Physical capital	Access to hand pumps for drinking water	Percent	91.05
	Access to NH, SH & District road network	km	354
	Household with access to radio or television	Percent	30.75
Natural capital	Cultivable Land (percent)—Net sown area/Total area	Percent	60.49
	Access to irrigation—Net Irrigated Area/Total Area ('000 ha)	Percent	45.85
	Livestock wealth[b]	000	953
Financial capital	Small savings in Post Offices & PPF	INR crores	48.9
	Households with access to banking services	Percent	43.3
	Per capita income	INR	15,601
Social capital	Villages with access to Agricultural Credit Societies/ Total villages	Percent	28.39
	Villages with access to town less than 15 km/Total villages	Percent	64.26
	Villages with access to pucca road/Total villages	Percent	57.49
	Villages with access to transport communications[c]	Percent	39.19

Source District Census Handbook 2011, Economic Survey of Bihar 2017–18, District Fact book
[a]*Number of Healthcare facilities per village (Average) = No of total healthcare facilities in district/ Number of villages in district*
[b]*Livestock wealth includes Cow, Buffalo, Sheep, Pig, Goat, Poultry*
[c]*Transport communications include bus service, railway facility and navigable waterways*

Conlusion

Generally, flood is defined as an unusually high stage in a river normally the level at which the river overflows its banks and inundates the adjoining area. The past study of flood trends shows that humans have gradually moved focus from flood control to adaption in order to limit the effects of flooding on various livelihood activities (Arman et al., 2010). This is in part due to the realization that flood control measures have constraints because of the fluctuating climatic changes, policy implementation challenges and limitations of flood prediction methods. Also some factors are often non-controllable like the fact that major tributaries originate from neighbouring countries, and as a result water flow is not always in control of the home state or nation. The main objective of this analysis is understanding flood risk and livelihood resilience in Begusarai district. For this analysis, we have used secondary source of data. After our analysis, we have observer that 2007 was major in the history of Bihar which has extremely impacted and affected the life and livelihood of the locals. And if we talk about livelihood activities, 70% people are directly engaged in agriculture activities. In the region, the percentage of labourers engaged in household industry is 4%, allied industry 0.2% and in other category the percentage is 26%.

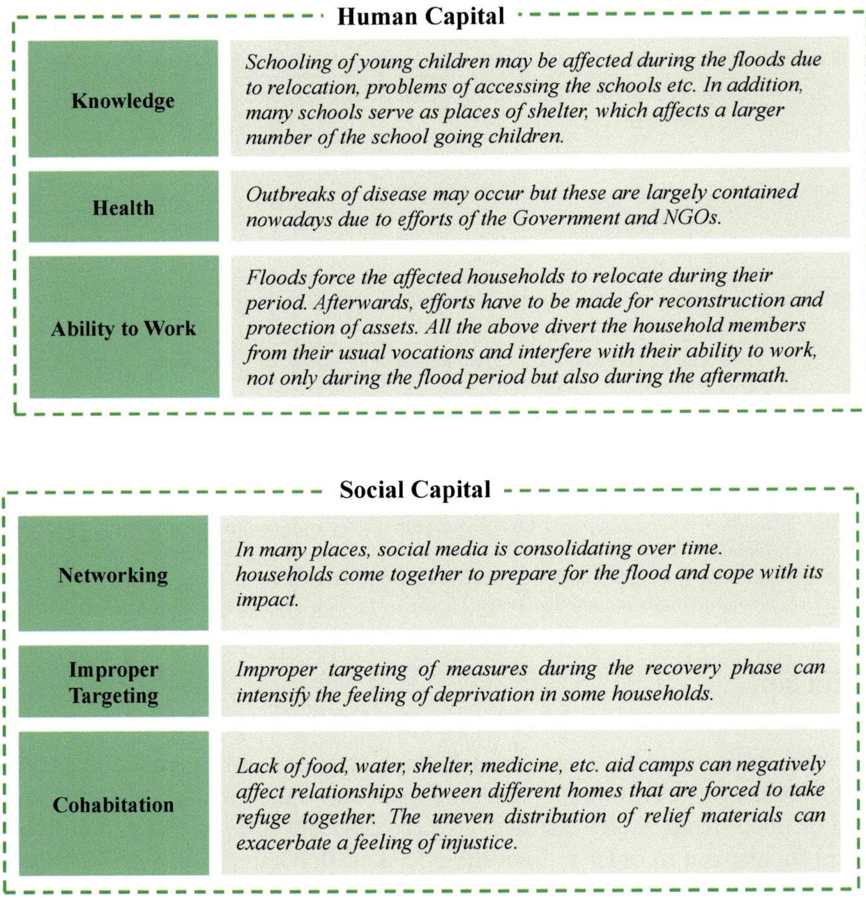

Fig. 3.4 Impact of floods on livelihood resources and options

Natural Capital

Diara Cultivation Uncertainty	*The diara land is found between the natural leeves of river formed due to its meandering.Cultivation of Kharif season crop is heavily impacted as majority of this fertile land is submerged when the river swells up.*
Water Logging	*Every year a section of the cultivable land suffers from extendted water logging even after the monsoon is over.*
Mono-cropping	*There is a significant area in Begusarai which falls under the Mokama group of Talls which is suitable only for single crop cultivation during Rabi season which impacts the overall yield.*
Soil Quality	*Floods also bring along sand and rocky materials that adversely affect the soil and render it unfit for cultivation of most crops.*
Course Change	*Rivers course change also affect life of the natives*
Ability to Work	*Floods force affected households to relocate during their period. Thereafter, efforts should be made to rebuild and protect the property. All of the above detracts from the household members of their usual vocations and interferes with their ability to work, not only in the flood period, but also in the next period.*

Fig. 3.4 (continued)

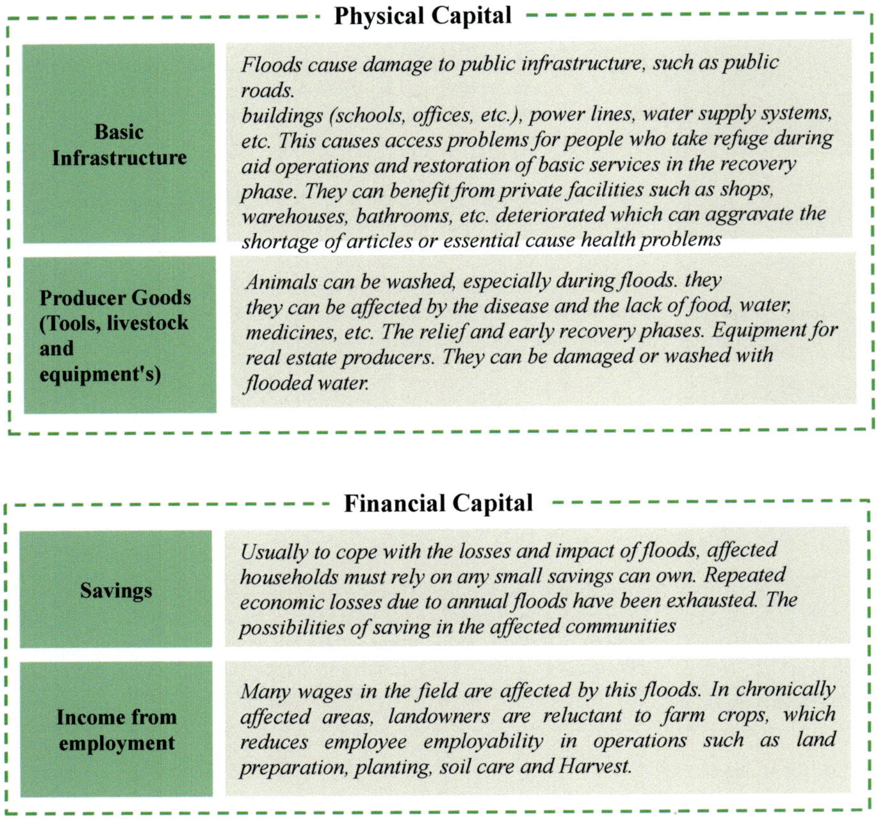

Fig. 3.5 Impact of flood on physical and financial capital

References

Agriculture and livelihood flood impact assessment in Myanmar. Ministry of Agriculture and irrigation, livestock, fisheries and rural development.

Arman, F. A., Yawson, D. O., Yengon, G. T., Odai., J. O., & Africa, E. R. A. (2010). Impact of floods on livelihoods and vulnerability of natural resources dependent communities in Northern Ghana. *Water, 2*, 130–139.

Ashley, C., & Carney, D. (1999). *Sustainable livelihoods: lessons from early experience*. DFID.

Francis, E. (1999). Learning from the local: rural livelihoods in Ditsobotla, North western Province, South Africa. *Journal of Contemporary African Studies, 17*(1), 49–73.

Islam, Md. S., Solaiman, Md., Islam, M. S., Tushar, T. R., & Kabir, M. H. (2015). Impact of flood on Char livelihoods and its adaptation techniques by the local people. *Journal of Science Research, 28*(2), 123–135.

Jelena, B., & Gourbesville, P. (2016). Resilience in flood risk management—a new communication tools. *Procedia Eng. 154*, 811–817.

Madhuri, D., Tewari., H. R., & Bhowmick, P. K. (2014). Livelihood vulnerability index analysis: an approach to study vulnerability in the context of Bihar: original research, Jamba. *Journalof Disaster Risk Studies, 6*(1), 1–13.

Mashebe, P., Jordaan, A., Zulu, A., & Kanjimba, A. (2016). The impact of flooding on the livelihood of people living in the Luhonono area in the Zambezi region, Namibia. *British Journal of Environmental Sciences, 4*(2), 1–9.

Rufat, S., Burton, C. G., & Maroof, A. S. (2015). Social vulnerability to floods: reviews of case studies and implications for measurement. *International Journal of Disaster Risk Reduction, 14*, 470–486.

Vis, M., Klijn, F., de Bruijn, K. M., & van Buuren, M. (2003). Resilience strategies for flood risk management in the Netherlands. *International Journal of River Basin Management, 2003*(1), 33–40.

Chapter 4
Recent Disasters in Kerala: Evidences from the Field

Varnav Somwal

Abstract The Western Ghats, (The Western Ghats are older than the Himalayan mountain chain and are internationally recognized as a 'hot-spot' of biological diversity. They run parallel to India's western coast and traverse Tamil Nadu, Kerala, Goa, Karnataka, Maharashtra and Gujarat.) a United Nations Educational, Scientific, Cultural Organization (UNESCO) World Heritage Site, is one of the 'hot-spots' of biological diversity in the world. The mountain range covers an area of 140,000 km^2 in a stretch of 1600 km parallel to the western coast of the Indian peninsula, traversing the states of Tamil Nadu, Goa, Kerala, Karnataka, Maharashtra and Gujarat. However, in recent times, the Western Ghats that were once covered in dense forests now has lost much of its natural beauty. Today, a large part of the range has been logged or converted to agriculture land for tea, coffee, rubber and oil palm or cleared for livestock grazing, reservoirs and roads. The growth of population around the protected areas and other forests has also led to habitat destruction, increased fragmentation, wildlife poaching and human–wildlife conflict. The biodiversity and ecosystem of the Western Ghats are threatened by pollution, mining and deforestation. Only one-third of the region is under natural vegetation, and much of this is degraded. A large part of the original natural vegetation was lost or converted to cultivated lands, coffee and tea plantations and hydroelectric reservoirs. Driven by economic development, population growth and the rising demand for power, agriculture commodities and minerals, pressures on the region's natural ecosystem are intensifying. Kerala, in particular, has had its fair share of disasters in recent times. God's own country is highly vulnerable to natural disasters and the changing climatic dynamics because of its location along the seacoast. Kerala is also one of the most densely populated Indian states making it more extremely vulnerable to damages and loses on account of disasters. Floods being the most common of natural hazard in the state. Nearly 14.5% of the state's land area is prone to floods. Between June 1 and August 18, 2018, Kerala experienced the worst floods in its history since 1924. During this period, the state received cumulative rainfall that was 42% in the excess of the normal average. According to the reports of the state government, 1259 out of 1664 villages spread across its 14 districts were affected. The devastating floods affected 5.4 million people, displaced

V. Somwal (✉)
Jindal Global Law School, O. P. Jindal Global University, Sonipat, Haryana, India
e-mail: varnav.somwal@gmail.com

V. R. Sharma and Chandrakanta (eds.), *Making India Disaster Resilient*,
https://doi.org/10.1007/978-3-031-50113-5_4

1.4 million people and took as many as 433 lives. Kerala is immensely vulnerable to disasters due to the fact that the study area is a tourist attraction as well as a home to numerous people. The area is densely populated. Bad dam management in the past has been one of a major reason for flooding in the area.

Keywords Western Ghats · Kerala disaster 2018 · Floods · Biodiversity · Cyclone · Dam management

Introduction

The world is moving at a very fast pace and there is a lot of chaos prevailing all over. This chaos is also leading to a simultaneous increase in different kinds of natural disasters which have claimed a lot of lives. A disaster is any actual event that may cause the loss of life or injury or property damage, socio-economic destruction or environmental destruction. Disaster management is the key to minimize such impacts. Disaster management is a continuous phenomenon of mitigating the impact of the disaster. It calls for collective and coordinated efforts. India is a fast-growing economy and has faced an increase in the number of disasters such as flood and cyclones. In India, according to the National Disaster Management Authority 1600 people are killed every year due to flooding.[1] In 2019, Extreme weather conditions killed more than 1500 people while floods alone claimed 850 lives.[2]

India's coast is extremely vulnerable to cyclones. With a coastline of 7517 km, the country is exposed to nearly 10% of the world's tropical cyclones.[3] Although cyclones affect the entire coast of India, the eastern coast is significantly more prone to cyclones as compared to the western coast. Among the cyclones that are formed in the Bay of Bengal, over 58% approach and cross the eastern coast. On the other hand, only 25% of the cyclones that develop over the Arabian Sea approach the western coast.[4]

[1] Chaitanya Mallapur, '*India Accounts for a Fifth of Global Deaths From Floods*', The Wire, 19 July 2018 (available at: https://thewire.in/environment/india-accounts-for-a-fifth-of-global-deaths-from-floods).

[2] Kunal Kambli, '*Top 5: Biggest Floods to Affect India in 2019*', The Weather Channel, 8 January 2020 (available at: https://weather.com/en-IN/india/news/news/2020-01-08-top-5-biggest-floods-affect-india-2019).

[3] '*Here's why cyclones hit eastern coast of India*', Deccan Herald, 29 April 2019 (available at: https://www.deccanherald.com/specials/heres-why-cyclones-hit-eastern-coast-of-india-731266.html).

[4] '*Cyclones and their impact in India*', National Cyclone Risk Mitigation Project (NCRMP).

Literature Review

Chandran 'On the ecological history of the Western Ghats' examines the great ecological interest of the Western Ghats landscape. The author traces over three millennia of forest utilization and management that have molded the Western Ghats landscape. The author presents a comprehensive history to understand the developments in the Western Ghats. Venkatesh and Rejimon Kuttappan (2018) presents a chilling overview of the loss of life and property in Kerala due to 11 straight days of tempestuous rainfall which resulted in loss of 15% of the State's GDP estimates for 2018–19. The authors analyzed data from a risk management agency, Care Rating, and found that over four million people have been affected by the floods in Kerala (most of them being laborers). Further, they analyzed data from IMD and found out that Kerala received 42% of excess rainfall in the year 2018. Venkatesh and Kuttappan (2018) observed that the Central Water Commission (CWC), India's only flood forecasting agency, doesn't have a flood forecasting system in Kerala. Not having a flood forecasting system in Kerala which is ecologically sensitive and prone to floods is depraving people of flood preparedness. Vishwa (2018) tried to explain the failure to stop degradation of Western Ghats worsened Kerala floods. He found that the Centre recognized the need to prevent further degradation of the fragile ecology of Western Ghats, it has failed to bring six states on board for urgent action. As a result, 56,825 km^2 of 'ecologically sensitive' area could not be earmarked as 'no go' zone for polluting activities and deforestation a prerequisite to save the region from constant environmental degradation.

'Kerala Post Disaster Needs Assessment: Flood and Landslides' (2018): The report recognizes Kerala's vulnerability to natural disasters and changing climatic dynamics because of its location. Floods are the most common of natural hazards in the State. The report also recognizes that 14.5% of the State's land is prone to floods. Landslides are also common in Wayanad, Kozhikode, Idukki and Kottayam districts. In 2018, Kerala experienced the worst ever floods in the history of the State since 1942. The report reviles that around 1259 out 1664 villages were affected. The floods and landslides affected 5.4 million people, displaced 1.4 million people and took 433 lives.

Objectives

The purpose of this study is to identify vulnerability to disasters from different perspectives and also to understand awareness and community preparedness in Western Ghats in general and Kerala flood 2018 in particular.

Study Area

Western Ghats in recent times has had its fair share of disasters. Cyclone Ockhi, the strongest North Indian Ocean cyclone of 2017, has wreaked havoc in Kerala, Tamil Nadu and Lakshadweep, raising questions about the role of disaster management agencies. Ockhi, termed as a 'very severe cyclonic storm' with wind speeds between 155 and 165 km per hour, hit the coastal areas of Tamil Nadu and Kerala.[5] Kerala again suffered its worst monsoon flooding in a century with more than one million people displaced, and more than 400 reported deaths.[76] Aid agencies and government groups set up more than 4000 relief camps, while rescue personnel made their way to submerged villages in helicopters and boats, bringing supplies, and evacuating those they could find.[87]

The purpose of this study is the assessment of vulnerability of cyclones, floods and landslides in Western Ghats (Fig. 4.1). Western Ghats is a mountain range that covers an area of 140,000 square km in a stretch of 1600 km parallel to the western coast of the Indian peninsula, traverse the states of Kerala, Tamil Nadu, Karnataka, Goa, Maharashtra and Gujarat.[8]

Location and Study Area

The Western Ghats region is highly subjected to natural catastrophe and chances of a Kerala like tragedy in the region cannot be ruled out. The Western Ghats' vulnerability to landslide is largely attributed to the deposits of overburden materials on the steep hill slopes. These deposits are primarily loose soil, tumbling stones and debris. The area witnesses intense rainfall, which triggers landslides and flash floods.

The floods that destroyed Kerala recently have brought into sharp focus the all-round ecological destruction.[9] The fact that there were 12 major landslides and hundreds of minor ones within a fortnight in the mountainous district of the state underscores how fragile the land has become over the decade.[10] Idukki and Wayanad

---

[5] Basudha Das, '*Cyclone Ockhi wreaks havoc on western coast: How India manages its disasters*', Business Today, 5 December 2017 (available at: https://www.businesstoday.in/current/economy-politics/cyclone-ockhi-wreaks-havoc-on-western-coast-how-india-manages-its-disasters/story/265335.html).

[6] Alan Tylor, '*Devastating Monsoon Floods in Kerala, India*', The Atlantic, 22 August 2018 (available at: https://www.theatlantic.com/photo/2018/08/devastating-monsoon-floods-in-kerala-india/568171/).

[7] *Ibid.*

[8] Western Ghats, UNESCO World Heritage Centre (available at: https://whc.unesco.org/en/list/1342/).

[9] Most of the floods occur during the monsoon season.

[10] '*Kerala floods: Man-made or nature's fury?*' The Hindu Business Line, 23 August 2018 (available at: https://www.thehindubusinessline.com/opinion/kerala-floods-man-made-or-natures-fury/article24762090.ece).

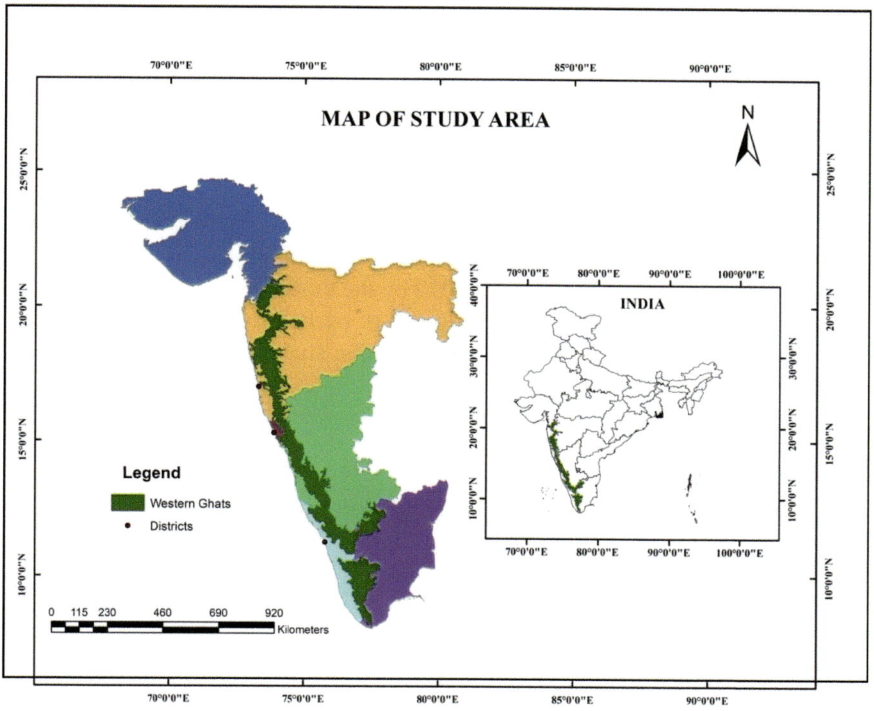

Fig. 4.1 Goa, Ratnagiri (Karnataka) and Kozhikode (Kerala). *Source by author*

districts, which fall on the Western Ghats, were completely cut off by the floods, as the roads connecting them with other. The flood killed more than 300 people.[11]

Western Ghats region is vulnerable to numerous natural disasters like floods, landslides and cyclones. According to the Cambridge Dictionary, vulnerability is the quality or state of being exposed to the possibility of being attacked or harmed, either physically or emotionally. In other words, vulnerability is the extent to which an individual, community or an area is exposed to the impact of a hazard.

Vulnerability of landslide in the monsoon season is a very common site in the Western Ghats. Apart from claiming human lives, they destroy hills and vast tracts of agriculture land. In 2018, Kerala suffered 5000 big and small landslides and landslips which lead to economic and life loss.[12]

[11] *Ibid.*

[12] '*Rebuild Kerala Development Programme*' A Resilient Recovery Policy Framework and Action Plan for Shaping Kerala's Resilient, Risk Informed Development and Recovery from 2018 Floods (available at: https://sdma.kerala.gov.in/wp-content/uploads/2020/08/RKDP-Plan-report.pdf).

Database and Methodology

The present study is based on both primary and secondary data sources. **Secondary data** and information have been collected from reports, articles, magazines and newspapers. Reports from various government and non-government organizations have been used.

Primary data has been collected through field survey including personal interviews with the help of structured questionnaire and field observations. This data was collected and compiled in March–April 2019 by the third year students from Geography department for the purpose of fulfillment of bachelor's degree from Shaheed Bhagat Singh College, University of Delhi. In this study, primary data has been collected through stratified random sampling.

The data collected from the field survey is tabulated and analyzed with suitable cartographic and statistical techniques.

Results and Discussion

The area is very prone to disasters as it poses the physical vulnerability potential for cyclone, floods, landslide and other associated climate related disasters. Understanding the socio- economic vulnerability of the region is also of paramount importance as it gives an in-depth understanding so that proper measures can be taken for management. Here in the succeeding section the discussion is on socio-economic, demographic and other related vulnerabilities.

Demographic Vulnerability

To assess the vulnerability data from 246 respondents have been collected where, 61% were male and the rest 39% were females. The ratio of females is much lower as compared to the males because the females were reluctant to communicate. **Sex composition** is an important factor while assessing disaster vulnerability because women are hard-hit by a disaster. Existing inequalities are the root cause of their vulnerability. Women are particularly vulnerable because they have fewer resources in their own right and under their own control. **Age** structure is a very important factor while assessing the social vulnerability. 45% of the total respondents belonged to the age group below 30 years, whereas 36% are from the age group 30–45 years. The remaining 14 and 4% belonged to the age group 45–60 years and above 60 years respectively. Keeping this in mind we can make an efficient and better management plan for disasters. By doing so we can reduce their vulnerability as well as the casualties caused by the disasters. **Family size** also gives us a comprehensive understanding of the socio-economic vulnerability. Out of the total families, 36%

had 2 to 4 members in their families, while 31% had 4 to 6 members. Whereas another 23% had less than 2 members and another 8% had 6 to 8 members in their families. 1%, on the other hand, had more than 8 members. The families who have more number of members will be more vulnerable than the families who have less members. To get a detailed understanding of the socio-economic vulnerability, it is crucial that we understand the composition of the respondents' family. If a respondent's family has **children and aged** persons, it will be fairly difficult for them to act promptly if a disaster occurs. Out of the total number of families, 38% were women, while 36% were males. The remaining 22 and 4% were children and aged persons respectively. **Disability** is an important actor while assessing the social vulnerability because people suffering from disability are more vulnerable to disasters as they are dependent on others for help. 97% of the respondents don't have any family member suffering from disability while the remaining 3% have a member suffering from the same.

Socio-cultural Vulnerability

Education plays a very important role in our lives as increases our awareness and makes us aware about our vulnerabilities. If a person isn't educated enough then he/she will not be much aware about anything and that will make them more vulnerable to disasters. 29% of the respondents have completed their education till the higher secondary level. 27% of them have studied till secondary level, whereas 21% of them are undergraduates. Fewer number of people have done their post-graduation or have studied higher than that. Visits to Primary School in Kozhikode (Fig. 4.2) was also done to assess the quality of education delivered.

According to the survey, 59% of the respondents fall in the income category of less than Rs. 25,000. 29% fall in the income category of Rs. 25,000 to 50,000. While 8% fall in the category of Rs. 50,000 to 75,000, followed by 5% in the category of Rs.75000 to 100,000. It is observed that majority of the people belong to low-income group. These people are more vulnerable than the people from middle- or high-income groups. As the majority of the people do not earn well or enough so they are living in houses belonging to the low- and middle-income group. Only the handful who are earning well have personal houses such as villas, mansions.

It can be observed that 47% of the respondents are living in houses belonging to the low-income group, whereas 44% are living in houses belonging to middle-income group. Only 8 per cent of the respondents live in houses belonging to high-income group. The remaining 1 per cent live is personal houses such as villas, mansions, etc. The respondents who live in houses belonging to low-income group will be the worst affected in case a landslide or a flood occurs. From the survey, it can be concluded that most of the people in the region fall in the income category of the less than 25,000, making them extremely vulnerable during the time of a disaster. Apart from that, in the study area we observed a lot of semi-pucca and kuccha houses. At the time of a disaster, these houses will be vulnerable. Most the families in the region

Fig. 4.2 School in Kozhikode, Kerala. *Source* Field survey, 2019

have children and aged people, who at the time of a disaster will be vulnerable as they require more attention and care. Only a mere 26% of the respondents have medical insurance, which is again a troubling sight. It is important that people have the necessary insurance, more importantly they are aware about the insurance policies.

Out of the total respondents, 83% are the **permanent residents** of the study area whereas the rest 10 and 7% are **tourists and migrants** respectively. Permanent residents are more vulnerable compared to tourists and migrants because during the time of a disaster it is possible that they can lose their livelihood. On the other hand, tourists and migrants can always go back to their native place.

The study area is dominated by the **Hindu** population. It is observed that 50% of the total respondents were Hindus. 44% of the respondents are Muslims, whereas the remaining 6% falls under the category of others which includes Christians and Parsis. Religion is not used as any important factor while assessing vulnerability but it sure is used for discrimination. During the time of rescue and relief measures, Muslim population is often neglected due to their religion. Hence, Muslims can be more vulnerable than any other religious groups.

Economic Vulnerability

The type of **construction** found in an area helps us in assessing the vulnerability of the area in case of a disaster. It can be seen that only 61% of the construction in the area is pucca whereas 28% is semi-pucca. The remaining 11% is kutcha. In case of a disaster, the most vulnerable construction will be kuccha as it can be destroyed easily during a flood or a landslide.

Transport facilities have always been capacity in case of a disaster. It helps in reaching the emergency facilities like hospitals, and police station faster. According to the survey, 57% of the respondents own a vehicle whereas the rest 43% don't. The people who don't own any vehicle are more vulnerable during a disaster as they will be always dependent on their neighbors or on the public transport in case of an emergency.

According to the survey, 54% of the respondents don't own any land in the area whereas the remaining 46% of the respondents have a **property** in the area. The people who use their land for farming and plantation are more vulnerable as it can be easily damaged in case a cyclone occurs. Also, the people that own land or property in the area more vulnerable as their property could be damaged during the time of a disaster.

Insurance, whether it is life insurance, medical insurance or property insurance, is always seen as a capacity in case of a disaster. According to the survey, 46 and 26% of the respondents have vehicle as well as medical insurance respectively. 12% of the respondents have general insurance such as LIC, whereas 11% have insured their houses. The remaining 2% have crop insurance. So, if a disaster occurs, then these people will be able to cover the losses through these insurances.

Domesticated **animals** such as livestock or pets are an important factor of assessing the social vulnerability. Animals are dependent on humans, so they are vulnerable in case of a disasters such as flood or cyclone. Out of the total respondents, majority of the respondents which is 76% do not have any domesticated animals. While the rest 24% have domesticated animals.

Flood is one of the disasters which is vulnerable in Western Ghats Flood can be caused by heavy rainfall, inadequate capacity of rivers to carry the high flood discharge, deficit drainage to carry away the rainwater quickly to streams/rivers. At present, mindless cutting of trees on a mammoth scale is taking place in Western Ghats. Because of this reason, the land of Western Ghats is unable to retain the water and losing its strength. Flood in Western Ghats are caused by climate change, and the disaster is exacerbated by bad developmental practices. Flood refers to overflow of water that can submerge a land causing infrastructural damage and also loss of life.

According to the survey, 88% of the total respondents are **aware** about flood as a disaster. The remaining 12% are not aware about it may be because the government has never conducted any training programs in the area, or this area has never been hit by a flood before.

Out of the total respondents, 80% of the respondents have witnessed or experienced a flood in their area and the other 20% of the respondents have never witnessed or experienced it. The majority of the respondents who experienced the flood belonged to the Kozhikode district of Kerala, and some were from Mandavi Beach, Ratnagiri.

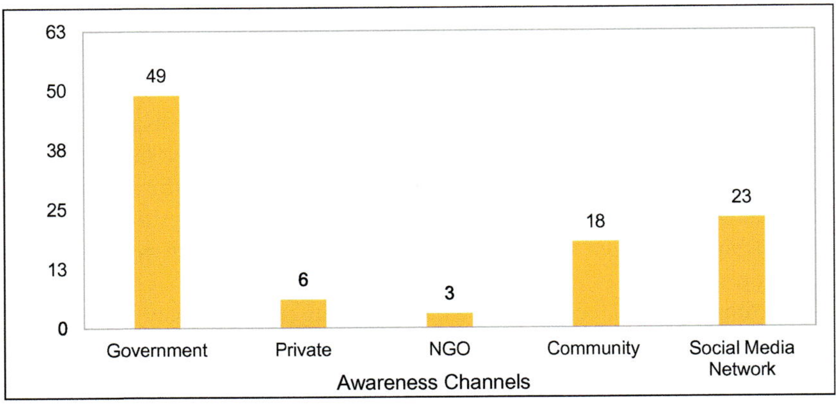

Fig. 4.3 Awareness channels. *Source* Field Survey, 2019

The major awareness channel (Fig. 4.3) which are available to spread awareness about flood are through the government agencies, private agencies, community, social media and many more. According to the survey, 49% of the respondents agreed that the government agencies are proved to be helpful during a disaster, whereas 23% of the respondents agreed social media network helps them being aware about disaster-related emergencies. 18% answered that community is more helpful. Very few of the respondents agreed private agencies and NGOs proved to be the most helpful.

Accessibility of **emergency facilities** such as hospitals, police stations, disaster relief, etc. is one of the capacities in case of a disaster. According to 54% of the respondents, the most accessible emergency facility is the police station, whereas for the other 35% hospital is the most accessible. While for the remaining disaster relief center is the more accessible. Police is the study area has a strong hold which is one of the reasons why it is accessible to a majority of people. Also, police stations are for every citizen and are easily accessible.

The major reasons because of which a flood occurs include heavy rainfall, overflowing of rivers, poor drainage system, failure of dams and many more (Fig. 4.4). Majority, which is 44%, of the respondents believe flood occurs due to heavy rainfall. The other 16 and 14% believes it is because of overflowing of rivers and failure of dams respectively. The remaining few of the respondents believe it is due to poor drainage, lack of vegetation and lack of initiatives by the administration.

The most vulnerable area in case of a flood is the area near the coastline and riverbanks. The coastline and riverbanks being closer to the sea, ocean or rivers make it more vulnerable than the other areas.

According to the survey, 29% of the respondents believe the children are the most vulnerable section of the society in case of a disaster whereas the other 26% believe the people suffering from disabilities are more vulnerable to disasters. The remaining 25 and 2% think the aged population and women are more vulnerable respectively. This information gives us an idea about the different sections of the society that are vulnerable during a disaster. It is difficult for old people and children to maneuver

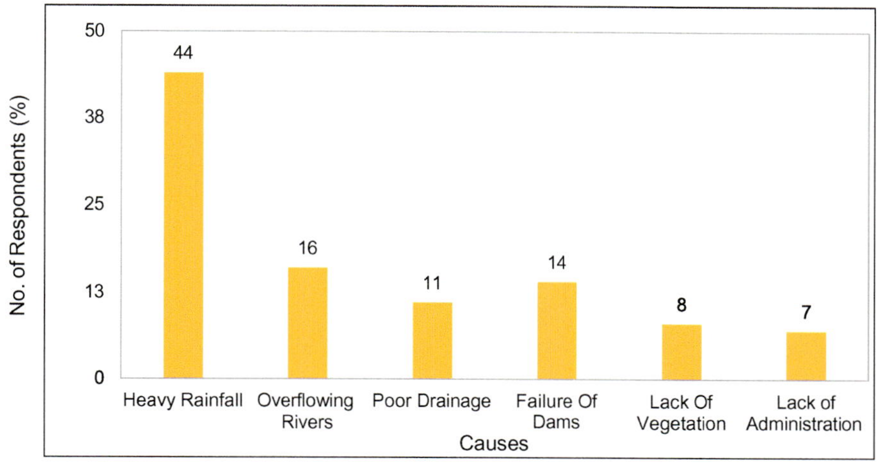

Fig. 4.4 Causes of flood. *Source* Field Survey, 2019

during the time of a disaster as they are dependent on other people for help. Hence, they become most vulnerable during a disaster. Figure 4.5 represents the Vulnerable Sections of the Society based on primary survey.

As we can see, 37 and 36% of the respondents believe low-income group and middle-**income** group are the most vulnerable economic group of the society, respectively. The major reason for this is the asset ownership, the rich have all the resources and assets that help them to recovery the losses caused by the disaster, on the other hand the poor rarely own any assets and if a disaster occurs the few assets they own might get destroyed thereby making them vulnerable.

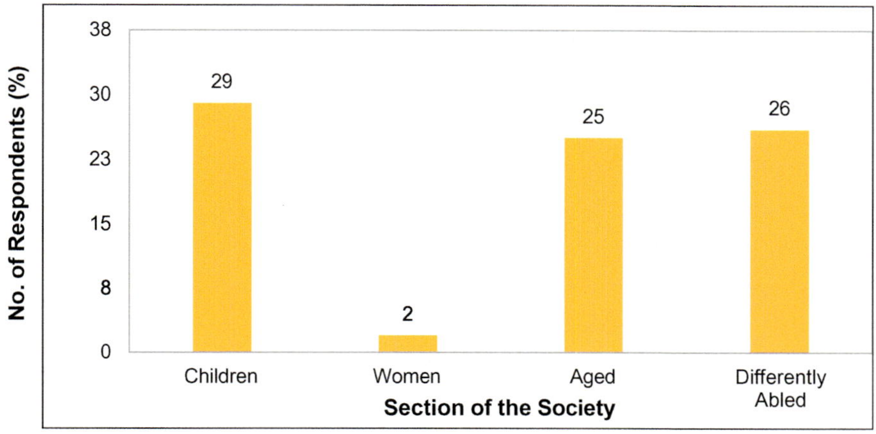

Fig. 4.5 Vulnerable section of the society. *Source* Field Survey, 2019

Another 22% people feel the high-income group is more vulnerable. The remaining 5% voted for the option none because they believe there isn't any one particular group which is more vulnerable than the other because a disaster won't affect a particular group rather it will affect everyone equally.

Recently, the state of Kerala was hit by a flood which badly affected more than half of the districts of the state. 38% of the respondents agreed that their area was highly affected by the flood whereas another 38% said their area was moderately affected. The rest of the respondents who denied their area being affected were from other parts of the study area.

The areas which were affected by the flood faced some serious damages in terms of loss of life, property and livestock. 36% of the respondents agreed they faced some damages to their property as well as there was a loss of their livestock. But due to God's grace, the study area didn't have any loss of life. In case of landslide, 64% of the respondents denied loss of any sort as the landslide never occurred in their area.

The government provides relief assistance to the people who were affected by the disaster. 72% of the respondents agreed that they were provided with relief assistance, whereas 28% of the respondents denied getting any relief assistance because most of them were not affected by the disaster as it did not occur in their area and few of them did not face any major losses, injuries or loss of life.

The relief assistance which was offered to the affected people included medical assistance, food, shelter, transport assistance, money, etc. (Fig. 4.6), majority of which is 28% of the respondents who were affected received food, whereas 27% received medical assistance. Another 21 and 18% of the respondents were given shelter and transport assistance respectively. The remaining 6% received money to recover their losses.

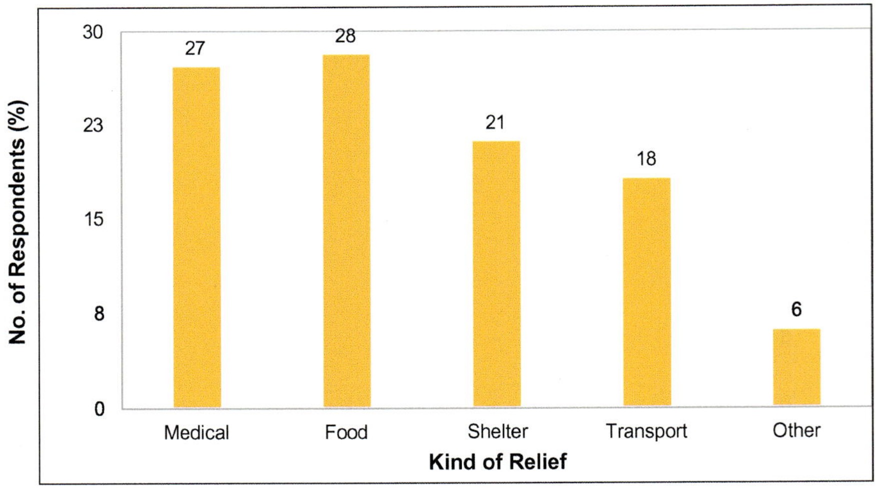

Fig. 4.6 Relief assistance offered to the respondents. *Source* Field Survey, 2019

The respondents were asked that during the time of the disaster were they offered any **compensation** or not. To which 66% of the respondents said yes that they were offered compensation for their losses; however, 34% of the people said they weren't offered any compensation. Some of the people in this category said no because they weren't affected, and hence, no compensation was offered.

Conclusion

Disaster vulnerability is a major issue that we sometimes purposely tend to ignore; however, this taken for granted attitude can cost lives and it has in the past. The vulnerability of a region increases if the people in the area are not aware and not prepared. Awareness level dictates to what extent a person can be vulnerable to a disaster. Majority of the people belonging to the study area were aware about flood and landslide as a disaster but most of them did not know what to do in case a flood or landslide occurs. Also, as we can see the respondents agreed that the most vulnerable section of the society includes aged population, children, disabled people and low-income group. However, if a flood and landslide with great intensity hits the area then straightaway everyone becomes vulnerable. We can conclude by saying the study area is highly vulnerable to floods and landslides due to the fact that the study area is a tourist destination as well as a home to numerous people. In the study area, it is clear that government is in the forefront of all the relief assistance that was provided to the victims. Also, in the study area community plays a vital role in case of a disaster. During the time of a disaster, communities' spaces like temples, mosques, schools and open fields are used by the people to gather there. The region has divided opinions on the role and genuineness of media reports. Some people feel that media is swayed by political agenda and acts as a spokesperson for some or the other political party. There are a few loopholes according to the respondents in disaster management in the region that should be corrected for better and efficient use of resources such as lack of awareness and ignorance.

Acknowledgements I would like to take this opportunity to express my heartfelt gratitude to all the teachers, who helped and guided me on the subject of my project, without whom this project would not have been completed. I am extremely grateful to Dr. Amrita Bajaj, Dr. Santosh Kumar, Dr. Ram Lal and students that accompanied me on my field trip. Further, I am grateful to all those people who spared their valuable time to fill the questionnaires based on the project topic. I would also like to thank the Geography Department of Shaheed Bhagat Singh College (M), who provided all the resources required for the project.

References

Kalyan Chakravarthy, Y. (2013). A study on incidence of heavy rainfall events over India during Summer Monsoon Season. wmo.int.

Kuriakose, S. L., Sankar, G. &Muraleedharan, C. (2008). History of landslide susceptibility and a chorology of landslide-prone areas in the Western Ghats of Kerala, India. *Environmental Geology, LVII*, 1553–1568.

Mandal, G. S., & Mohapatra, M. (2010). Cyclone Hazard Prone Districts of India: A Report. *National Disaster Management Authority, Government of India, Indian Meteorological Department Mausam Bhawan New Delhi*, 10–28 (2010). Indian Metrological Department.

Mangesh, G., et al. (2016). Cyclone Phyan-induced Plankton Community succession in the coastal waters off Goa, India. *National Institute of Oceanography, CXI*, 1091–1097 Cyclone Ockhi-report.

Rajendran, K., Kitoh, A., Srinivasan, J., Mizuta, R., & Krishnan, R. (2012). Monsoon circulation interaction with Western Ghats orography under changing climate. *Theoretical and Applied Climatology, CX*, 555–571.

https://ndma.gov.in/en/

https://www.timesnownews.com/india/article/cyclone-ockhi-600-fishermen-rescued-from-south-india-alert-for-fishermen-along-maharashtra-gujarat-coast/138525

https://sandrp.in/2013/07/26/climate-change-in-western-ghats-4x4-report-and-beyond/

https://whc.unesco.org/en/list/1342

https://www.gktoday.in/gk/western-ghats/

Chapter 5
Floodplain Mapping Using HECRAS Model and Geospatial techniques—A Case Study of Varanasi City

Vishal Mishra, Anuj Tiwari, and Prabuddh K. Mishra

"Never trust the Rivers-an Ancient Sanskrit Couplet"

Abstract Varanasi, an ancient city of India, lies on banks of River Ganga, one of the largest rivers of the world. Varanasi is proposed to be developed into a Smart City in the first phase of Indian Smart City Mission. The resilience of a city is its ability to persevere in the face of an emergency, so it can continue functioning despite serious challenges. Flood is a frequent phenomenon for Varanasi, but in recent years, it was deluged. The flood-risk zones mapping makes the first step of flood control measures. In this chapter, a one-dimensional hydraulic model, i.e. HEC-RAS has been integrated with Remote sensing and GIS techniques to map the floodplain zones of Ganga River located around the city of Varanasi. The pre-processing includes geometry setup such as digitizing the stream centerline, right bank and the left bank of the river. SRTM DEM is used in ArcGIS software extension HEC-GeoRAS 4.3 for pre-processing of GIS data, i.e. derivation of the channel geometry for input to HEC-RAS. Next, HEC-RAS simulations were performed to generate a water surface profile for a given design flood condition. Finally, simulated results have been imported in ArcMap 10.1 and being overlaid with the DEM for obtaining flood risk zones and flood inundation maps. Floodplain map analysis shows that more than 300% area has a probability of inundation as compared to the normal flow of the river. The results

V. Mishra
Department of Civil Engineering, Indian Institute of Technology, Roorkee, Uttarakhand, India

A. Tiwari
Geomatics Section, Department of Civil Engineering, Indian Institute of Technology, Roorkee, Uttrakhand, India

P. K. Mishra (✉)
Department of Geography, Shivaji College, University of Delhi, New Delhi, India
e-mail: prab19@gmail.com

V. R. Sharma and Chandrakanta (eds.), *Making India Disaster Resilient*,
https://doi.org/10.1007/978-3-031-50113-5_5

of this research will be helpful in developing a smart Varanasi, which will offer a safe habitation to its residents.

Keywords Floods · HEC-RAS · Hydrological model · Flood inundation

Introduction

A flood is defined as an incident when waters from a river, lake, ocean, or other surface water feature rise above regular boundaries as per ASCE Handbook of Hydrology, 1996. It is a general phenomenon that in the rainy season the discharge in rivers increases with the excess rainfall and the river water crosses its boundary. Floodplain is the relatively flat and low-lying area bordering the river, which floodwater inundates when it breaks or spills over the riverbanks (Kumar, 2016). In the context of floods, the lower Gangetic Plains of Northern India are very vulnerable. According to 11th plan working group, the total flood-prone areas in India are 45.640 million hectares out of that about 7.340 million hectares lie in Uttar Pradesh covering 30.16% of its total geographical area (Planning Commission, 2011). In U.P., annual estimated loss due to floods is 432 crores (Uttar Pradesh State Disaster Management Authority). Alone in 2016 eight districts of eastern UP witnessed floods, whose estimated loss of crop was over 1216 lakh (The Indian Express, 2016).

Varanasi (also known as Benaras), an ancient city, lying on the banks of river Ganga sacred to millions of Hindus boasts of the existence of nearly twenty-five hundred years (Wood, 2015; Eck, 1999). Varanasi is a populated city and tourist hub owing to its popularity, sacredness and religious importance (Kumar et al., 2021). Varanasi has been shortlisted Ministry of Urban Development (MoUD) of India as one of the 98 smart cities for the 'Smart City Challenge'. A smart city must have some core infrastructure elements; some of them are the sustainable environment, safety and security of citizens, and sanitation (Smart City: Mission Statement and Guidelines, 2015). These are severely affected by floods in case of Varanasi. Even the land-use pattern of the Varanasi gets affected by the floods. Varanasi is periodically flooded in monsoon season. Some of the recent events are summarized in graphs shown in Fig. 5.3. Varanasi has seen the worst floods in 2013 and 2016 (Figs. 5.1 and 5.2). Also, in 2014 and 2015, the city was gripped by a flood which resulted in some inundation problem, but it did not have any adverse effects. The four graphs plotted for the years 2013 to 2016 for daily water level, in between 29 July and 5 September each year. The danger flood level for Varanasi is 71.26 m, and the river crossed it and reached to a maximum of 72.63 m and 72.56 in 2013 and 2016 respectively (UP Irrigation).

Fig. 5.1 Submerged buildings due to floods in different years at Varanasi

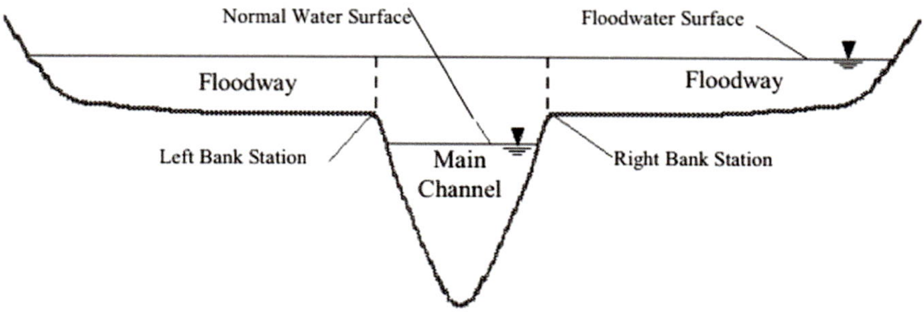

Fig. 5.2 A HEC-RAS cross section

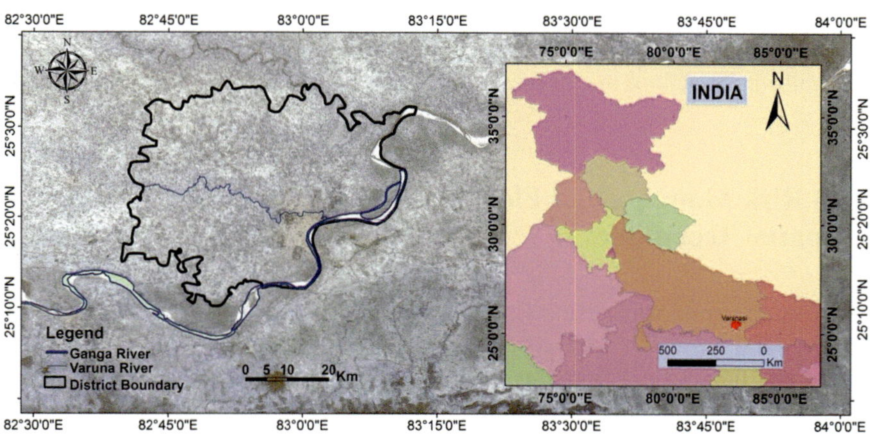

Fig. 5.3 Map of the study area

Need of Study

The Government of India has announced the **SMART** cities project for over 100 cities and **AMRUT** project for 500 cities. Flood protection is one of the thrust areas for these cities. Varanasi as lying along the confluence of Varuna and Ganga is very much prone to flood.

A detailed study of floodplains is not economically feasible for developing a nation of India. According to FEMA (2006), "a riverine study typically costs $5000 to $10,000 per mile of the stream that is to be mapped". For developing nation, cost factor makes the development of traditional flood inundation maps virtually unaffordable for these regions. Thus, there is a need for the development of an inexpensive, fast approach to flood inundation mapping, especially for developing countries such as India.

The Inter-Governmental Panel on Climate Change (IPCC) has identified that there will be an increase in the number of extreme flood events in the future as a consequence of climate change (Pachauri et al., 2014). It mentions the risk of inland flooding, which is of greater risk to human life and assets in urban areas. It is needed to be ensured that our urban areas do not get disrupted due to natural hydrological disasters like cyclones and flooding caused by severe rainfall as was experienced by Chennai, Gurugram, Bengaluru, Hyderabad, etc. during 2015–16.

Floodplain mapping is of substantial usage for land developers, urban planners and engineers. It also forms the basis for flood management programs and flood-risk mapping (U.S. Army Corps of Engineers, 2016). In developed countries, for example, in the European Union, The European Union Floods Directive has made the establishment of flood maps necessary for high-risk areas (Civil Protection, 2007).

Flood Risk Mapping and Floodplain Mapping Using Geospatial Techniques

Floodplain mapping using remote sensing data integrated with the hydrodynamic model is not a very old concept though the satellite remote sensing has been used for mapping flood extent from the 1970s. Application of physical models was the first step in the development of hydraulic floodplain modelling. These small-scale models of the river systems simulate different scenarios for predicting the outcome of an event. Starting from the 1940s, the need for faster, easier methods of hydraulic modelling were felt. Hand computations were too time-consuming, and physical models were costly, impractical, and too case specific. By the early 1960s, the first hydraulic modelling programs were developed which answered the problems of existing methods. The GIS also developed independently until the late 1980s. Remote sensing technology and GIS has become the key tool for flood monitoring and management from a few decades. Beavers (1994) pioneered some of the first work which was based on the integration of hydraulic modelling and GIS. Goodchild et al. (1996) gave a good collection of articles for performing environmental modelling, which included hydrologic modelling using GIS. Philip et al. (1998) used different GIS-derived models to derive potential inundation surfaces and assessed their prediction using SAR imagery. The integration of remote sensing and GIS has made a substantial contribution as compared to the ground-based observations for identification and monitoring flood extents which helps in disaster response and mitigation.

While dealing with numerical flood modelling, the bottleneck that one encounters is the flow-complexity in the floodplains, and especially regarding the depiction of small-scale topographic features and hydrodynamic control exerted over the flow. Hydrodynamic modelling requires meticulous topographic features for describing the inherent complexity of the river system within urban areas. Recent advances in Geospatial technologies widened the range of available terrain data, and now higher-resolution digital elevation models (DEMs) are available for hydraulic modelling. Accuracy of flood estimation depends primarily on the resolution of DEM used in flat terrain (Sanyal & Lu, 2004; Casas et al., 2006) and the steepness of terrain slopes perpendicular to the river flow direction (Brandt, 2016). Brandt and Lim (2012), Saksena (2015) have analysed the effect of DEM resolution accuracy on flood inundation mapping. Cook and Merwade (2009) studied the effect of different types of topographic (DEM) data, geometric configuration and modelling approach of flood inundation mapping.

The various flood risk-mapping programs around the world use a variety of river modelling methods to estimate flood depths, velocities and extents. Real or imaginary rainfall scenarios are input into a network of rainfall-runoff models representing hydrological sub-units, whose outputs provide the inputs to a model for the river network and significant features such as floodplains and reservoirs. The model detail may incorporate all significant river controls, such as bridges, culverts, gates, defences and other features, as well as the main details of the floodplain, using construction and topographic information obtained from conventional survey and remote sensing techniques (Optical Satellite Imagery, Synthetic Aperture Radar, or Light Detection and Ranging) (Sene, 2008).

Hutti et al. (2014) conducted their study on Bennihalla River. Hutti et al., 2014 created a flood hazard zonation map. Madhusudhan et al. 2016 used HEC-RAS with SWAT model for estimating flood submergence area. Saini et al. (2016) performed the flood-risk assessment in urban areas of Ambala city using HEC-RAS.

Multi-criteria decision analysis is also a Geospatial Technique other than GIS integrated numerical modelling which is used for Urban Flood Hazard Zonation. It is better for the first stage for the studying floods (Fernández & Lutz, 2010). Kia et al. (2012) used artificial neural network for flood modelling.

Types of River Models

There are many ways in which flood inundation modelling can be performed. On the basis of dimensionality, these models traverse a vast spectrum ranging from zero-dimension models which utilizes a water level versus flow rate rating curve; one-dimensional (1D) models such as the well-known HEC-RAS code, two-dimensional (2D) models which have their roots in the shallow-water equations; Navier–Stokes equations based three-dimensional (3D) models and hybrids approaches (Sanders, 2007). For this study, we have selected a 1D approach to modelling.

1-D Models

HEC-RAS and MIKE 11 are widely applied, well-proven 1D River modelling packages.

2-D Models

Two-dimensional (2D) flood inundation models have been a hot spot of research. The basic assumption of the 2D modelling is that there is hydrostatic vertical pressure distribution, and furthermore that turbulent momentum dissipation can be modelled with bottom stresses as per a fourth power drag law. There are many 2-D models such as Flo-2D, TUFLOW, RMA4 and CCHE2D-FLOOD (Qi et al., 2013), which can be used for flood modelling.

HEC-RAS

With the starting of computer era in the twentieth century, complex mathematical modelling of physical phenomenon such as flood became possible. Programs initially used geometric profile of selected cross sections of the river and floodplain and the analysis of bridge effects was computed manually.

In 1966, U.S. Army Corps of Engineers (USACE) released a FORTRAN program named "Hydrologic Engineering Center (HEC) program". In 1968, this program was revised, expanded, and rereleased as HEC-2, Water Surface Profiles.

In 1991, the HEC started working on the development of a replacement program since there were difficulties in inputting and outputting the data in HEC-2. In 1995, it released the very first version of HEC-RAS (River Analysis System). Since then HEC-RAS has been updated from time to time with current version HEC-RAS 5.0.3 released in 2015.

HEC-RAS is a 1-D steady flow model, intended for computing the water surface profile. The HEC-RAS system can model supercritical, subcritical, and mixed-flow regimes for drainage pattern consisting of a full network of channels, a dendritic system, or a single river reach. HEC-RAS is used to simulate floodplain, sediment and sediment transport, water quality, dam break, blocking bridges, scour phenomenon, and ice-covered river (ShahiriParsa et al., 2016). The results of these models are typically applied in floodplain management and flood insurance studies to evaluate the effects of flood inundation (Handbook, 1996).

Many of these software tools follow a similar theme for linking hydraulic modelling and GIS. Cross-sectional parameters are extracted from a terrain model and imported into a hydraulic model. After the user executes the hydraulic model, the output is processed for display and analysis in a GIS. However, this technique requires a high-resolution DTM as the source of input cross-sectional descriptions.

Many aspects of hydrologic engineering, including rainfall-runoff analysis; river hydraulics; simulation of reservoir system; flood damage analysis; and real-time river forecasting for reservoir operations are covered by the HEC-RAS.

The primary advantage of one-dimensional models is that they are very accurate in their approach of describing the water levels and discharge along with the longitudinal profile of river at the selected cross section of the river (Thakur & Sumangala, 2006) The one-dimensional hydrodynamic modelling is based on the theory of open channel flow. St. Venant equations for unsteady open channel flow form the base for the 1D hydrodynamic model. Saint Venant is given by the expression

$$\frac{\delta Q}{\delta t} + \frac{\delta\left(\frac{\alpha Q^2}{A}\right)}{\delta x} + gA\frac{\delta h}{\delta x} + \frac{gQ|Q|}{C^2AR} = 0$$

$$\frac{\delta Q}{\delta x} + \frac{\delta A}{\delta t} = q$$

where

- Q stands for discharge (m^3/s),
- q lateral inflow, (m^2/s),
- A flow area, (m^2),
- h stage above datum, (m),
- R resistance/hydraulic radius, (m),
- α coefficient of the momentum distribution.
- C coefficient of Chezy resistance, ($m^{1/2}/s$),

There are general assumptions associated with Saint Venant equations.

- It considers fluid incompressible and homogeneous.
- It is applicable for one-dimensional flow and considers uniform depth across the cross section.
- The bottom slope is a small and small variation in longitudinal geometry.
- Hydrostatic pressure is assumed.

To perform the floodway analysis, the user must first determine the existing natural conditions (100-year) floodwater surface profile for the river reach. The HEC-RAS has the ability to compute water profiles rapidly for several different types of hydraulic systems.

HEC-RAS, for hydraulic analysis, utilizes several parameters as the input of the stream channel geometry and water flow. These parameters are used to establish a series of cross sections along the stream. Each cross section is divided into three segments, which are left floodway, main channel, and right floodway, as illustrated in Fig. 5.3.

Study Area

Varanasi is located in the Northern Gangetic Plains of Uttar Pradesh (India) (Fig. 5.4). The district boundaries extend between 25° 10′ N to 25° 37′ N latitude and 82° 39′ E to 83° 10′ E longitude. The area is covered by the Survey of India toposheet No.63 K/ 5,63 k/15. The city of Varanasi is situated on the left bank of the river. Geographically, the Ganga defines the borders of Varanasi through its floodplain. Varanasi is plagued by floods annually, and it is located on the high ground above the dangers of the roaring rivers. On the opposite side of Varanasi from the Ganges lies the Varuna River, this also defines the shape of the city by a floodplain. The Gomati River, a meandering river flows along its northern borders. A small local rain-fed stream Assi also flows in the southern side of Varanasi. The Ganga, Varuna and Assi are the three natural streams, which receive the stormwater of Varanasi city. The east bank is a floodplain flooded with water as the Ganga swells in monsoon leaving layers of silt and sand deposition along the convex shoreline.

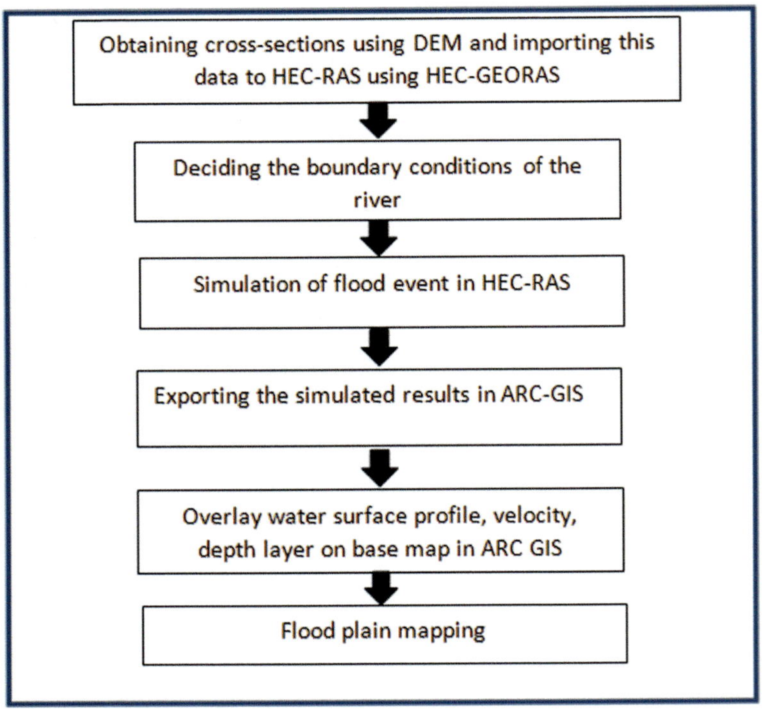

Fig. 5.4 Flowchart of hydraulic modelling of a river

Data

The following input data are required to conduct a flood routing and water level simulation: channel geometry; boundary conditions; tributary inflows; and channel resistance. In addition, it is desirable to have discharge and water level data at key points along the modelled reach to assess HEC-RAS model performance.

SRTM DEM of 30 m resolution has been used for the study. Baral et al. (2016) studied different DEM for Indian region. The inference drawn from their study is that the mean and standard deviation of SRTM-DEM is almost close to the values derived from the topographical map as compared to ASTER GDEM and CARTOSAT DEM. Topographic maps and Google Earth have been used for digitizing the channel banks. Manning's Roughness Coefficient, i.e. Manning's *n,* which represents the roughness or friction of the channel/floodplains offered to flow, has been referred from literature (Te Chow, 1959).

Methodology

The flowchart of the methodology adopted is shown in Fig. 5.5. Bank lines are digitized along with the approximate location of the top-of-bank on both sides of all of the streams following the topographic map and Google Earth. The geometric data required to define in HEC-RAS includes: Cross-sectional data, Reach lengths which are measured between cross sections) and the Stream junction information (Reach lengths across junctions and tributary angles). HEC-RAS requires the bank stations to be specified for each cross section. By drawing the bank lines that intersect the cross sections, the Geo-RAS utility can determine where that bank station falls on each cross section. The HEC-RAS model was setup using 53 cross sections to provide the channel width and bed elevation. These sections are extended over both sides of the channel using the SRTM DEM to provide for floodplain topography. The channels so generated in ArcMap are shown in Fig. 5.6. The cross sections are generated at an interval of 500 m. In order to move into the GIS environment, the output data from HEC-RAS is extracted. HEC-GeoRAS is the bridge between GIS and hydraulic model HEC-RAS. The approach for floodplain mapping in ArcView generates a perspective of the floodplain far superior to that generated by the limited visualization tools offered in HEC-RAS.

Entering Boundary Conditions

The type of flow entered depends upon the type of analysis to be performed. In our case, the steady flow analysis is simulated. The discharge of 12,000 cu m/s is taken for the worst flood condition. It includes the flow data, number of profiles computed

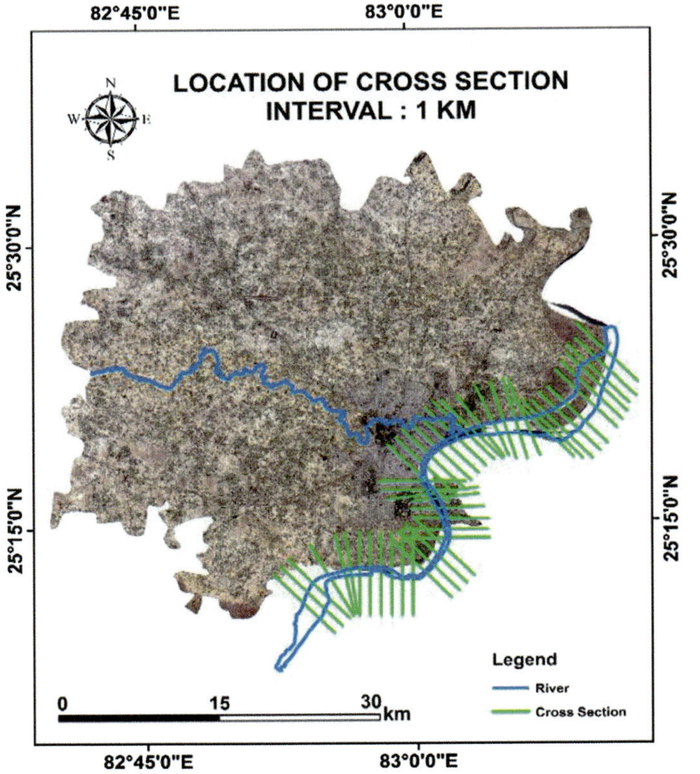

Fig. 5.5 Location of cross sections generated in Arc-map

and river system boundary conditions. Boundary conditions are required to perform the calculations. In this study, the normal depth is used as a boundary condition and steady flow data.

Results

The objective of the study was to obtain the flood-risk zones and the flood inundation maps of Ganga river in Varanasi using a hydraulic model, i.e. HEC-RAS. After giving all the input parameters to the software for the computation, the output in terms of the table and the graphs are obtained which includes: The cross-sectional output the value of ground elevation, velocity head, water surface elevation, total velocity, max channel depth, losses, average velocity, wetted perimeter, etc. This output is presented in Figs. 5.7 and 5.8 which contains flood inundation map, cross-sectional profile and rating curve respectively. The maximum water level which has been observed is 44.79 m.

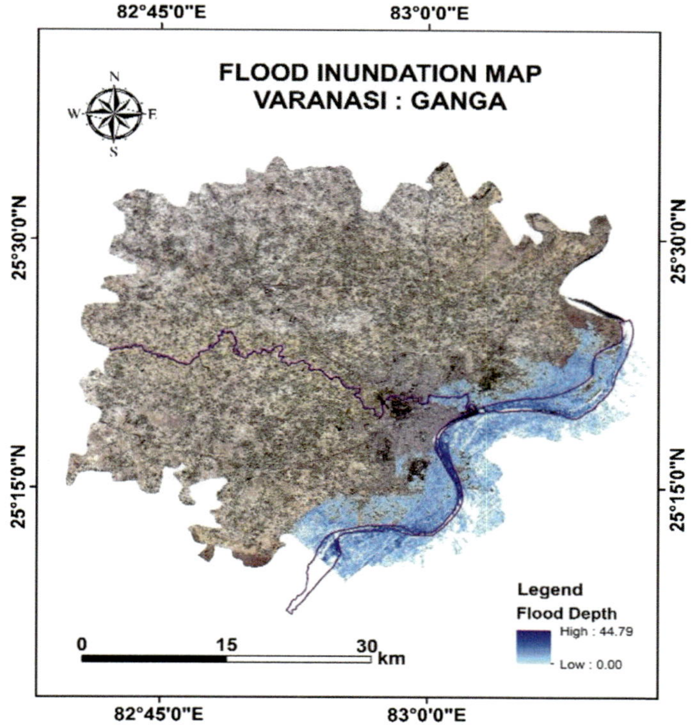

Fig. 5.6 Flood inundation map of Varanasi

Conclusion and Future Work

We draw the following conclusion from our study:

1. The HEC-RAS provides the flood profile for the worst flood intensity. This profile will help in shaping the necessary flood disaster mitigation policies and plans.
2. The accuracy of plotting the flood profile is dependent upon the DEM used.
3. Flood modelling using HEC-RAS is effective tool for hydraulic study, handling of disaster management measures.

The accuracy of hydraulic modelling depends upon the DEM resolution; hence this study can be carried out with more accurate elevation data obtained from LiDAR technology. Also, consideration of unsteady flow in the model will improve the water level and inundation results. HEC-RAS simulates different water surface profiles of different T-year floods scenario, which can be used for further flood zone analysis.

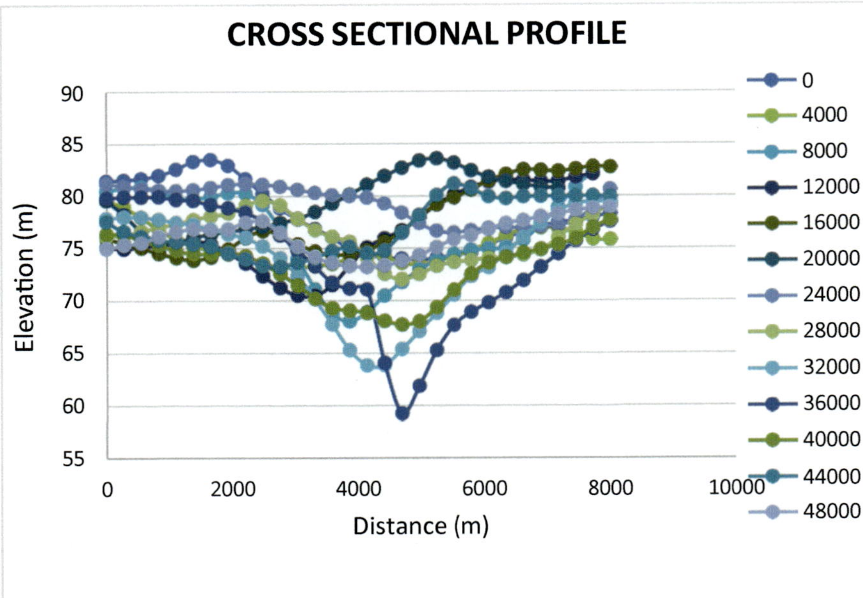

Fig. 5.7 Cross-sectional elevation profile generated by HEC-RAS using SRTM DEM

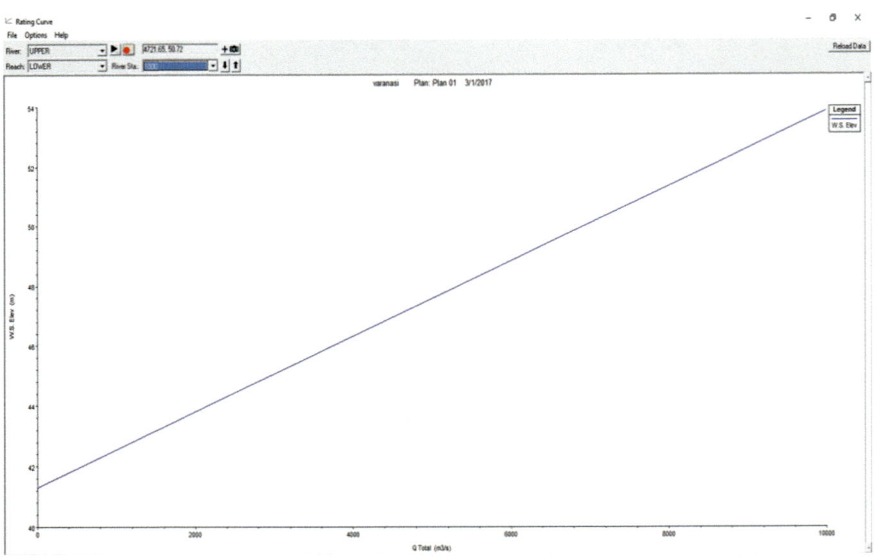

Fig. 5.8 Rating curve of a section

References

Baral, S. S., Das, J., Saraf, A. K., Borgohain, S., & Singh, G. (2016). Comparison of Cartosat, ASTER and SRTM DEMs of different terrains. *Asian Journal of Geoinformatics, 16*(1).

Beavers, M. A. (1994). Floodplain determination using HEC-2 and geographic information systems. Doctoral dissertation, University of Texas at Austin.

Brandt, S. A., Lim, N. J. (2012, September 5–7). Importance of river bank and floodplain slopes on the accuracy of flood inundation mapping. In: *International conference on fluvial hydraulics. River flow 2012* (pp. 1015–1020). CRC Press/Balkema (Taylor & Francis).

Brandt, S. A. (2016). Modelling and visualizing uncertainties of flood boundary delineation: Algorithm for slope and DEM resolution dependencies of 1D hydraulic models. *Stochastic Environmental Research and Risk Assessment, 30*(6), 1677–1690.

Casas, A., Benito, G., Thorndycraft, V. R., & Rico, M. (2006). The topographic data source of digital terrain models as a key element in the accuracy of hydraulic flood modelling. *Earth Surface Processes and Landforms: the Journal of the British Geomorphological Research Group, 31*(4), 444–456.

Civil Protection. (2007). Directive 2007/60/EC of the European Parliament and of the Council of 23 October 2007 on the assessment and management of flood risks.

Cook, A., & Merwade, V. (2009). Effect of topographic data, geometric configuration and modeling approach on flood inundation mapping. *Journal of Hydrology, 377*(1–2), 131–142.

Eck, D. L. (1999). *Banaras, city of light.* Columbia University Press.

FEMA. (2006). Floodplain management requirements: A study guide and desk reference for local officials. Available at http://www.fema.gov/medialibrary-data/20130726-1539-20490-0241/nfip_sg_unit_3.pdf. Accessed on 10/2/2017.

Fernández, D. S. & Lutz, M. A. (2010). Urban flood hazard zoning in Tucumán province Argentina using GIS and multicriteria decision analysis. *Engineering Geology, 111*(1–4) 90–98. 10.1016/j.enggeo.2009.12.006.

Goodchild, M. F., Steyaert, L. T., & Parks, B. O. et al. (Eds.) (1996). *GIS and environmental modeling: progress and research issues.* Wiley.

Handbook, H. (1996). *ASCE manuals on engineering practice No. 28.*

Hutti, B., Noor Monsoor, C. M., Mahesh Bilwa, L. (2014). Flood hazard zonation mapping using geoinformatics technology; Bennihalla Basin, Gadag and Dharwad District, Karnataka, India. *International Journal of Engineering Research & Technology*, ISSSN, 2278-0181.

Kevin, S. (2010). *Flood warning forecasting and emergency response.* Heidelberg, Berlin: Springer Berlin.

Kia, M. B., et al. (2012). An artificial neural network model for flood simulation using GIS: Johor River Basin, Malaysia. *Environmental Earth Sciences, 67*(1): 251–264.

Kumar, R. (2016). Flood hazard assessment of 2014 floods in Sonawari sub-district of Bandipore district (Jammu&Kashmir): An application of geoinformatics. *Remote Sensing Applications: Society and Environment*, 4188–203. 10.1016/j.rsase.2016.10.002.

Kumar, A., Agarwal, V., Pal, L., Chandniha, S. K., &Mishra, V. (2021). Effect of land surface temperature on urban heat Island in Varanasi city India. *J, 4*(3), 420–429. 10.3390/j4030032.

Pachauri, R. K., Allen, M. R., Barros, V. R., Broome, J., Cramer, W., Christ, R., ... & van Ypserle, J. P. (2014). Climate change 2014: Synthesis report. *Contribution of working groups I, II and III to the fifth assessment report of the intergovernmental panel on climate change* (p. 151). Ipcc.

Philip, A., Townsend, S., & Walsh, J. (1998). Modeling floodplain inundation using an integrated GIS with radar and optical remote sensing. *Geomorphology, 21*(3–4), 295–312. 10.1016/S0169-555X(97)00069-X.

Qi, H., Qi, P., & Altinakar, M. S. (2013). GIS-Based spatial Monte Carlo analysis for integrated flood management with two dimensional flood simulation. *Water Resources Management, 27*(10), 3631–3645. 10.1007/s11269-013-0370-8.

Saini, S. S., Kaushik, S. P., & Jangra, R. (2016). Flood-risk assessment in urban environment by geospatial approach: A case study of Ambala City, India. *Appl Geomat, 8*, 163–190. https://doi.org/10.1007/s12518-016-0174-7.

Saksena, S. (2015). Investigating the role of DEM resolution and accuracy on flood inundation mapping. In *World environmental and water resources congress 2015* (pp. 2236–2243).

Sanders, B. F. (2007). Evaluation of on-line DEMs for flood inundation modeling. *Advances in Water Resources, 30*(8), 1831–1843. 10.1016/j.advwatres.2007.02.005.

Sanyal, J., & Lu, X. X. (2004). *Application of remote sensing in flood management with special reference to monsoon Asia: A review. Natural Hazards, 33*(2), 283–301. 10.1023/B:NHAZ.0000037035.65105.95.

ShahiriParsa, A., Noori, M., Heydari, M., & Rashidi, M. (2016, January 9). Floodplain zoning simulation by using HEC-RAS and CCHE2D models in the Sungai Maka river. *Air, Soil and Water Research*. ASWR-S36089.

Smart City: Mission Statement and Guidelines. (2015, June). Govt. of India, Ministry of urban development downloaded from http://164.100.161.224/upload/uploadfiles/files/SmartCityGuidelines(1).pdf.

Te Chow, Ven.(1959). *Open channel hydraulics*. New York: McGraw-Hill Book Company, Inc.

Thakur, P. K., & Sumangala, A. (2006). Flood inundation mapping and 1-D hydrodynamic modeling using remote sensing and GIS techniques. In: *ISPRS orange book publications during ISPRS/ISRS commission IV symposium on: "Geospatial database for sustainable development"* (pp. 27–30). Goa.

The Indian Express. (2016, August 23). Uttar Pradesh Floods: Schools Closed in Varanasi, Allahabad until Thursday. *Express News Service*. Web. February 2, 2017.

U.S. Army Corps of Engineers. (2016). *Hydrologic engineering center HEC-RAS river analysis system: Application guide, version 5.0*.

Wood, M. (2015). *The story of India*. Random House.

Part II
Climate Change and Land Use

Chapter 6
Contending Global Warming by Popularising Environment-Friendly-Fuel Compressed Natural Gas (CNG)

Soma Sengupta and Anjan Sen

Abstract The release of greenhouse gases into the air leads to depletion of ozone layer, which in turn causes global warming and several associated natural disasters. Vehicular emissions due to use of fossil fuels is the principal cause of atmospheric pollution, and Compressed Natural Gas (CNG) is a cheaper eco-friendly alternative, as it produces fewer undesirable gases. However, high price of CNG vehicles is a major barrier to wider and quicker adoption as a fuel. Public mass-transportation vehicles swiftly adopted CNG since they could easily recover the installation cost. The purpose of the present study is to explore how social marketing approach can be used to make CNG a success among private vehicle owners. This will go a long way in reducing hazards related to air pollution and aid in managing possible atmospheric disasters. The research comprises three case studies—CNG in public transport buses in Delhi; Maruti Udyog Limited; and the CNG kit market. The study found that though legal notification can forcibly engineer a shift to CNG, a voluntary shift would require convincing the public about the benefits of CNG, and removing any doubts associated with it. CNG vehicles are not popular among customers due to several perceived fears like, price of fitment, concerns of safety, reduced boot space, and waiting time at CNG filling stations. This was revealed during analysis of data collected from 250 households in Delhi and NCR, which prevented people to switch to CNG. Designing a people-oriented campaign or social marketing strategy can cause social acceptance of green fuels, by building awareness and addressing perceived fears, making the CNG kits affordable, and opening more CNG fueling stations. Partnership among all the stakeholders should be encouraged. The government should have a vibrant policy to phase out fossil-fuel-operated vehicles, and promote CNG, to reduce the occurrence of atmospheric hazards and disasters.

S. Sengupta (✉)
Department of Commerce, Kamala Nehru College, University of Delhi, New Delhi, India
e-mail: dr.somasengupta@gmail.com

A. Sen
Department of Geography, Delhi School of Economics, University of Delhi, Delhi, India

V. R. Sharma and Chandrakanta (eds.), *Making India Disaster Resilient*,
https://doi.org/10.1007/978-3-031-50113-5_6

Keywords Compressed natural gas · Eco-friendly fuel · Psychological fear · Social marketing strategy · Attitudinal change

Acronyms

ANOVA	Analysis of variance
CFC	Chloro fluoro carbon
CNG	Compressed natural gas
CO2	Carbon di oxide
CSE	Centre for science and environment
CSR	Corporate social responsibility
DTC	Delhi transport corporation
IGL	Indraprastha gas limited
i-GPI	Intelligent-gas port injection
NCR	National capital region
NGO	Non-governmental organization
ORT	Oral rehydration therapy
PIL	Public interest litigation
RWA	Resident welfare association
SD	Standard deviation

Introduction

Global warming and consequent climate change is today a real environmental problem and the primary ecological challenge confronting humanity. It has now been firmly established that humans themselves through several 'un-friendly' activities are the prime cause and source of global warming, which in turn leads to several 'man-made' disasters. The ten principal human causes of global warming, ranked in descending order by significance in 2018, are travel and transportation, industrialization, deforestation, livestock production, factory farming, consumerism, overuse of electricity, overfishing, use of aerosols, and inability to change. If these issues are not addressed within the next decade, the planet will face global environmental catastrophe of gigantic proportions, which will cause large-scale war, dire poverty, immense water shortages, and colossal die-offs of species.

With travel and transportation being the number one cause of global warming and associated disaster, the present paper addresses the various dimensions of this key cause. The vast majority of vehicles on different media (land, air or water) are powered by a variety of fossil fuels. As the fossil fuel is burnt to power the engines, the vehicles release carbon and other pollutants, impairing the quality of both air and water. In fact, transportation has been a huge world-wide contributor of greenhouse gas emissions since 2016.

Vehicular emissions is one of the major causes of pollution, specifically in cities. Compressed Natural Gas (CNG) or methane stored at high pressure is a cheaper and eco-friendly alternative to fossil fuel. On the other hand, since it is a natural gas that is lighter than air and disperses quickly when released, it is safer than other fuels in the event of a spill.

CNG vehicles are of different types. One is getting a CNG kit fitted in traditional automobiles which were running on petrol or diesel. The other one is CNG vehicles which are manufactured for CNG use, either alone, i.e., 'dedicated' or with a dual fuel or bi-fuel mode. The market for CNG vehicles is ever growing since its inception in India as well as all over world.

Public transportation vehicles were the early adopters of CNG in India, because of court orders and since they could quickly recover the installation cost. But private vehicle owners are still reluctant to switch to the green fuel.

Significance of the Study

CNG is a boon to the mankind with merits like, huge cost saving, less polluting, and help to mitigate greenhouse emissions, thereby preventing any atmospheric disasters. CNG vehicles have lower maintenance cost in comparison to other hydrocarbon fuel powered vehicles.

However, there are a few limitations too. CNG tank occupies a large space in the trunk of the car or bed of a truck. Absence of standardization of CNG kits coupled with high installation cost acts as a barrier to its acceptability. Moreover, in India, CNG filling stations are neither in sufficient number nor they are located at all possible convenient locations, and hence, long queues could be seen at the filling stations.

In order to popularize CNG, besides developing the infrastructural support, there is a need for developing an effective social marketing campaign to change the attitudes and behavior of people. This study will help in designing and implementing a social marketing strategy to make CNG acceptable among the target audience.

Review of Literature

According to Philip Kotler (2004), 'marketing is a societal process by which individuals and groups obtain what they need and want through creating, offering, and freely exchanging products and services of value with others'.

Kotler and Andreasen define social marketing as 'differing from other areas of marketing only with respect to the objectives of the marketer and his or her organization. Social marketing seeks to influence social behaviors not to benefit the marketer, but to benefit the target audience and the general society'. This technique has been used extensively in international health programs, especially for contraceptives and oral rehydration therapy (ORT), and is being used with more frequency in the USA for such diverse topics as drug abuse, heart disease and organ donation (Weinreich, 1999). It utilizes concepts of market segmentation, consumer research, product concept development and testing, directed communication, facilitation, incentives, and exchange theory to maximize the target adopters' response (Kotler & Roberto, 1990).

Demographic profile of the target audience decides the success of a social marketing campaign, as was shown by a study by Burnett's (1981), who identified variables (behavioral and demographic) that distinguish donors from non-donors. The results indicated that donors tend to be male, married, having children, have rare blood types, and low self-esteem, to be low risk takers, very concerned with health, better educated, religious, and quite conservative. And the non-donors tend to be the opposite on all these dimensions. Mayer (1976) emphasizes that the market segment of the socially and ecologically conscious population is large and marketers must make serious efforts to reach it despite the limitations of doing so.

Bloom and Novelli (1981) identified eight basic decision-making areas in social marketing: market analysis, market segmentation, product strategy development, pricing strategy development, channel strategy development, communications strategy development, organizational design and planning, and evaluation.

The impact of social marketing and social advertising was explored by researchers for different campaigns like, anti-tobacco campaign (Andrews et al., 2004); alcoholic beverage advertising (Franke & Wilcox, 1987); cigarettes advertising or ban on cigarette advertisements (Holak & Reddy, 1986; Pechmann & Shih, 1999; Teel et al., 1979; Henke, 1995); and AIDS campaign (Bush & Boller, 1991).

Green products or environmentally favorable products are generally expensive and faces difficulty in being popular among the masses, and hence, does not remain profitable for the firms. The green product innovation and green process innovation were positively correlated to the corporate competitive advantage (Chen, et al. 2008; Oyewole, 2001). Marketers sometimes deliberately make false or exaggerated 'green' claims. Critics refer to this practice as 'green washing' (Ottman, 2002).

Hence, social marketing of green product like, CNG is needed to popularize it among the masses by disseminating information, promoting it, and hence, introducing a behavioral change among the target audience.

Objectives of the Study

Delhi urban area with a population of 25 million has more vehicles than the aggregate of three next metro cities—Mumbai, Kolkata, and Chennai. A silent disaster is in the making. The greenhouse gases coupled with construction, transforms the city into a gigantic urban heat island, with the air quality being 'severe' at least in ten of the twelve months in a year, which is a cause for several ailments. Delhi happens to be among the top five polluted cities in the world.

The purpose of the study was to explore how social marketing approach can be used to make CNG a success in the market, which though a small step, but in the long-run will contribute in making cities resilient to impending disaster, and will help in its management.

The specific objectives of the research are as follows:

- To critically evaluate the measures undertaken by different stakeholders in promoting CNG in Delhi.
- To analyze the applicability of social marketing strategies, and design an effective campaign for acceptance of environmental-friendly CNG in Delhi.

Hypothesis of the Study

The hypothesis of the study is that there is no significant difference in the opinions of users and non-users regarding factors preventing the use of CNG.

Research Methodology

The study includes three case studies based on secondary data sources, i.e., CNG in public transport buses in Delhi; Maruti Udyog Limited; and the CNG kit market. Primary data was collected from 250 households in Delhi and NCR with the help of a questionnaire which was statistically analyzed.

Findings of the Case Studies

The outcomes of the case studies are as follows:

Case Study 1: Public Transport Buses Switched to CNG in Delhi

Delhi was considered as the third most polluted country of the world in early 1990s. The pollutant level in the city's air was so high that it was leading to a rise in respiratory diseases. The leaves of the trees and plants near major roads darkened due to the high pollution level. An environment activist, lawyer M. C. Mehta filed a Public Interest Litigation (PIL) in the Supreme Court of India calling for measures to improve air quality in Delhi. The combustion of adulterated diesel in vehicles was the reason behind a very high level emission of pollutants in the environment. The

Environment Pollution (Emission and Control) Authority recommended the emission standards which were then enforced in the city. The Supreme Court issued orders in 1998 for public transport buses in Delhi to switch over to single fuel mode of CNG by April 1, 2001. Several hurdles were there in the path of a swift conversion from diesel to CNG run buses which were as follows:

- The politicians, bureaucrats, chassis manufacturers, gas suppliers and government blamed each other for missing the deadline for implementation of the Court orders.
- Issues related to the construction of bus bodies were taken up with the placement of orders for bus chassis. Automobile giants Ashok Leyland and others were not in favor of converting the existing fleet into CNG vehicles.
- There was lack of perception about the issues involved—the Transport Ministry took time to notify the emission norms; the testing agencies were unclear about the certification procedures; the DTC was not clear about the safety of the technology; and the government was not willing to fund an unproven technology since public funds were involved.
- The idea of targeting the public transport buses was criticized by a section of public as well as by the experts like, Tata Energy Research Institute and Indian Institute of Technology Delhi. They were of the view that orders were given without doing enough research on the sources of pollution.
- CNG was not available outside Delhi, and hence, it posed a problem for the DTC buses plying between Delhi and neighboring States as there were no gas filling stations in their route.
- There was a dual problem—the inadequate number of CNG refueling stations was a barrier to acceptance of CNG vehicles by consumers; on the other hand, the less number of CNG vehicles made investment in CNG refueling stations unprofitable. Hence, there was a major problem with the vehicle refueling infrastructure in Delhi.
- The Delhi Transport Corporation (DTC) bought two thousand CNG buses. With a fleet of 2120 CNG buses, DTC became the largest CNG city bus fleet operator in the world.
- By July 2001, the queues at CNG filling stations became longer and bus operators including DTC could not get gas after waiting for eight to ten hours. The compressors of CNG buses broke down frequently; there were cylinder bursts and a few buses caught fire, leading to questions regarding safety of such buses.
- The Delhi government under the order of Supreme Court in 2004 made it mandatory for all city buses and auto rickshaws to run on CNG with the intention of reducing air pollution.

Despite, these issues the public transport buses in Delhi are running successfully on CNG. Beside that the autos were also forced to become CNG enabled. Convincing the auto owners was easier as then the price difference between CNG and petrol made it viable for scooters and taxis to switch to CNG.

Steps Taken by the Authorities to Popularize CNG

The following steps were taken by the government to popularize CNG in Delhi.

- Financial incentives from the government by way of cheap loans from the Delhi Finance Corporation for replacement of old taxis and scooters hastened the process of induction of CNG vehicles.
- There were several myths associated with CNG like, it causes cancer; it is not safe; CNG emits more greenhouse gases; there is not enough CNG, and so on. These myths when removed through advertisements in different media by CSE, the inhibition of owners of autos, scooters, and taxis were reduced.
- The number of CNG filling stations was increased so that they can meet the growing demand of CNG.

Steps Needed to Be Taken up by the Authorities to Popularize CNG in Private Vehicles

There is a strong need to give impetus to strengthen the CNG infrastructure substantially. This is not possible if CNG continues to be dispensed by a public sector monopoly supplier like IGL. There is need to allow other players including private players to set up the CNG infrastructure in Delhi. Encouraging competition in the setting up of CNG supply stations is required. This will ensure more number of filling stations and better services and encourage private vehicle owners to switch to this green fuel.

Though legal notification can automatically lead to a shift to CNG, a rather willful shift requires continuously convincing the public about the benefits of CNG and removing the myths associated with it.

Case Study 2: Maruti Udyog Limited—CNG Models

By August, 2010, Maruti launched five CNG cars in Delhi, Mumbai, and Gujarat. Marketing strategies used to popularize the CNG models are as follows:

New technology used to develop the CNG models A newly developed i-GPI or 'intelligent-Gas Port Injection' technology was used for such models. This technology is claimed to be safe, reliable, efficient, clean, and responsive to the environment or in other words, environment friendly.

Benefits of Maruti Factory Fitted CNG Vehicles Maruti highlighted the benefits like vehicle body designed for CNG system; performance with gasoline powered engine; safety through high-quality components, CNG system leak-proofing; high fuel efficiency; dual ECU system; lower running costs by sixty percent; peace of mind for the customers as the vehicles were backed by full warranty coverage and service support across the country.

Price of Maruti CNG Cars Price of all the CNG cars are approximately fifty to fifty-five thousand rupees more compared to their petrol versions.

Promotion of Maruti CNG Cars Beside advertisement in newspaper, all the products are extensively covered in their website. The website of Maruti is vibrant and gives a comparative analysis of its CNG and petrol versions of the models.

Distribution of Maruti CNG Cars The CNG variant is distributed in the same manner as the petrol or diesel versions through their showrooms, though the demand for such cars is less than those of petrol or diesel models.

Target Segment Since the CNG models are not very popular among the households for private use, Maruti targets the small firms into travel and tourism, to popularize such models.

However, there are a number of shortcomings in the company's approach to popularize their CNG models, like the models are not promoted well and the advertisement in newspaper and radio is rarely seen; the promotion did not highlight the need for a shift to such models; the customers are reluctant to pay fifty thousand rupees extra for such models. Moreover, the average effective life of such car is five years and the saving will not be able to recover the initial high price within that period.

CNG vehicles as such are not popular among the customers because of the psychological fears like, higher price, safety concern, lesser boot space, and waiting time at CNG filling stations. Maruti is not addressing those issues.

Case Study 3: CNG Kit Market

The CNG kit and the cylinders are available in most part of the city. The stores or fitment centers are nowadays, advertising the product through their websites. But the customers have to explore the vehicle repairing market of an area to find out the stores which are providing such facilities. Standardization of the product and services offered by stores is a big issue. The specific characteristics of CNG kit market are as follows:

- Government-approved CNG kit fitment centers are there in every city. Most of the renowned outlets or centers have their websites as well.
- Variants of kit include sequential and conventional kits. The kits installed are either imported from Italy, Argentina, and Poland; or the Indian brands. The most popular brands selling such kits are Lovato and Tomaset to. They are available in both, Argentina and Italy versions and the cost of the kits range between twenty-two thousand rupees to sixty thousand rupees.
- A smooth and efficient transition of petrol based vehicle to CNG fuel source is ensured by such fitment centers. Though the loss of pickup after installation and enhancing mileage are two important issues, the per kilometer running cost after installation of CNG kit is quite less.

The CNG kit market is all confusing for the private vehicle owners because of the reasons like, non-availability of a comprehensive list of authorized fitment

Table 6.1 Number of users and non-users

Respondents	Number of respondents	Percentage (%)
Users	21	08.4
Non-users	229	91.6
Total	250	100

Source Primary Survey

centers; lack of perfection in work; eat-up boot space; lack of standardization and quality maintenance; the CNG kit brands and different country-of-origin poses more confusion for the customers.

The authorized fitment shops, the government, and the transport authority should address these issues. The product should be promoted well. The target audience should be made to understand the need to shift to CNG mode by getting the CNG kit fitted in their existing petrol or diesel vehicles by explaining the environment friendliness of the product which also reduces the fuel bill of the individuals.

Analysis of Data Collected from Households

For the purpose of the study, data was collected from 250 respondents from Delhi and its neighboring areas, and then the data was statistically analyzed. The results of the analysis are presented in the following paragraphs.

Number of Users and Non-users

Out of 250 respondents, only 21 (8.4%) have CNG in their vehicle.152 respondents (91.6%) were not having a CNG vehicle (Table 6.1). Hence, CNG is yet not very popular.

Factors Motivating Switch to CNG Private Vehicles

Most of the respondents opined that the increasing price of petrol or diesel and the cost saving resulting from switching to CNG vehicle is the strongest motivating factor (Table 6.2). Good experiences of others who are using CNG vehicles also motivate people to switch to such vehicles, and hence, this factor got the second rank. People who travel long distances every day are motivated more to switch to CNG vehicles. There is a low concern for nature or environment.

Table 6.2 Ranking of motivating factors

Factors	Mean	S.D.	Ranks
Increasing prices of petrol or diesel and cost saving by using CNG	4.86	0.44	I
Experiences of others	4.79	0.48	II
Suggested by mechanic of a garage	4.12	0.53	III
Concern for environment or nature	3.21	0.72	VII
Have to travel long distances	4.08	0.49	IV
Availability of factory fitted CNG kit in vehicle models	3.66	0.65	VI
Advertisements and government notices	0.83	0.76	IX
Lower maintenance cost	2.47	0.78	VIII
Low cost of getting a CNG kit fitted	3.72	0.74	V

Source Primary Survey

Factors Preventing Use of CNG in Private Vehicles

The most important deterrent is the long waiting time at CNG fueling stations (Table 6.3). This is closely followed by the factor 'very few CNG fueling stations at convenient locations'. The third rank went to the factor 'safety concern' followed by the 'high cost of installation of CNG kit' and 'takes up additional space in the trunk of a car'.

Table 6.3 Ranking of factors preventing use of CNG in private vehicles

Factors	Mean	S.D.	Ranks
High cost of installation of CNG kit	4.66	0.55	IV
Branded CNG kits are not available	4.54	0.58	V
Safety concern	4.79	0.56	III
No quality assurance	4.37	0.67	VII
Takes up additional space in the trunk of a car	4.52	0.62	VI
Lack of information/ not promoted well	4.23	0.64	VIII
High prices of factory fitted cars	3.96	0.68	IX
Long waiting time at CNG fueling stations	4.94	0.42	I
Very few CNG fueling stations at convenient locations	4.90	0.46	II
Cannot relate it as being eco-friendly	3.09	0.73	XII
Not aware of the harmful effect of petrol or diesel	3.25	0.68	XI
Not aware of outlets selling or installing CNG kits	3.51	0.72	X

Source Primary Survey

Table 6.4 Ranking of ways to make CNG in private vehicles popular

Ways	Mean	S.D.	Rank
By reducing the installation cost	4.57	0.52	III
By encouraging automobile manufacturers to come out with low-priced CNG vehicles	4.28	0.53	V
By increasing the price of petrol and diesel and providing CNG at a subsidized price	4.34	0.49	IV
By having more number of CNG filling stations at accessible places	4.76	0.61	I
By assuring the customers that CNG is safe to use	4.69	0.68	II
By branding the CNG kits	4.18	0.70	VII
By reducing the size of the cylinder so that they don't take up a large space	4.22	0.68	VI
By celebrity endorsement	3.32	0.69	XII
By educating the potential customers about the significance of environment friendly CNG	3.88	0.64	IX
By making the customers understand the cost saving and lower maintenance cost	4.06	0.67	VIII
By subsidizing the product by the government	3.56	0.71	X
By promotion of the product by government sponsored advertisements creating public awareness	3.48	0.59	XI
By strictly enforcing the switch to CNG in private vehicles through legal orders	3.21	0.71	XIII

Source Primary Survey

Efforts to Make CNG in Private Vehicles Popular

Most of the respondents opined that CNG in private vehicles can be made popular by having more number of CNG filling stations at accessible places (Table 6.4). This will reduce the waiting time at CNG fueling stations. Assurance regarding CNG being safe and affordable will encourage adoption of the green fuel.

Difference of Opinions of Users and Non-Users of CNG in Private Vehicles

In order to test the hypothesis, stating that there is no significant difference in the opinions of users and non-users regarding factors preventing the usage of CNG in private vehicles, both t-test and one-way ANOVA were conducted for twelve factors (Table 6.5). The t-values and the F-ratios were found to be significant for most of the factors like 'high cost of installation of CNG kit', 'safety concern', 'no quality assurance', 'takes up additional space in the trunk of the car', 'high prices of factory

Table 6.5 Difference of opinions of users and non-users of CNG in private vehicles

Factors	Users $N = 21$		Non-Users $N = 229$		t-value	F-Ratio
	Mean	S.D.	Mean	S.D.		
High cost of installation of CNG kit	4.34	0.53	4.69	0.62	2.50*	6.27*
Branded CNG kits are not available	4.33	0.61	4.56	0.64	1.58	2.50
Safety concern	4.47	0.54	4.82	0.58	2.66**	7.08**
No quality assurance	4.15	0.61	4.39	0.52	1.99*	3.98*
Takes up additional space in the trunk of a car	4.20	0.64	4.55	0.60	2.54*	6.47*
Lack of information/ Not promoted well	4.12	0.61	4.24	0.53	0.98	0.96
High prices of factory fitted cars	3.64	0.62	3.99	0.59	2.59*	6.71*
Long waiting time at CNG fueling stations	4.73	0.41	4.96	0.51	2.01*	4.03*
Very few CNG fueling stations at convenient locations	4.60	0.42	4.92	0.41	3.42*	11.67**
Cannot relate it as being eco-friendly	2.98	0.72	3.10	0.67	0.78	0.61
Not aware of the harmful effect of petrol or diesel	3.15	0.71	3.26	0.69	0.70	0.49
Not aware of outlets selling or installing CNG kits	3.20	0.70	3.54	0.68	2.19*	4.79*

**Significant at 0.05 level of significance.* ***Significant at 0.01 level of significance*
Source Primary Survey

fitted cars', 'long waiting time at CNG filling stations', 'very few CNG filling stations at convenient locations', and 'not aware of outlets selling or installing CNG kits'.

On the basis of these results, the hypothesis is rejected. There is a significant difference in the opinions of the users and the non-users of the CNG in private vehicles regarding the factors preventing the usage of such vehicles.

Conclusions of the Study

The following conclusions can be drawn on the basis of the present study:

Lack of Awareness People are still not aware about the environment friendliness of CNG and about the harmful effect of petrol and diesel on the environment.

High Psychological Costs A number of psychological fears or costs were found among the people regarding their decision to switch to this product, such as high cost of installation of CNG kit; long waiting time at CNG fueling stations; takes up additional space in the trunk of a car; branded CNG kits are not available; safety concerns, and so on.

Not Applying a Social Marketing Strategy to Deal with the Issue The government and the automobile companies are not attempting on bringing an attitudinal change among the target audience by applying the principles of social marketing.

Social Marketing Implications to Popularize CNG in Private Vehicles

The implications for the automobile companies and the government are as follows:

Developing a Social Marketing Mix An integrated marketing mix by blending the elements of marketing will help in bringing an attitudinal and behavioral change among the owners of private vehicles so that they willfully shift to CNG from petrol or diesel.

Designing the Product While designing a CNG vehicle, the automobile companies should keep in mind that minimum space should be occupied by the CNG storage tank.

Emphasis on Actual Price and Psychological Price The CNG vehicles should be made affordable for the target audience. The authorities and the automobile companies should strive to reduce the psychological fears or non-monetary price associated with CNG vehicles. This can be done by opening up more number of CNG fueling stations, redesigning the vehicles, making available standardized and branded CNG kits.

Promotional Strategy The promotional campaigns should strive to remove the myths associated with CNG that it is not safe and emits more greenhouse gases; educate and inform people about the benefits of CNG and the harmful effect of the other types of gasoline. The promotions should also reduce the psychological costs or fears. Informative websites of the government and automobile companies may be helpful in educating the target audiences. An integrated promotion mix, i.e., advertising (along with a celebrity endorsement), personal-selling, and sales promotion should be adopted to disseminate information and persuade target audience.

Distribution Strategy The authorities should open up more number of CNG fueling stations. The automobile companies should ensure that the CNG models are made available in every part of the country. The government should come out with a list of CNG kit fitment centers.

Public Seminars and workshops can be conducted by the automobile companies and the Government or authorities inviting the general public. Their opinions and fears should be heard and resolved.

Partnership Partnership should be encouraged between the stakeholders, i.e., the automobile companies, politicians, bureaucrats, chassis manufacturers, gas suppliers, and the government to popularize CNG. Frequent meetings and discussions between the stakeholders are imperative.

Policy The government should lay down a clear-cut policy to phase-out the petrol or diesel operated vehicles. To build up a critical mass of CNG vehicles in the country, cues can be taken from other countries. In Singapore, an incentive for using this eco-friendly fuel was initiated for two years under which a green vehicle rebate (GVR) of forty percent was given on the newly registered green passenger vehicles.

Purse-Strings The government may create a fund to promote CNG in the country and for that purpose the Non-Governmental Organizations (NGOs) can be roped in.

Hence, social marketing can ease out the process of shifting from petrol and diesel to CNG by the private vehicle owners. This will not only reduce the air pollution levels, it will also reduce the fuel bills of private vehicle owners. The perils of global warming is looming over the world and its wrath is demonstrated quite frequently in the form of different forms of disasters striking different cities of the world every now and then. The adoption of CNG will reduce the emission of harmful gases, global warming, and occurrence of natural disasters.

References

Andrews, J. C., Netemeyer, R. G., Burton, S., Moberg, D. P., & Christiansen, A. (2004). Understanding adolescent intentions to smoke: An examination of relationships among social influence, prior trial behavior, and anti-tobacco campaign advertising. *Journal of Marketing, 68*(3), 110–123.

Bloom, P. N., & Novelli, W. D. (1981). Problems and challenges in social marketing. *Journal of Marketing, 45*(2), 79–88.

Burnett, J. J. (1981). Psychographic and demographic characteristics of blood donors. *Journal of Consumer Research, 8*(1), 62–66.

Bush, A. J., & Boller, G. W. (1991). Rethinking the role of television advertising during health crises: A rhetorical analysis of the federal AIDS campaigns. *Journal of Advertising, 20*(1), 28–37.

Chen, Y. S. (2008). The driver of green innovation and green image: green core competence. *Journal of Business Ethics, 81*(3), 531–543.

Franke, G., & Wilcox, G. (1987). Alcoholic beverage advertising and consumption in the United States, 1964–1984. *Journal of Advertising, 16*(3), 22–30.

Henke, L. L. (1995). Young children's perceptions of cigarette brand advertising symbols: Awareness, affect, and target market identification. *Journal of Advertising, 24*(4), 13–28.

Holak, S. L., & Reddy, S. K. (1986). Effects of a television and radio advertising ban: A study of the cigarette industry. *Journal of Marketing, 50*(4), 219–227.

Kotler, P., & Roberto, E. (1990). *Social marketing: Strategies for changing public behavior*. Free Press.

Kotler, P. (2004). *Marketing management: Analysis, planning, implementation, and control* (11th ed.). Prentice-Hall of India Pvt. Ltd.

Mayer, R. N. (1976). The socially conscious consumer: Another look at the data. *Journal of Consumer Research, 3*(2), 113–115.

Ottman, J. (2002). The real news about green consuming. http://www.greenmarketing.com/index.php/articles/. Accessed 22 February 2013.

Oyewole, P. (2001). Social costs of environmental justice associated with the practice of green marketing. *Journal of Business Ethics, 29*(3), 239–251.

Pechmann, C., Shih, C. F. (1999). Smoking scenes in movies and antismoking advertisements before movies: effects on youth. *Journal of Marketing, 63*(3), 1–13.

Teel, S.J., Teel, J.E., Bearden, W.O. (1979). Lessons learned from the broadcast cigarette advertising ban. *Journal of Mar*keting, *43*(1), 45–50.

Weinreich, N. K. (1999). What is social marketing? webmaster@socialmarketing.com. Accessed 22 February 2013.

Chapter 7
Increasing Vulnerability of Arabian Sea Towards Cyclonic Storms

A. Athul and Sushma Gulria

Abstract India has a coastline of about 7516 km. Even though the North Indian Ocean (NIO) only spawns 7% of cyclones, its devastating effect is comparatively very high in India. Cyclone genesis in Arabian Sea was very less compared to Bay of Bengal due to its unfavourable climatology for cyclone genesis. Changes in the climatology of Arabian Sea are increasing its vulnerability to tropical cyclones. This study reveals the impact of climate change in the increasing vulnerability of Arabian Sea to cyclone genesis.

Keywords Vulnerability of Arabian Sea · Tropical cyclone · Arabian Sea climatology · Bimodal system

Introduction

India is highly vulnerable to natural hazards especially earthquakes, incessant rains, high winds, etc. being the common threats. High wind speeds form depressions and subsequently turn into cyclones of different intensities. Cyclones in Indian Ocean are known as tropical cyclones. These are weather systems in which winds equal or exceed gale force (minimum of 34 knot, i.e. 62 kmph). Indian sub-continent is exposed to nearly 10% of tropical cyclones in the world because of its 7516 km-long coastline. There are 13 coastal States/Union Territories (UTs) comprising of 84 coastal districts which are affected by cyclones during different seasons. With respect to cyclone vulnerability, Bay of Bengal is known to be more sensitive as compared to the Arabian Sea. This is because of the presence of remnants of typhoon and relative temperature difference between Arabian Sea and Bay of Bengal. Arabian Sea is colder than Bay of Bengal; moreover, the communities living in west coast are

A. Athul (✉)
Department of Disaster Management, Pondicherry University, Pondicherry, India
e-mail: athulperul@gmail.com

S. Gulria
National Institute of Disaster Management (NIDM), New Delhi, India

V. R. Sharma and Chandrakanta (eds.), *Making India Disaster Resilient*,
https://doi.org/10.1007/978-3-031-50113-5_7

less exposed and hence, become less vulnerable towards cyclones. Recently and as noticed, the cyclone activity in Arabian Sea is increasing because of climate change and anthropogenic emissions leading to increase in pollution, which in turn is making the west coast more vulnerable towards cyclones. Although only couple of cyclones have been generated in Arabian sea each year, which are not severe in nature and therefore, making less disaster prone towards cyclones. The strongest storm in the recent time was Very Severe Cyclonic Storm **'Ockhi'** which was formed in last week of November 2017 and affected the coast of Tamil Nadu and Kerala. Ockhi caused an estimated US$5.07 billion economic loss, whereby nearly 282 fishermen lost lives from the State of Kerala and Tamil Nadu. One of the most destructive cyclonic storms in Arabian sea, in terms of human life loss, was Very Severe Cyclonic Storm **ARB02** (03A designated by JTWC) which made land fall in western coast in 1998 with a maximum sustained wind speed of 105 knot (194 km/h). More than 1100 deaths were attributed (Aldinger, 1998). The cyclones Ockhi and ARB02 clearly corroborate that Arabian Sea cyclones can be intense and have the potential to become severe as well.

The characteristics of tropical cyclones are changing. Generally for a depression to become a cyclone it takes nearly 72 h. But in the case of cyclone Ockhi, it took less than 24 h only. The Intergovernmental Panel for Climate Change (IPCC) Third Assessment Report (2001) tells that there is scope for some increase in tropical cyclones maximum intensities in a warmer conditions are 'likely', in some regions. This change in characteristics of tropical cyclones may be because of climate change. Kelvin Wash, Jinhua Yu and Yuquing Wang who worked on 'Climate change and Tropical cyclones' suggest that future projections based on theory and high-resolution dynamical models consistently plead that greenhouse warming will cause the globally averaged intensity of tropical cyclones to shift towards stronger storms, with intensity increase of 2–11% by 2100. Existing modelling studies also consistently project decreases in the globally averaged frequency of tropical cyclones by 6–34%. Sea Surface Temperature (SST) in all regions which is prone to tropical cyclone genesis is increasing during the past several decades. Recently, the economic damage and disruption due to cyclones are rapidly increasing. This is primarily because of agglomeration of increased infrastructure and population rise in coastal regions. Moreover, in developing countries, populations are gradually relocating towards coastal areas. Hence, climate change can be attributed as one of the several factors likely to affect future evolution of damage from tropical cyclones.

Tropical cyclones are different from other meteorological hazards because they can trigger other coastal hazards such as coastal flooding, coastal erosion, storm surges, salinity ingress, landslides in continental slope. IPCCs' Fifth Assessment report (2013) states that 'climate change exacerbates the vulnerability of coastal regions to extreme and impulsive physical process, such as storm surge and storm waves'. Future changes in storm surge and ocean wave climates are a dynamic side issue of how climate change influences coastal regions (Nobuhito Mori, 2016). It has also been reiterated that climate change due to emission of greenhouse gases is accelerating cyclone genesis both in frequency and intensity; further to this, climate change is also causing sea level rise due to thermal expansion of water and polar ice cap melting. If these two scenarios are combined together, it can create high-amplitude

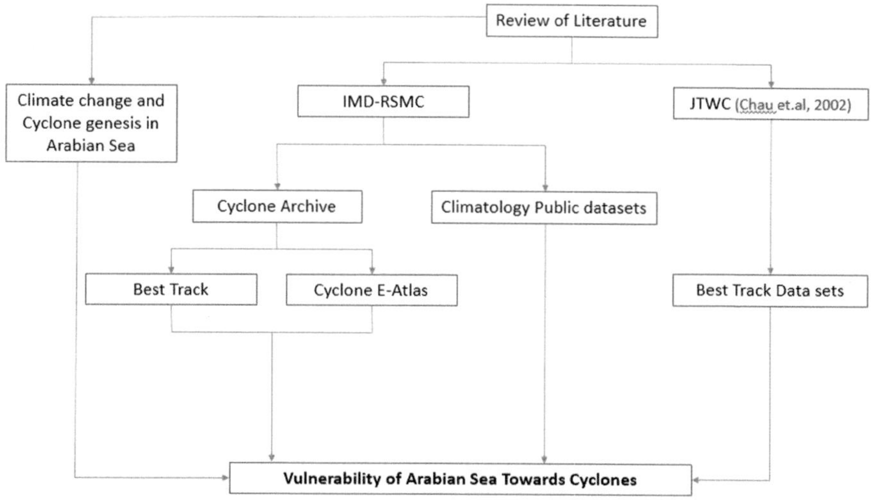

Fig. 7.1 Flow chart for methodology

storm surges. These storm surges can cause coastal erosion and can alter whole shoreline and subsequently and inadvertently change coastal dynamics. Henceforth, the occurrence of such intense cyclones can change the whole coastal dynamics which will invariably have a detrimental impact on the coastal communities and heighten their vulnerabilities.

Secondary Data

Data were collected for the study period 1981–2015 from different sources. Data regarding the cyclone activity in Arabian Sea Basin were collected from Regional Specialised Metrological Centre, Indian Metrological Department (IMD) and Joint Typhoon Warning Centre (JTWC). RSMC Cyclone archive is having accurate cyclonic data so maximum data were collected from cyclone archive. Cyclone track and its deviations were collected from E-Atlas which is published by IMD. With the help of best track data extracted from IMD RSMC and JTWC, statistical analysis is carried out.

Methodology (Fig. 7.1)

Cyclones in India

India has a coastline of about 7516 km. Even though the North Indian Ocean (NIO) is only creating 7% of cyclones, yet, the devastating effect is comparatively very high in India. North Indian Ocean basin is having a bimodal frequency of cyclone formation since North Indian Ocean is having 2 cyclone seasons, i.e. May–June and October–November respectively with primary peak in November and secondary peak in May. Eastern side of the North Indian Ocean basin is the major area for cyclone genesis. Cyclone is created in Eastern side, and initially, it moves in a west-northwesterly direction. Arabian Sea was less prone to cyclone than Bay of Bengal with a ratio 1:4. Between 1891 and 2006, 306 cyclones have crossed east coast, out of which 103 were of severe nature. In the same period, 48 cyclones were generated in west coast, out of which 24 were severe. Bay of Bengal is more vulnerable towards cyclones because of many reasons. The primary reason being the temperature, as Bay of Bengal is warmer than Arabian Sea. This temperature helps to sustain moisture and warmer sea attracts formation of cyclones. Bay of Bengal is having adjoining Andaman Sea and Pacific Ocean and the remnants of Typhoon can produce cyclones in Bay of Bengal. Majority of cyclones from Bay of Bengal won't reach Arabian Sea because of the presence of large land mass, further, after landfall, Tropical cyclones from Bay of Bengal generally dissipates. The presence of Western Ghats is also responsible for making Arabian Sea less vulnerable to the impacts of heavy winds (Fig. 7.2).

Cyclones are known by their destructive behaviour like destructive winds, storm surges and torrential rainfalls. Winds during cyclone can have detrimental impact on infrastructures like housing, lifeline services such as hospitals, schools, food storage facilities, roads, bridges, culverts, livestock and crops. Torrential rainfall can also trigger flooding and landslides. The most destructive cyclone faced by India in the past has been the Orissa Super Cyclone of 1999. Nearly 9893 people lost their lives, and 15,681,072 people were affected by this cyclone.

Coastal lines of India again and again affected by numerous cyclone where the list starts from Laila in 2010 to Fani in 2019. Laila was the first cyclonic storm to affect southeastern India in May month. Laila developed on May 17 in the Bay of Bengal and made landfall in Andhra Pradesh. It was the worst storm to hit Andhra Pradesh in the last 14 years. In the same year, another cyclone was developed in the same basin on October, and it was named as Nilam which effected the coast of Tamil Nadu. In 2013 November, a weak system was developed in Bay of Bengal which caused heavy rainfall in eastern India. In the same month, another cyclone is formed in the coast of Andaman and Nicobar Islands which is known as Lehar. Lehar caused damages in Andaman Island, Andhra Pradesh and Orissa. Hud Hud was developed on 2014 October, and it caused extensive damages in Vizag and moves through a northwards track towards Uttar Pradesh. Ockhi is categorized as Very Severe Cyclonic Storm which is developed from Arabian Sea and affected the coasts of Tamil Nadu, Kerala and Lakshadweep in the month of November 2017. Recently, Orissa is affected by a

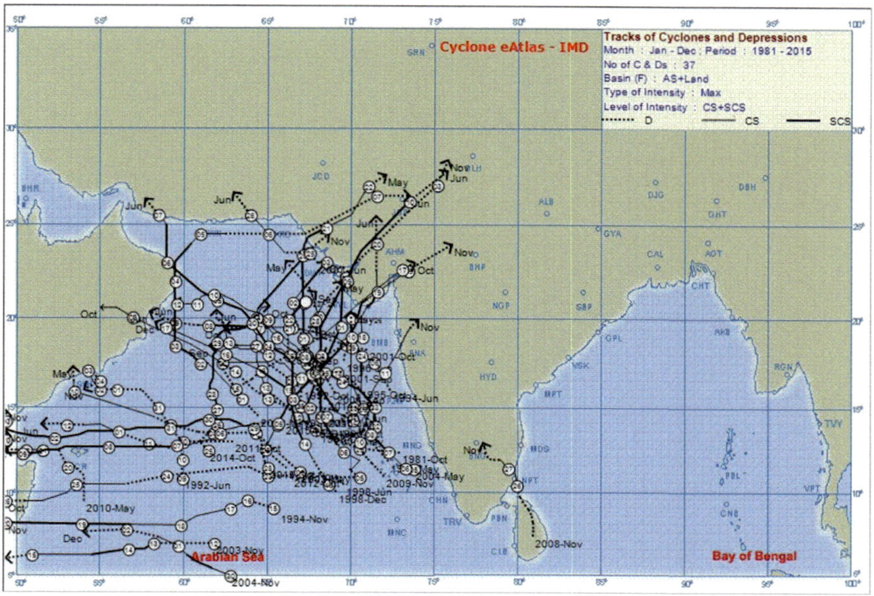

Fig. 7.2 Trajectories of tropical cyclones formed in Arabian Sea basin during 1981–2015 (*Source* IMD Cyclone e-Atlas)

cyclone known as Fani which lies in the Extreme Severe Cyclonic storm. 89 people are known to have killed by this cyclone.

Tropical cyclone genesis in Arabian Sea during 1981 to 2015 was considered for this study. Data were collected from Regional Specialized Meteorological Centre for Tropical Cyclones over North Indian Ocean, IMD. Arabian Sea cyclones are having a bimodal system. The cyclonic seasons are May–June and October–November. Records suggest that postmonsoon cyclones are more severe in nature than premonsoon cyclones.

Best track genesis of Arabian Sea cyclones as a storm has occurred in North Indian Ocean and west of 77°E. Most Arabian Sea cyclones form near to the west coast, in close vicinity to the State of Kerala, Karnataka and Goa coast and move northerly and northeasterly track or near to UAE and Gujarat. Storms forming near to centre of the basin and closer to the equator move more towards easterly direction and hence, these kinds of cyclones are more dangerous in Indian context. Very less cyclones move in westward direction, but most of the depressions are having a westward movement. Cyclonic storm developed in Arabian Sea is not making landfall in South Indian coast, but these regions were affected by intense rain and strong winds during maturity of the cyclone. Kerala coast is getting rain in non-monsoon season due to the development of low-pressure depressions, and this plays a role in the variation of weather in the region. Continuous rains due to depression are causing floods in low-lying areas. Cyclones formed in Bay of Bengal which progress towards Arabian Sea are not included.

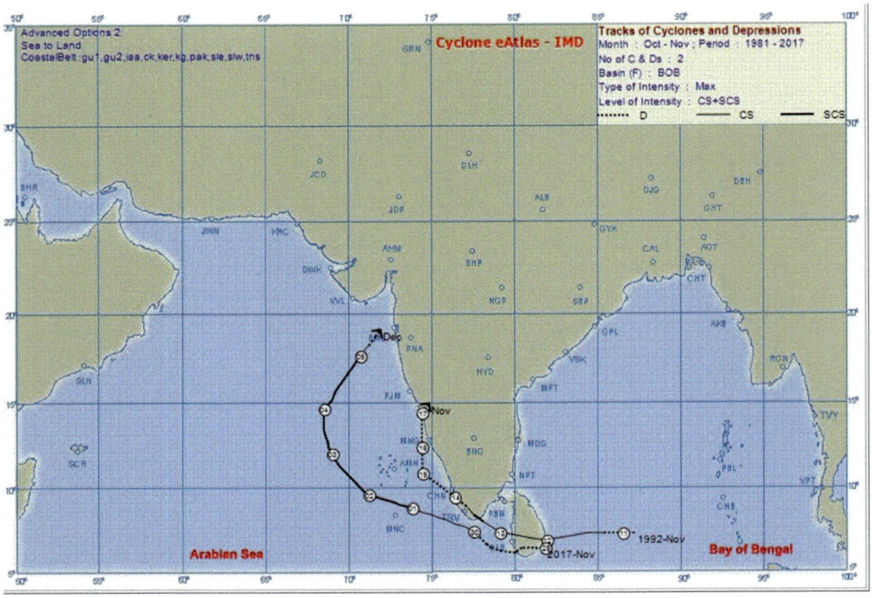

Fig. 7.3 Trajectory of tropical cyclone formed from Bay of Bengal and moved through Arabian Sea during 1981–2017 (*Source* IMD Cyclone e-Atlas)

Only two cyclones were developed from Bay of Bengal (BOB) basin which moves through Arabian Sea (AS) and caused landfall in west coast during 1981–2017 (Figs. 7.2, 7.3).

Ockhi is the only one severe cyclonic storm which formed over Bay of Bengal and intensified by Arabian Sea. In 1992 November, one cyclonic storm had developed over Bay of Bengal, but it decayed into a depression by Arabian Sea. These two cyclones were formed in the month of November so evidently it is clear that second cyclonic season is vulnerable for the movement and intensification of cyclone from Bay of Bengal to Arabian Sea.

Annual frequency of cyclone in Arabian Sea basin is plotted in a bar diagram with a five-year interval time during a period from 1981 to 2015. Linear trend line analysis of depression (DP), cyclonic storm (CS) and severe cyclonic storms (SCS) are showing an increase in frequency. Fifteen depressions were formed in 2011–2015 period which is three times than 1986–1990. A steady increase in the number of depression formed is observed from 1986 to 2015. 8 cyclonic storms were formed during 2001–2005 and 2011–2015. But only 3 cyclonic storms were formed between 1981–1985 and the frequency of cyclonic storms was increasing with an undulating behaviour (Fig. 7.4).

Only 2 severe cyclonic storms (SCS) were formed during 1981–1985, but between 1996–2000 and 2001–2005 six severe cyclonic storms (SCS) had formed. Even though it is not showing a steady increase, it is evident that SCS are increasing in an undulating behaviour.

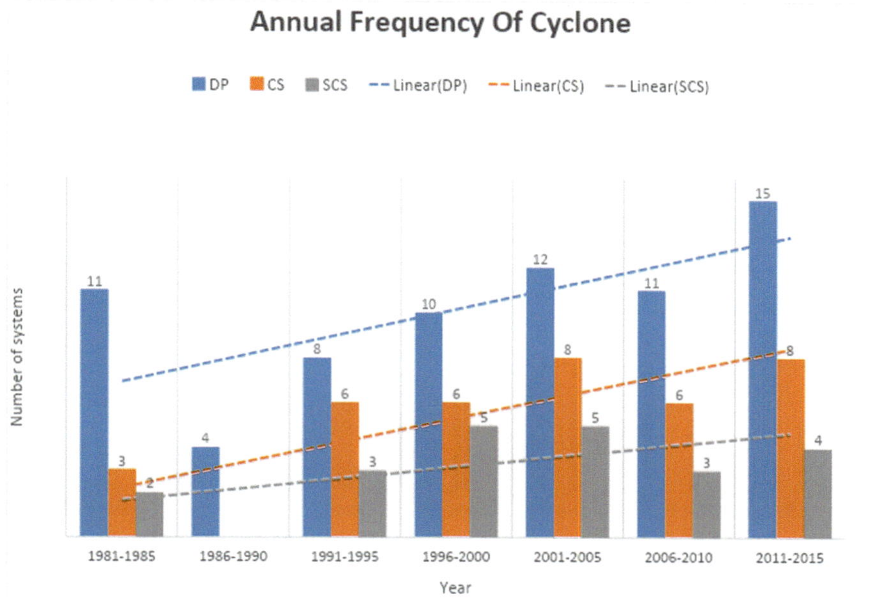

Fig. 7.4 Annual frequency of cyclone in Arabian Sea basin during 1981–2015

The most observed calm time was from 1986 to 1990. No cyclonic storms (CS) or severe cyclonic storms (SCS) were formed and only 4 depressions (DP) were observed during this period. So, average annual statistics will be less than one depression per year and exactly 0.8 depressions per year.

Figure 7.5 shows month-wise annual frequency of cyclone and gives clear month-wise dependence of cyclone during each five-year period. During 1986–1990, two cyclones had formed; first cyclone season in July and another two were in second cyclonic season of November. Between June 2006–2010 and October 2011–2015, 15 systems were developed which is marked as the most active months in the study period. 11 systems were developed in second cyclone season during 2011–2015.

Month-wise frequency of cyclonic disturbance during study period (Fig. 7.6) clearly shows that second cyclonic season is responsible for the formation of maximum number of cyclones. 32 cyclones were formed during second cyclonic season and 25 cyclones in first cyclone season. The first four months (January, February, March and April) are not generating any cyclonic system.

During January and February, the sea surface temperature (SST) will be less than 26.5 °C which is the primary condition required for cyclone genesis. But during March and April, humidity plays the role. Humidity will be less during summer season. During May, temperature soars and also the arrival of Monsoon increases the humidity and it facilitates the cyclone genesis. Figure 7.7 depicts the Annual cycle of monthly mean Arabian Sea SST (°C; 7.5°–20°N, 50–80°E) for the period 1979 to 2008. Comparatively, second cyclonic season is having less SST than first

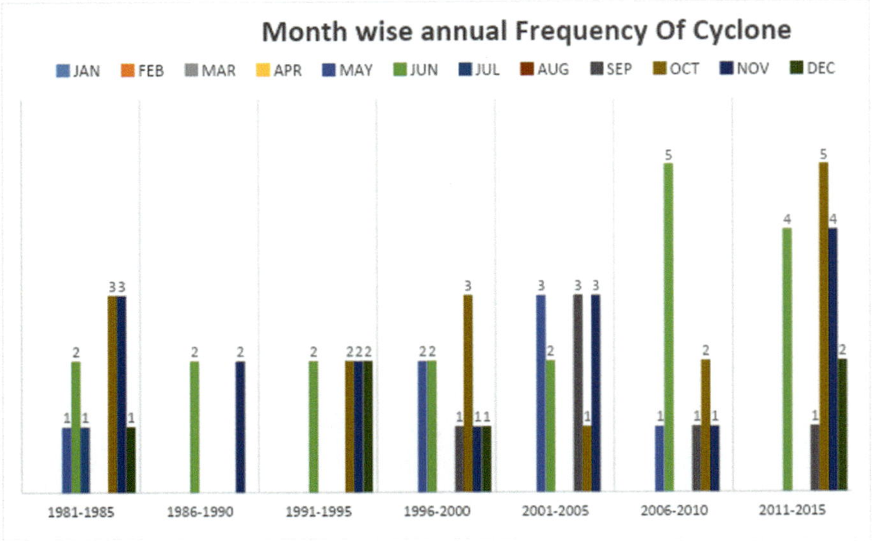

Fig. 7.5 Month-wise annual frequency of cyclone in Arabian Sea basin during 1981–2015

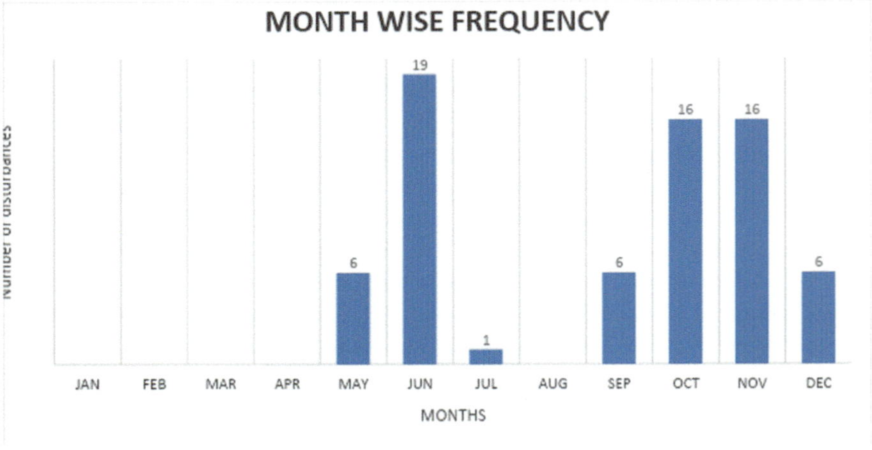

Fig. 7.6 Month-wise frequency of cyclone in Arabian Sea basin during 1981–2015

cyclonic season yet, cyclone genesis is more in the later season. This may be because of the presence of high humidity and low-pressure behaviour. In this season, relative humidity plays the key role. Primary parameters also play sufficient role in cyclone genesis, but these have not been included as part of this study.

The total life period of cyclonic storms in Arabian Sea during 26 years starting from 1991 is about 87 days (Fig. 7.8). Life period of super cyclonic storms (SUCS) is about 1.3 days (33 h), very severe cyclonic storms (VSCS) were there for 22 days 12 h

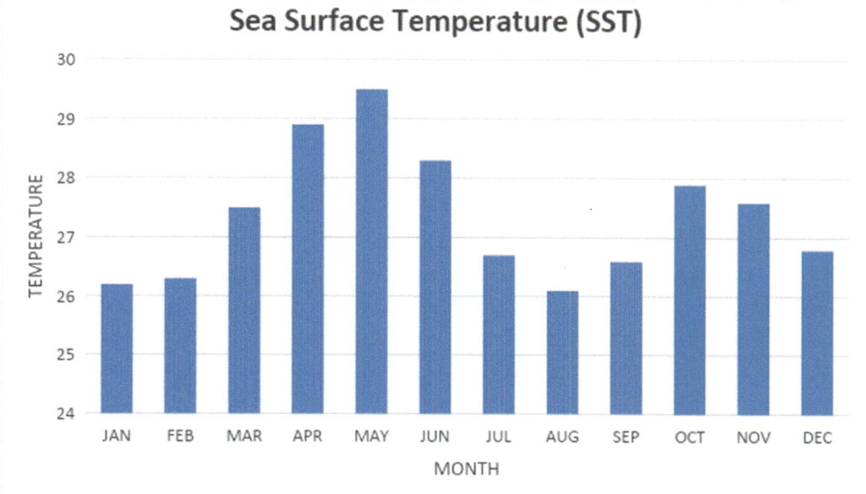

Fig. 7.7 Annual cycle of monthly mean Arabian Sea SST (°C; 7.5°–20°N, 50–80°E) for the period 1979 to 2008 2011 (*Source* Evan)

(540 h) and severe cyclonic storm (SCS) life period have been 16 days 15 h (399 h). Postmonsoon is more prone for cyclone origins. But here monsoon season also shows cyclones because June is added along with monsoon season. Super cyclone numbers got raised during postmonsoon season.

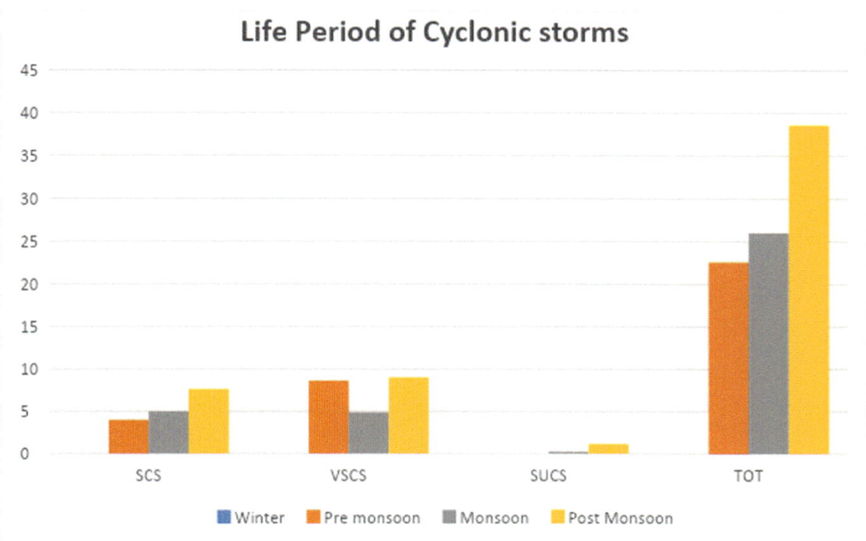

Fig. 7.8 Life period of cyclonic storms in Arabian Sea during different seasons 1991–2015

41 cyclonic storms formed in the Arabian Sea during 1979–2008; among these, 8 storms were classified as severe cyclonic storms and one is classified as super cyclone. These cyclones had an average lifetime of 3.4 days with a range of lifetimes from 1 to 9. This histogram (Fig. 5.8) proves the bimodal system and decreasing tropical cyclone activity during the peak of South West Monsoon. During premonsoon, storms occur during May or June. A study by Evan (2011) hypothesized that the preference of one month over the other is a function of the south-west monsoon onset date. Early onset will favour cyclone genesis in May and vice versa (Evan and Camargo, 2010).

Arabian Sea cyclone system develops mostly in two seasons which is already discussed. South-west monsoon onset depends on the cyclone genesis in May and June. But November storms were depended with low-pressure anomalies over the Bay of Bengal. The formation of Ockhi cyclone is more relevant during this as the low-pressure system over Bay of Bengal was responsible for the system generation. A general trend is observed in November storm formation in Bay of Bengal and Arabian Sea. A November storm formation happens either in the Arabian Sea or the Bay of Bengal but will not happen in both at the same time. The November storm formations in both basins are tabulated in Table 7.1.The exception cases for these formations were 1986, 2013 and 2015. Evan et al. hypothesized this behaviour up to 2008, and in the present study, this hypothesis has been re-verified by projecting the fact up to 2017.

Table 7.1 Tropical depressions over North Indian Ocean during postmonsoon season 1891–2017

Arabian Sea (AS)	Bay of Bengal (BOB)
1979	1981
1980	1983–1988
1982	1992
1986	1995
1993	1998
1994	2002
1997	2005
2003	2007
2004	2008
2009	2010
2011	2012
	2014
	2016–2017

Results and Discussion

In this study, an attempt was made to understand the increase in vulnerability of Arabian Sea towards cyclonic storms. Variation of tropical cyclones genesis in North Indian Ocean is discussed with latitudes. Tropical cyclones are formed in area between latitudes 8 and 15°N, but in the month of June to September, little activity is observed to the north of 15°N. By observing the tracks of cyclones, Bay of Bengal is the area of highest incidence but it is not usual to move across Southern India and intensify in Arabian Sea. Potential impact of cyclones towards mankind is increasing because of the increase in global population and property values in tropical cyclone-prone area.

Northern Indian Ocean is having two cyclonic season, May–June and October–November with primary peak in November and secondary peak in May. Arabian Sea is less prone to cyclones compared to Bay of Bengal in the ratio 1:4. The potential loss done by tropical cyclones is because of their destructive winds, storm surges and torrential rainfall. In this context, storm surge is more disastrous to coastal communities. Storm surges are also occurred in two cyclone seasons. When a cyclone moves towards a coast, then the right forward sector of the cyclone experiences on-shore wind which pushes sea water towards coast and will cause storm surges. By analysing tropical cyclones over Arabian Sea, it consists of bimodal system of cyclone genesis and postmonsoon cyclones are more severe than pre monsoon cyclones. Most of the Arabian Sea cyclones formed near to west coast and move northerly and northeasterly track. November season is vulnerable for the movement and intensification of cyclones from Bay of Bengal to Arabian Sea. Annual frequency of cyclones in Arabian Sea basin is showing an increase in frequency (Fig. 7.3). Second cyclonic season is responsible for the formation of maximum number of cyclones. 32 cyclones are formed in second cyclone season, and 25 cyclones formed in first cyclone season (Figs. 7.5 and 7.6). Life period of SUCS is about 1.3 days (33 h), VSCS is 22 days & 12 h (540 h) and SCS was active for 16 days &15 h (399 h) (Fig. 7.8). The hypothesis proposed by Evan, preference of one month over the other is a function of the south west monsoon onset date. Early onset will favour cyclone genesis in May and vice versa and is verified with data.

Conclusion

The present study reveals that Arabian Sea is now becoming more and more vulnerable towards cyclonic storms. The cyclone activity is not distributed all over the year, so it has a bimodal system. Bimodal system of cyclone genesis in Arabian Sea is having two cyclonic seasons, one is during May and other is at November. Postmonsoon cyclones are more dangerous than premonsoon cyclones. During monsoon, no cyclones are observed over Arabian Sea.

Coastal communities are becoming more and more vulnerable to a wide range of coastal hazards including severe storm events, tsunamis, shoreline changes, and resource degradation. This increase in vulnerability is mainly due to increasing coastal population. 23% of world's population (nearly 1.2 billion people) and more than 50% of the Indian Ocean region's population live within 100 km of shoreline and 100 m of sea level. Coastal communities are mostly living in rural areas and a small fraction in medium cities. So the living standard of coastal communities is poor as compared to urban living standards. In Indian context, most of the coastal communities are below poverty line and they are partially separated from the mainstream. So, basic last mile services and impending hazard warning and response systems become less effective. Increase in population density will increase the intensity of disasters because more and more people and infrastructure get exposed. This increase in population coupled with frequency and duration of storms, sea level rise and other coastal hazards cause severely impact and slow the disaster recovery processes. Climate change is also becoming a big threat to existence of these coastal communities. The impact of climate change is having primary impacts on Oceans. It is because the land area is less compared to ocean area. As we have already discussed that climate change is having effects in cyclone genesis and cyclones are also affecting coastal communities even before they make land fall. Therefore, it becomes imperative to make these communities adapt and resilient towards climate change. Hence, for building a resilient coastal community, it is required to build their copping capacities. Resilience is the capacity of an individual, community and a natural system to adapt to and recover from undesirable changes.

References

Aldinger, W. T. (1998). *1998 Annual Tropical Cyclone Report.* JTWC.

Evan, A. T. (2011). Arabian Sea tropical cyclones intensified by emissions of black carbon and other aerosols. *Nature, 479*, 94–97. https://doi.org/10.1038/nature10552

Evan, A. T., & Camargo, S. J. (2010). A climatology of Arabian sea cyclonic storms. *Journal of Climate, 24*, 140–158. doi:https://doi.org/10.1175/2010JCLI3611.1

Nobuhito Mori, T. T. (2016). Impact assesment of coastal hazards due to future changes of tropical cyclones in the North Pacific Ocean. *Weather and Climate Extremes, 11*, 53–69.

Chapter 8
Mapping Agricultural Drought Vulnerability at a Regional Level Using GIS—A Case Study

C. Prakasam, R. Saravanan, and Varinder S. Kanwar

Abstract Drought is a highly complicated yet least understood of all the other natural disasters. It is due to the lack of rain and shortfall of humidity. The existence of the farming drought's early warning framework can be remarkably valuable in a call for enhancing the drought readiness and to reduce the impacts of drought. During this examination, a GIS-based method was used for assessment of horticultural drought susceptibility for the hydropower undertakings of Larji watershed of Kullu region. Larji hydropower project (126 MW) undertaking was planned to be built in June 1984 on Beas River as a power advancement plot and in a tender to outfit the massive hydropower potential available in the State. Binwa watershed, located in Kangra area of Himachal Pradesh, comprises of a stroke of Lesser Himalayas and Shivalikrough inclines. The key indices that are to be assessed to describe the farming drought susceptibility are Normalized Difference Vegetation Index (NDVI), Normalized Difference Water Index (NDWI), Land Use Land Cover (LULC), and Land Surface Temperature (LST). According to the drought's impact aspect due to these indices, the weight is specified. On account of the weights, overlay analysis is set up to provide drought susceptibility zones in the investigation. Thematic maps of every index will be formed using GIS software based on allowed loads. The results of the indices are such as NDVI ranges from − 0.1968 to + 0.625, LST varies from − 9.88 to 44.18, NDWI ranges between − 0.224 and + 0.278, and the Land Use Land Cover classification classified the forest cover, degraded forest, fallow land, barren land, and glacier. Based on the weights given, the study shows drought susceptibility's steps in the region with a reasonable range of values. This is a startling issue because it is in the perennial river's downstream locale and Larji dam. The guide that arises out of this local drought susceptibility can facilitate leaders to generate valuable farming drought mitigation tactics.

C. Prakasam · R. Saravanan (✉) · V. S. Kanwar
Department of Civil Engineering, Chitkara University, Himachal Pradesh, India
e-mail: saravana8689@gmail.com

C. Prakasam
Department of Geography, Assam University, Diphu Campus, Diphu, India

R. Saravanan
Ecofirst Services Ltd, Tata Consulting Engineers Limited, Bangalore, India

V. R. Sharma and Chandrakanta (eds.), *Making India Disaster Resilient*,
https://doi.org/10.1007/978-3-031-50113-5_8

Keywords Agricultural drought · Vulnerability mapping · Normalized difference vegetation index · Himachal Pradesh · Geographic information systems

Introduction

The definition of drought differs based on the causal factor. Agricultural drought is the result of a lack of availability of water for the agricultural practices followed by either the hydrological or the meteorological drought. It is not only the natural phenomena that can be the cause for this, but also the man-made activities that contribute toward the agricultural drought. Drought susceptibility is a concept which indicates the likelihood of harms from peril in a particular place by concentrating on the framework status former to the catastrophe. Drought susceptibility has been rendered as a probable misfortune in the district because of the lack of water during the drought season. The collection of sufficient temporal and spatial data is tremendously disturbing, mainly in areas with hard topography. The implementation of GIS, therefore, is of primary importance. Using the GIS tool Normalized Difference Vegetation Index (NDVI), Normalized Difference Water Index (NDWI), Land Use Land cover (LULC), and Land Surface Temperature (LST) have been analyzed in this research work.

Lei, Yongdeng, and Lili Luo built up a quantitative drought risk assessment demonstrated for the mid-season paddy crop type and evaluated that the drought vulnerability also incorporated a progression of vulnerability mapping of drought. The vulnerability was mapped for northeast China with the most astounding vulnerability index pursued by South China, North China, and Northwest China in the order of their risk. Nazareth, Ethiopia, surveyed the spatial-worldly difference in farming drought period pattern, and seriousness of this was assessed using NDVI, Standard Precipitation Index (SPI), and Water Requirement Satisfaction Index (WRSI), which were unusual. The net results indicate that the cropping periods from 2000 to 2005 experienced an improved horticultural drought and a reduction of grain yield with the spatial difference in the level of significance. Padhee S. K. made an attempt to differentiate affectability of downscaled soil humidity to meteorological drought for estimation of agricultural drought period. Irregularity in NDVI and soil humidity (multi-day composites) for Rabi periods of 2000–01 to 2009–10 were managed. Fatemi, Mehran, and Mahdi Narangifard emphasized that for quick and long haul change discovery and observing in terms of spatial and temporal changes utilizing (RS) remote sensing and GIS data is of utmost significance in producing data about the latest LULC, LST, and NDVI showing ERDAS Imagine 9.2 and results show a massive urban growth reflected as degraded vegetation. Le Page, Michel, and MehrezZribi broke down the prospective for utilizing the indices such as NDVI, (SMI) Soil Moisture Index, and LST for Northwest Africa in the measurement and consistency of drought. The result illustrated a high correlation between them and a drought period of (2007–2017). Wilhelmi, Olga V., and Donald A. Wilhite presented a methodology for spatial, GIS-based agricultural drought vulnerability assessment for Nebraska hypothesizing that the vital social and biophysical aspects were soils, climate, soils, land use, and irrigation. The outcomes demonstrate that the most susceptible regions to agricultural drought period are the non-watered cropland and rangeland on sandy soils situated in territories with an exceptionally high likelihood of lack of dampness in regular yield. Wu, JianJun, GuangPoGeng, HongKui Zhou, JingHui Liu, QianFeng Wang, and JianHua Yang selected six main crops as the agriculturally drought affected and assessed vulnerability for global scale at a 0.5° resolution and mapped spatially. The outcomes showed that the zone rates of various evaluations of worldwide vulnerability to agricultural drought were in the order of high to low, i.e., 38.96% to 7.26%, respectively.

Objectives

The Larji watershed study area is completely influenced by the outcome of construction and operation of the hydropower dam which indirectly influences the agricultural practices. The present research calculates the indices to identify the level of drought impact in the Larji watershed. This research also agitates the administrators and the local farmers regarding the rising drought susceptibility in their region in a spatial and sequential form and aids them in drawing out appropriate mitigation measure.

Study Area

Larji hydropower project has an installed capacity of 126 MW. It is located on Beas River, which is in Kullu district of Himachal Pradesh and possessed by Himachal Pradesh state (Fig. 8.1).

Electricity Board Limited (HPSEB). The catchment area of the H.E project is extended over an area of 4921 km^2. The Larji dam site is situated at an altitude of 2299 m mean sea level. The project was accomplished in September 2007. The catchment experiences precipitation due to the southwest monsoons and the western disturbances that bypass the northwest part of the country at the time of winter as well. The region's flora is mainly constituted of various broadleaves species such as Kosh, Khanor, Walnut, and Kharusu. This comprehension of medicinal plants can also be seen in this range. Due to the intrusion of expansion activities like road construction, water supply scheme, etc., there is a high level of soil erosion affecting soil integrity. The livelihood of the people in the region is cultivation, agriculture, horticulture, and rearing of sheep and goat. The storage capacity of the Larji reservoir is 230 ha-m, which is sufficient for operating the power station at full installed capacity for more than four hours in a day during lean periods. The project will facilitate energy generation of 587 Gigawatt-hours in 90% reliable year pattern of flows. Since the project is on the River Beas, its standard slope is calculated when the river emerges

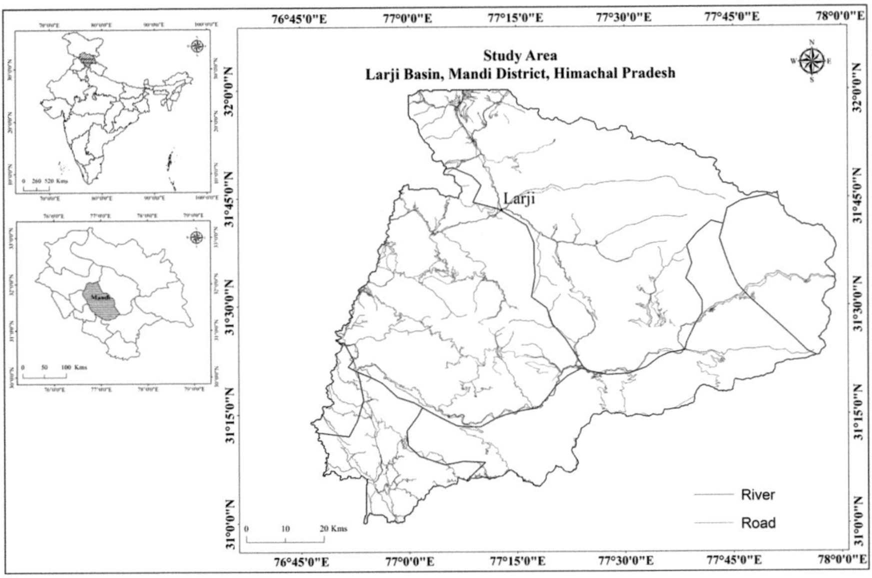

Fig. 8.1 Study area: Larji Basin, Mandi district, Himachal Pradesh

from the Rohtang, to cover a distance of 28.9 km up to Manali; there is a fall of 77 m/km.

The major portion of the catchment falls in the district of Kullu and a part in Mandi, i.e., Seraj Forest division, Parvati forest division, Kullu forest division, Great Himalayan National Park, Kanawar Sanctuary, Khokhan Sanctuary, and a part of Panarasa Forest Range of Mandi Forest Division. It covers a distance of 38.6 km from Manali to Kullu, and there is a fall of 15.6 m/km. Soil erosion is the disparaging process for the soils, which may be defined as the wearing away of the earth's surface by breakdown and transportation of the soil by water, ice, and wind. Catchment area for Larji hydropower project is cut off from heavy rainfall region around Dharmashala in the west by a high ridge, running north–south, and Uhl River in the east.

Methodology

The Landsat 8 OLI/TIRS C1 Level-1 (https://earthexplorer.usgs.gov/) has been utilized for this research to calculate the indices. Each index is calculated using the computation between the different reflectance bands. After the outcomes are inferred, a separate data segment is created in the GIS tools to enter the weights. The weight that is given depends on the properties or the nature of the index that would support the event of the drought helplessness in the examination area. The weighted overlay analysis is finished regarding the weights given, and the mapping is generated with the classification that ranges from low to exceptionally high defenselessness to drought.

Normalized Difference Vegetation Index (NDVI)

Sruthi S. and MA Mohammed Aslam emphasized that NDVI is one of the most commonly used indices to estimate the measure of radiation being consumed by plants. The visible and near-infrared bands are used to calculate this index. NDVI ranges between (− 1 and + 1) to depict vegetation density and health. The λred and λNIR represent the reflectance of the red and NIR bands in the below-mentioned formula.

$$\text{NDVI} = \frac{\lambda\text{NIR} - \lambda\text{red}}{\lambda\text{NIR} + \lambda\text{red}}$$

Alshaikh and AmalYahya surveyed NDVI for June 2008 using the data obtained from the EgySat-1 satellite to calculate four NDVI-derived water indices. There were seven factors and two constraints that were executed in the view of delineating the rice cultivation suitability map using the Spatial Decision Support System (SDSS), and the inputs involved two constraints and seven factors.

Land Surface Temperature (LST)

It is the combination of the temperature of bare soil and vegetation. Sruthi S. and MA Mohammed Aslam also underlined that the LST can be estimated using the infrared spectral channels to calculate the upper brightness temperatures in the atmosphere.

$$\text{LST}_{ij} = \frac{\text{BT}_{\max} - \text{BT}_j}{\text{BT}_{\max} - \text{BT}_{\min}}$$

where the BT_j is the average brightness temperature (BT) value for a composite period of interest and $\text{BT}_{\min}$ and $\text{BT}_{\max}$ are minimum and maximum pixel-specific temperatures logged over the period of long term.

Gidey, Eskinder, OagileDikinya, Reuben Sebego, EagilweSegosebe, and AmanuelZenebe made utilization of the MOD11A2 Terra and calculated the Land Surface Temperature (LST) by 250 m spatial goals alongside the half-breed TAMSAT month-to-month rainfall resulting in imperative ascent by 0.52–1.08 °C showing drought period level and additionally proposed that the impact of drought period could be diminished through involving the smallholder ranchers in a wide scope of on- and off-ranch rehearses.

Normalized Difference Water Index (NDWI)

Gu, Yingxin, Eric Hunt, Brian Wardlow, Jeffrey B. Basara, JesslynF. Brown, and James P. Verdin highlighted that the Normalized Difference Water Index (NDWI) is an alternate for assessing and also an additional index that is strongly linked to the plant water content.

Hazaymeh, Khaled, and Quazi K. Hassan used the MODIS and Landsat 8 data to outline agricultural drought indicator (ADI) by assessing the indices Normalized Difference Water Index (NDWI), Standardized Contrast Vegetation Index (NDVI), VSDI, NMDI, MSI, and Land Surface Temperature (LST).

$$\mathrm{NDWI} = \frac{\lambda\mathrm{Green} - \lambda\mathrm{NIR}}{\lambda\mathrm{Green} + \lambda\mathrm{NIR}}$$

The NDWI is calculated using the green reflectance band of water and NIR.

The outcomes establish that this technique would be significant for managing and monitoring agrarian drought environments in semi-parched areas.

Land Use Land Cover Mapping

The LULC classification was done in order to obtain a wide level of cataloging to infer different LULC classes. As this work focuses on agriculture, the classification will be in accordance with the vegetation components and water resources. Raghavswamy V., R. Nagaraja, and N. C. Gautam monitored the land use/land spread to delineate drought inclined area during a period from 1980 to 1982 as far as rural use is concerned. It was categorized into 12 Level-II and 5 Level-I classes. The pattern of progress in LULC was featured as the drought inclined area.

Analysis and Discussions

The Landsat 8 data was operated for the purpose of examining the drought indices. The components chosen for the vulnerability study are NDVI (Fig. 8.2), LST (Fig. 8.3), and NDWI (Fig. 8.4). The first index, NDVI, ranges from -0.1968 to + 0.625 between the titular ranges of − 1 and + 1. The area covers the maximum degree of the low range representing drought-prone to a maximum degree. Even the highest range of value does not point to prosperity in the vegetation which indirectly shows the susceptibility in the study area. The second index LST varies from -9.88 to 44.18, representing that the surface temperature is high and that supplies much to the evapotranspiration leaving no dampness for the agriculture. The higher the LST, the higher is the area prone to drought. The third index, NDWI, in the study area ranges between − 0.224 and + 0.278 with an extensive negative range. The LULC analysis (Fig. 8.5) was carried out with vegetation cover as primary key, in that aspect the Landsat data was classified as water, forest cover, fallow land, degraded forest, barren land, and glacier. The results show that the extent of degraded forest is comparably high followed by the forest cover. Along the poor vegetation, the fallow land is present. The river flows as the glacier melts are present as considerable land cover. The barren land is present in the least.

The LULC examination results into forest cover, degraded forest, fallow land, barren land, glacier, etc., upon which generally covered by the poor vegetation, fallow land, and glacier. Based on the distinctiveness of the land use, suitable weight is given, trailed by the re-classification of the range of the indices with appropriate weights that are given.

The weighted overlay analysis was carried out to demarcate the drought-vulnerable zones (Fig. 8.6). Based on the literature studies and expert's opinion, the weights were given. The drought vulnerability here is focused on the vegetation, and hence weights are given focusing on the factor that might the vegetation in the region. The weights were given based on indices that show sign of vulnerability, and NDVI ranging between 0.62 and − 0.19 is categorized into four raster layers. The values ranging in negative were given value of 4 weights since the negative value represents the poor vegetation in the region and in the same manner up to 1. Likely the LST, the highest value 44.18 and its subsequent category are given a weight of 4 as the Land Surface Temperature is high and the contribution to the drought vulnerability is high. The NDWI also follows the NDVI weight generation and is provided accordingly. The fourth layer is that the weight for the Land Use Land Cover categorized as barren land and poor vegetation is given as 4, glacier's contribution is marked as 3, forest cover and fallow land are given as 2 and water is marked as 1 accordingly that contributes to the drought vulnerability.

Based on the weight given, the analysis is carried out providing with drought vulnerability categorized as low, moderate, high, and very high. About 41.08% of the study area is very highly vulnerable to drought in the near future, 45.63% is of highly vulnerable, 13.02% as moderate, and 0.26% as low vulnerability. The vegetation in this study area is supported by the perennial river, and hence, the drought vulnerability of about 47.2% of the vegetation cover is quite alarming. The natural free-flowing water is dammed for the production of hydropower in the upstream side, and hence, the downstream vegetation is highly vulnerable. The drought vulnerability is very high in near the place where the water is controlled by the hydropower project. This study is primarily focused on the lean months, and hence, the lean month (November to February) was chosen. The glacier part of the region is shown as less vulnerable since the vegetation is the primary focus in the study. If this condition prevails in the region, the drought percentage will increase and also the glacier contributes much to the study area which remains ice most of the year not paving way for the source of water in the region. Hence, the stored water in the dam is the main source of water, which is controlled for hydropower production. This study helps us to view the alarming nature of the drought vulnerability of which the construction and

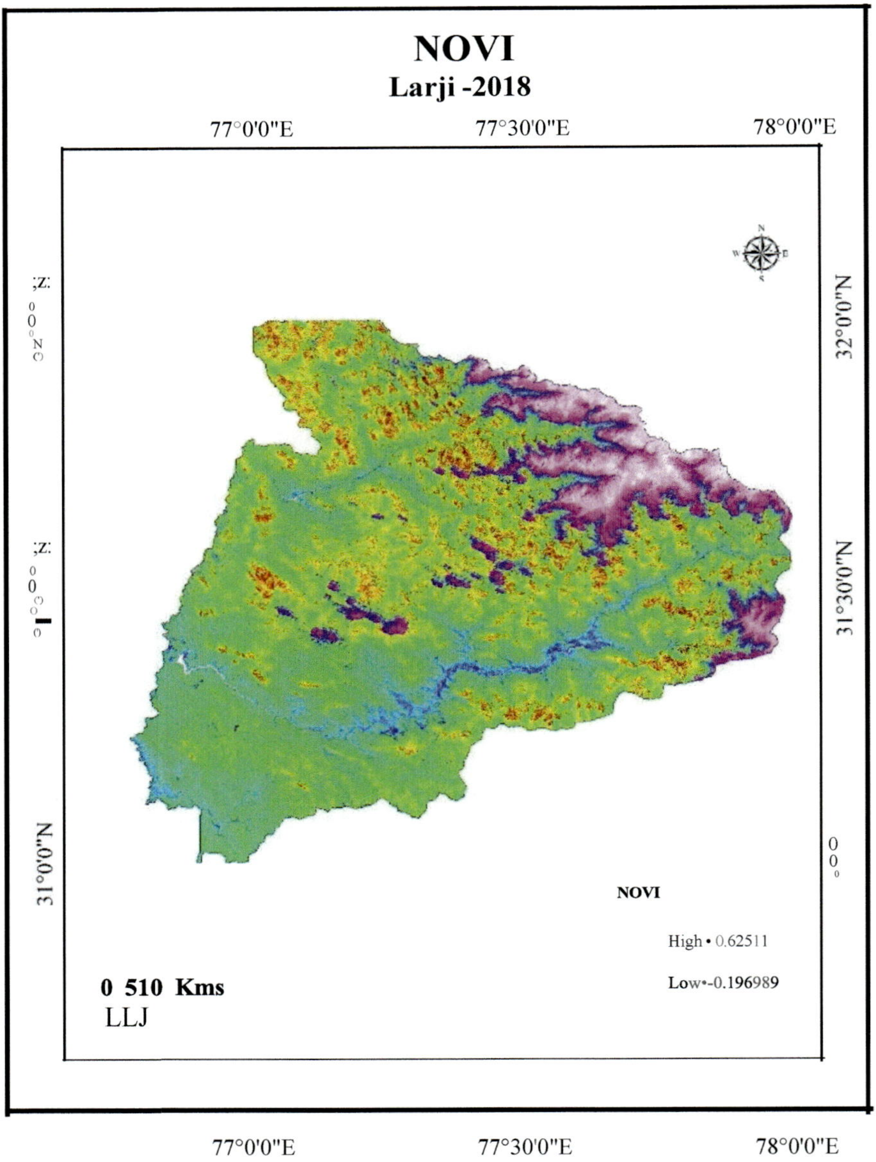

Fig. 8.2 NDVI analysis

operation of the Larji hydropower dam could be a reason. Hence, the maintenance of the minimal flow is emphasized much as a part of the analysis.

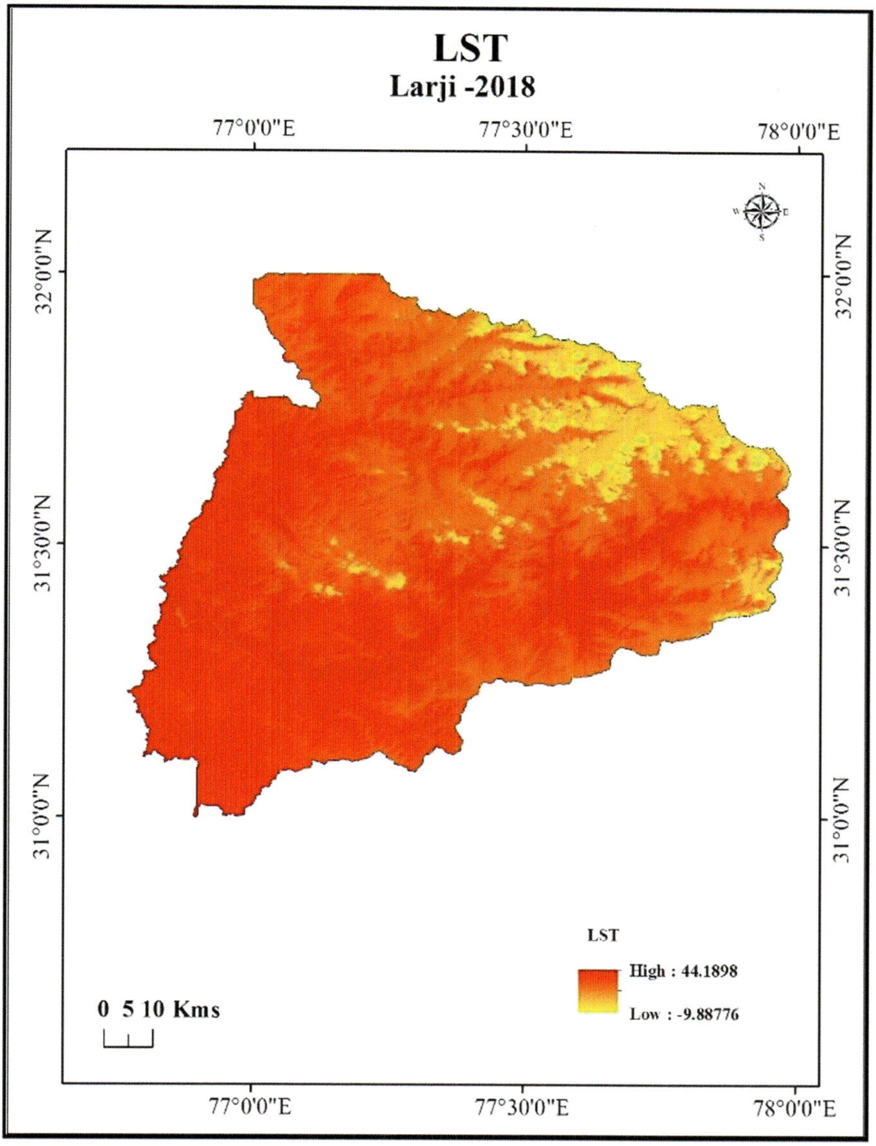

Fig. 8.3 LST analysis

Validation from the Field Data

The questionnaire survey was carried out in the Larji downstream (Fig. 8.7), and the following analysis was concluded. The data on the Larji hydropower project's ecosystem states that the flora in the region is occupied by various broadleaves species

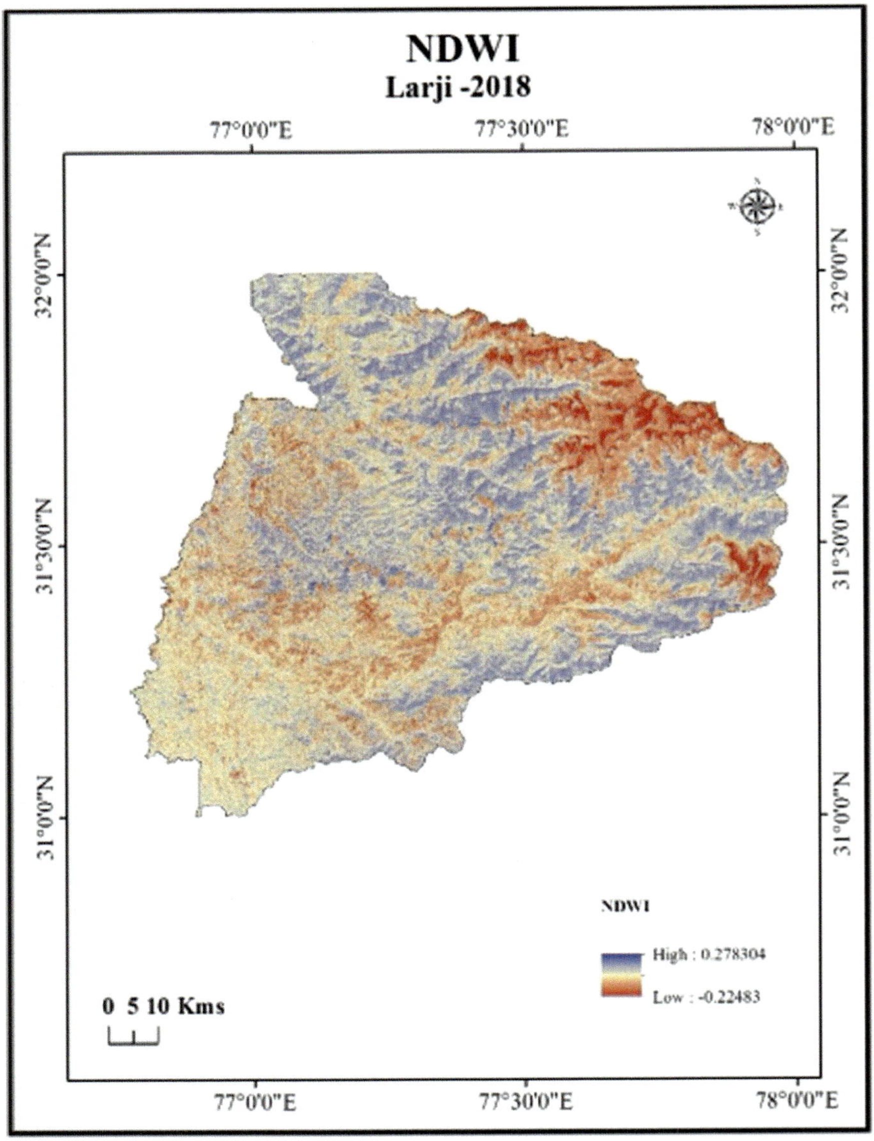

Fig. 8.4 NDWI analysis

like Kosh, Khanor, Walnut, and Kharusu. The long range of medicinal plants also can be seen in this range. Due to various development activities like construction of the road, water supply schemes, etc., the soil erosion takes place at a higher level causing the soil integrity. The occupation of the people in the region is cultivation, agriculture as well as horticulture and rearing of sheep and goat to meet out their cash requirements for their livelihood.

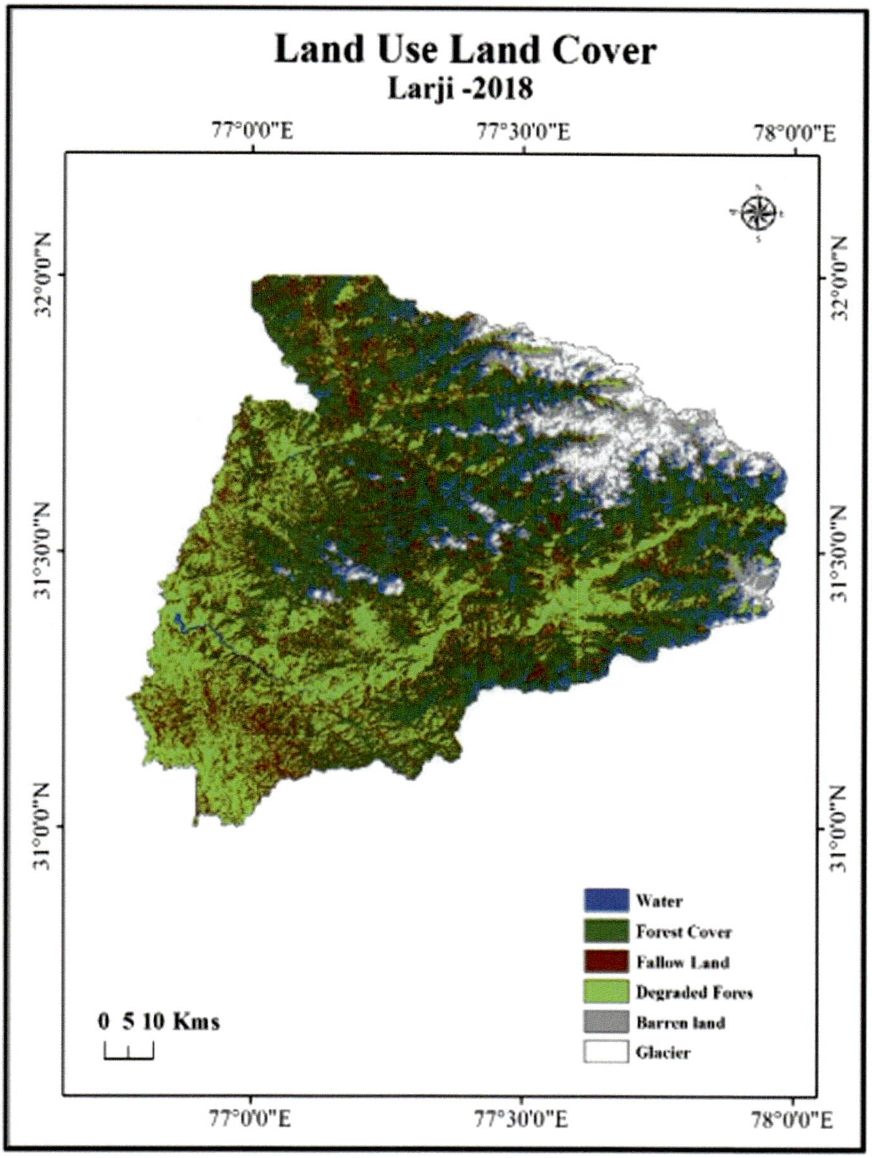

Fig. 8.5 LULC analysis

The basic perception about the impact of hydropower project was surveyed, and descriptive analysis was carried out. The null hypothesis is that the environmental flow is maintained in the dam, and there is no significant impact on the ecology considering the availability of water for the livelihood purpose. These questionnaires were surveyed among the farmers, locales, fishing occupation, etc. The significant

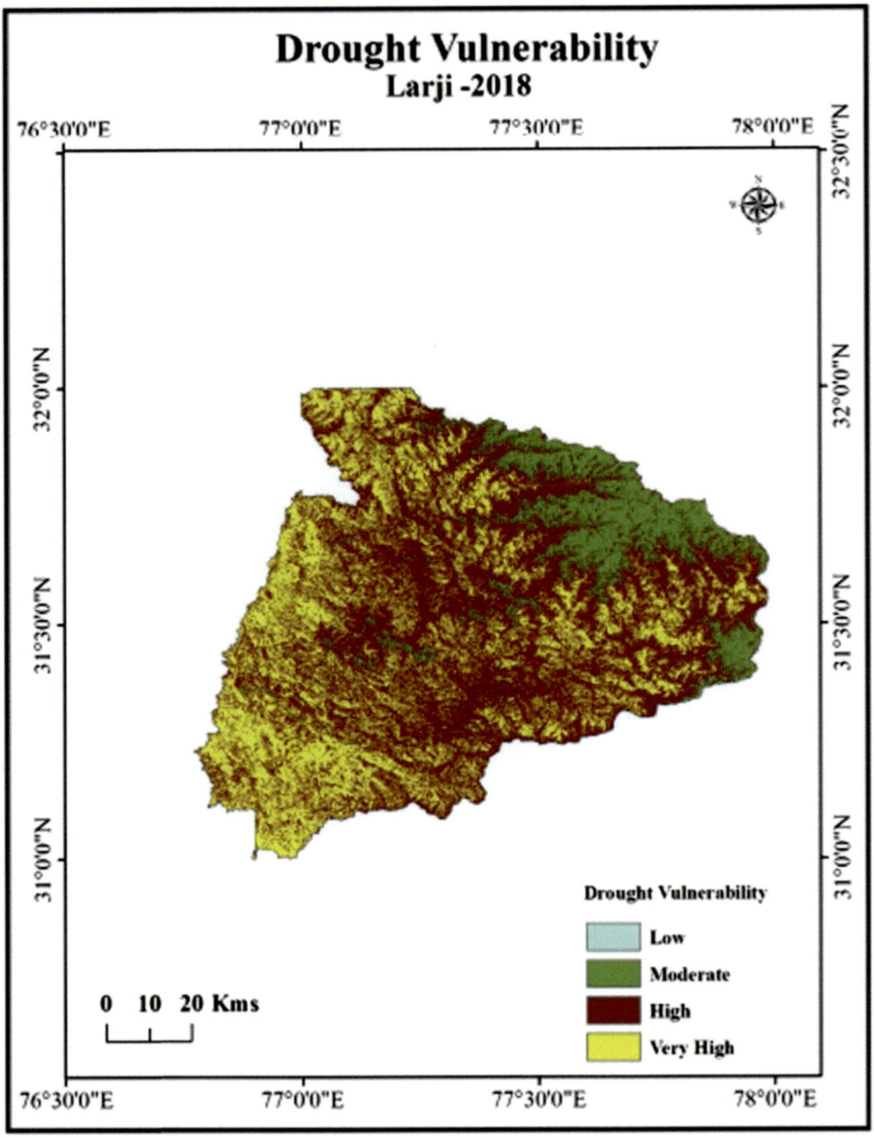

Fig. 8.6 Drought vulnerability mapping

P-value for the variables supporting the maintenance of environmental flow is less than 0.05; hence, the hypothesis cannot be validated. In the selected study area, the agriculture is heavily affected, and one of the main reasons stated by the locale is the operation of the hydropower project. The water required for the survival of the fish ecosystem is also at stake as the water during the lean period is much less than the required. One of the main sources of water for the livelihood in these region is the

Fig. 8.7 Questionnaire survey

dam water, and during the lean period, the hypothesis of sufficiency is insignificant; hence, there is deficiency of water for the basic amenities during the lean period. On the other hand, the administers of the Larji hydropower projects were also questioned of which the analysis are the Larji hydropower project is located at 969 m elevation. It is primarily focused on producing hydropower, and hence, the maintenance of the minimal flow is the matter of many factors to be considered. The 42 cumecs is the total discharge during the lean period of which 15% is 7 cumecs. The lean period is here considered from January to February. For the better functioning of the dam, the gate height cannot be less than 10 cm. The total storage of the dam is of about 343 hec/m, of which 250 cumec for each 3 machine constitutes together to produce 126 MW.

Conclusion and Recommendations

Drought is a very complex phenomenon yet most understudied concept. Effective measures need to be adopted in order to deal with it in a lot better way. This has to be understood so that we are better prepared in advance to act against it at times of the occurrence of droughts.

In order to achieve this, a GIS-based assessment of horticultural drought susceptibility methodology has been harnessed to the best of its capability for the hydropower undertakings of Larji watershed of Kullu region. The main indices utilized in the assessment are NDVI, NDWI, LULC, and LST. The final result of the amalgamation of drought susceptibility factors was calculated by the simple addition of the weights GIS. Four maps which were previously created were added together by using Raster Calculator. Geographic areas with large numeric values of the sum of weights are believed to be relatively more susceptible to agricultural drought than areas with lesser values. The resulting map was re-classified into a class of six identifying areas with "low susceptibility", "moderate", "high", and "very high vulnerability". Using this drought susceptibility map, the drought-prone areas are recognized and are classified as very high, high, moderate, low to moderate, and low drought susceptible zones. The outcomes of various factors on the resulting map were also analyzed. The decision-makers can make use of these maps to supply useful drought mitigation tactics in the essential area. Here, it is expected that the drought is the result of the man-made activities such as construction and operation of the hydropower dam in the region. The minimal ecological flow maintenance is a question mark in terms of this study area.

Finally, it is recommended that the administration must genuinely look into the drought issue through encouraging resource management practices and enhance the drought assessment and implementation unit to facilitate in minimizing the unfavorable impacts of drought. The NGT order to maintain 15% of the minimal flow is the prime solution that the dam administers could follow as a primary mitigation measure. Concerned experts must work closely with partners who may be exclusively or indirectly influenced by the drought period in doing this. The socio-economic data ought to be similarly examined while assessing drought risk to understand the helpless groups in a much better manner.

Acknowledgements The research work done is a part of DOES & T, Himachal Pradesh, funded research project. We would like to express our sincerest gratitude to DOES &T, Himachal Pradesh, for funding this research project.

References

Fatemi, M., & Narangifard, M. (2019). Monitoring LULC changes and its impact on the LST and NDVI in District 1 of Shiraz City. *Arabian Journal of Geosciences, 12*(4), 127.

Gidey, E., Dikinya, O., Sebego, R., Segosebe, E., & Zenebe, A. (2018). Analysis of the long-term agricultural drought onset, cessation, duration, frequency, severity and spatial extent using Vegetation Health Index (VHI) in Raya and its environs, Northern Ethiopia. *Environmental Systems Research, 7*(1), 13.

Gu, Y., Hunt, E., Wardlow, B., Basara, J.B., Brown, J.F., & Verdin, J.P. (2008). Evaluation of MODIS NDVI and NDWI for vegetation drought monitoring using Oklahoma Mesonet soil moisture data. *Geophysical Research Letters, 35*(22).

Hazaymeh, K., & Hassan, Q. K. (2017). A remote sensing-based agricultural drought indicator and its implementation over a semi-arid region, Jordan. *Journal of Arid Land, 9*(3), 319–330.

Le Page, M., & Zribi, M. (2019). Analysis and predictability of drought in northwest africa using optical and microwave satellite remote sensing products. *Scientific reports, 9*(1), 1466.

Lei, Y., & Luo, L. (2011). Drought risk assessment of China's mid-season paddy. *International Journal of Disaster Risk Science, 2*(2), 32–40.

Nazareth, E. (2014). Remote sensing and GIS based agricultural drought assessment in East Shewa Zone, Ethiopia. *Tropical Ecology, 55*(3), 349–363.

Padhee, S. K. (2013). Agricultural drought assessment under irrigated and rainfed conditions. PhD diss., Master Thesis. Andhra University, Visakhapatnam, India.

Raghavswamy, V., Nagaraja, R., & Gautam, N. C. (1983). Application of multidate satellite imagery for land use planning and management in the drought prone areas of Karnataka. *Journal of the Indian Society of Photo-Interpretation and Remote Sensing, 11*(3), 47–54.

Sruthi, S., & Mohammed Aslam, M. A. (2015). Agricultural drought analysis using the NDVI and land surface temperature data; a case study of Raichur district. *Aquatic Procedia, 4*, 1258–1264.

Wilhelmi, O. V., & Wilhite, D. A. (2002). Assessing vulnerability to agricultural drought: A Nebraska case study. *Natural Hazards, 25*(1), 37–58.

Wu, J., Geng, G., Zhou, H., Liu, J., Wang, Q., & Yang, J. (2017). Global vulnerability to agricultural drought and its spatial characteristics. *Science China Earth Sciences, 60*(5), 910–920.

Part III
Fire/Smog

Chapter 9
HRVC Assessment of Urban Fire Hazard: A Case Study of Malviya Nagar, Delhi

Vishwa Raj Sharma, Kavita Arora, and Kamal Bisht

Abstract The cities in India have the poorest fire safety records of any of the developing countries. Homes and public buildings are generally built and expended for requirements without considering the safety measures. Commercial outlets within the residential areas are common phenomena in urban areas of India. There is a critical need for awareness and urgent debate about the growing vulnerability of urban dwellers in the context of unregulated spatial growth and development. In last few years fire hazard became a major problem for urban dwellers in India and caused huge number of death, loss of property and environmental degradation. Delhi is also known as heavy fire-risk city because of its ever-growing population, unplanned construction, crowded places, high-rise buildings and commercial and industrial growth in NCR. The city is also known for ignoring the safety laws to favor the economic activities. The recent incident happened in Khirki Extension a crowded residential area of Malviya Nagar was one such example. In this context the study analyzed the spatial distribution of fire incident occurred in different areas of Delhi in last few years, their impacts on residents and causes which constitute and increase vulnerabilities. The study also suggested some measures to manage the fire in urban areas.

Keywords Urban fire · Hazard · Vulnerability · Warehouse

V. R. Sharma
Department of Geography, University of Delhi, New Delhi, India

K. Arora
Department of Geography, University of Delhi, New Delhi, India

K. Bisht (✉)
Shaheed Bhagat Singh College, University of Delhi, New Delhi, India
e-mail: kamal_bisht24@hotmail.com

V. R. Sharma and Chandrakanta (eds.), *Making India Disaster Resilient*,
https://doi.org/10.1007/978-3-031-50113-5_9

Introduction

Urban fire can be defined as fire predominantly occurs in the cities or metropolitan as well as urban area with the prospective of quickly extent to nearby buildings that may be cause of loss of human life and valuable property such as residential area, commercial offices, vehicles and schools.

In modern world of rapid urbanization, vulnerabilities for fire incidents are increased in both developed and developing nations. In developed nation such as London Grenfell tower, Lancaster West Estate, 24-story residential building was under fire about 13 days and more than 150 firefighters remain engaged to control that, 65 people were rescued, and 74 people lost their life. Breckland lodge fire was controlled by more than 70 firefighters. Chiles port city, Valparaiso city face such an incident where more than ten thousand people evacuated from the fire, which cause of death at least 12 local resident and vanished number of homes. This phase of rapid urbanization in India makes the city more complex in structure and vulnerable to fire hazard. Delhi is continuous sufferer of fire hazard from many decades, a number of people loss there life and wealth, and destruction of resource as well as environment degradation also occurred. The Uphaar Cinema fire is one of the worst fire tragedies in the history of Delhi. In 1997 when fire broke in Cinema 59 people were died and 103 were seriously injured due to stamped during the fire. Lack of safety instruments and unacquaintedness about them were the main reasons of that accident. People were not aware about the tackling of this type of mishap, and most of the deaths occurred due to stamped. Few years back another massive fire broke in the building of National Museum of Natural History, and precious historical and heritage collections were burnt and turned into ashes. Some of the newspaper reported that "Fire safety mechanism was not functioning".

Methodology

Hazard, risk, vulnerability and capacity analysis (HRVC) is one of the prominent tools in disaster management. In the present study fire hazard incident in Delhi was analyzed and risk and vulnerability of local community were depicted through Geographical Information System (GIS) and remote sensing. The study tried to correlate the frequency of fire incidents and increasing population density along with built-up expansion from 2003 to 2016. Tables and graphs are prepared to show the number of fire accidents that occurred in the last few years. Remote sensing technique is used for analyzing urban expansion which became a cause of urban fire so frequently in high-density population areas. High-density built-up areas that are highly vulnerable and have less capacity to cope or resist hazards are identified, using remote sensing and GIS application. Secondary sources like Delhi Fire Service reports, different articles and research papers are used to acquire information about the fire incidents that occurred in Delhi.

Hazard, Risk, Vulnerability and Capacity Analysis (HRVC)

The HRVC analysis is a compound of different dimensions of disaster management. Firstly, understand the components of this model systematically. Hazards can be defined as events that create threats to people and infrastructures, which may cause a disaster, and there are two classifications of these: manmade (chemical accident, road and railway accidents, fire, building collapse) and natural (floods, droughts, earthquake cyclones, tsunami, landslides). Vulnerability is the extent to which a community and geographic area are likely to be deformed by the cause of a particular hazard on account of its proximity to hazardous terrain or a disaster-prone area. It is the likely extent of damage due to a hazard. It also has classifications such as physical vulnerability (depending on physical location of people), economic vulnerability (economically weak people are more vulnerable) and social vulnerability (some sections of the population such as old age people and children). According to the Disaster Management Act, 2004, disaster is defined as a "catastrophe, mishap, calamity or a grave occurrence in any area arising out of natural or manmade causes, or by accident or negligence, which results in substantial loss of life and human suffering or damage to, or destruction of, or degradation of environment, and is of such a nature, or magnitude as to be beyond, the coping capacity of the affected community of the affected area". Capacity is the abilities for understanding, skills, strength and resources of individuals or communities, which empower them to prevent as well as prepare or stand against a disaster. Risk is the probability of harmful consequences such as deaths, property or livelihoods loss as well as economic or environmental damage of community which, resulting from interaction between natural or human-induced hazards and vulnerable conditions (HPC Report, 2001).

Analysis of Past Fire Incidents in Delhi

Fire, cause by human or nature, can pose hazard to people, properties and environment, possibly resulting in psychological damage, physical injuries, even death and significant economic losses (Yao & Zhang, 2016). Delhi is highly prone to hazards like earthquake, flood, fire accident as well as building collapse and epidemics. Rapid urbanization makes the city more vulnerable about fire accident. NCT of Delhi total area is 1483 km^2, which holds a population of about 1, 67, 87, 941 in number. The density of the city is about 11,320 population per sq. km (Census of India, 2011). The substantial growth of population and industrialization make the city more vulnerable for different disasters. In the compact areas of the city fire hazard is mostly happened due to chemicals, LPG explosives as well as short circuit of electrical systems. In the city, areas which gradually congested because of influx of population become more vulnerable for the fire hazard. According to Delhi Fire Service statistics, Delhi had more than 234,291 fire incidents during 2003–2016 that results 3100 deaths and 20,239 person's injuries. In this period 102 medium and 11 serious fire accidents were occurred. The Uphaar Cinema tragedy and Lalkuan fire are the well-known episodes. Fire safety provisions were introduced in the building bye-laws in 1983. The Delhi Fire Prevention and Fire Safety Act, 1986, was notified by the Government of India in a gazette on December 12, 1986, after a fire incident in Sidharth Hotel, New Delhi.

Table 9.1 highlights that in the last fifteen years fire accident increased in the city, which caused loss of property and human life.

Figure 9.1 shows the number of call to report fire incidents received by the fire department in different years. Frequency of calls to report fire incidents was increased in North West zone in last few years, but it always remained high in South zone. Frequent calls itself show less preparedness and less resilience regarding fire accidents.

Figure 9.2 clearly shows that every year Delhi faces the challenge of many fire incidents. The figure showed number of deaths occurred in Delhi due to fire from 2003 to 2016. In 2005 to 2006 more than 450 people were died; after that the average of 200 people loss there life every year due to fire. While analyzing the reasons of these incidents in the city it was found that the densely populated areas and the places where built-up expansion happened at large scale in the last few years are more vulnerable.

Table 9.1 Total call and deaths during 2003–04 and 2015–16

S. No. calls	Year	No. of calls	Approx. loss in lakhs	Property saved in lakhs	Injured	Deaths	Medium	Serious	Major
1	2003–04	14,595	5874	8750	1334	235	17	05	–
2	2004–05	14,208	4681	6629	1687	272	27	05	–
3	2005–06	16,340	4720	6457	2191	470	16	01	–
4	2006–07	14,291	5587	14,903	1743	303	16	03	–
5	2007–08	15,718	5922	29,369	2057	351	9	02	–
6	2008–09	16,452	5902	29,471	2225	380	06	02	–
7	2009–10	21,314	–	–	2598	423	10	2	–
8	2010–11	22,187	–	–	243	447	10	3	–
9	2011–12	18,143	–	–	2132	357	13	1	–
10	2012–13	22,581	–	–	1979	285	9	2	–
11	2013–14	22,726	–	–	2299	372	16	1	–
12	2014–15	23,242	–	–	2068	291	7	2	–
13	2015–16	27,089	–	–	2099	339	11	Nil	–

Source Delhi Fire Service

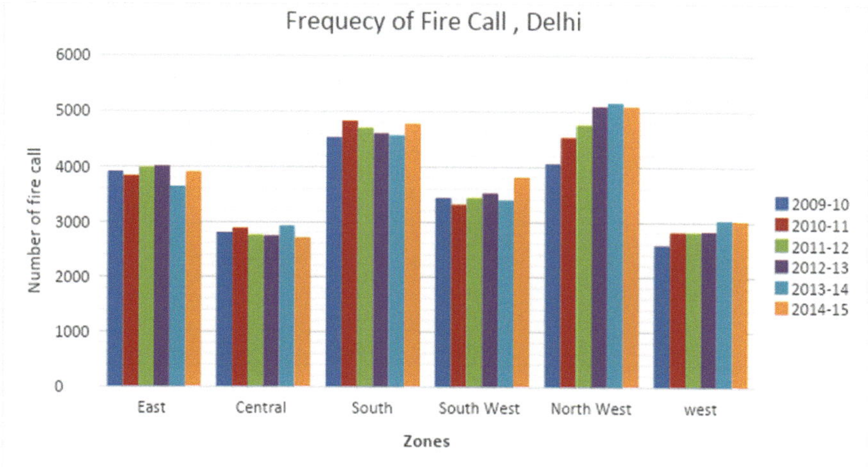

Fig. 9.1 Frequency of call received by the fire department to report the incidents. *Source* Delhi Fire Service, 2015

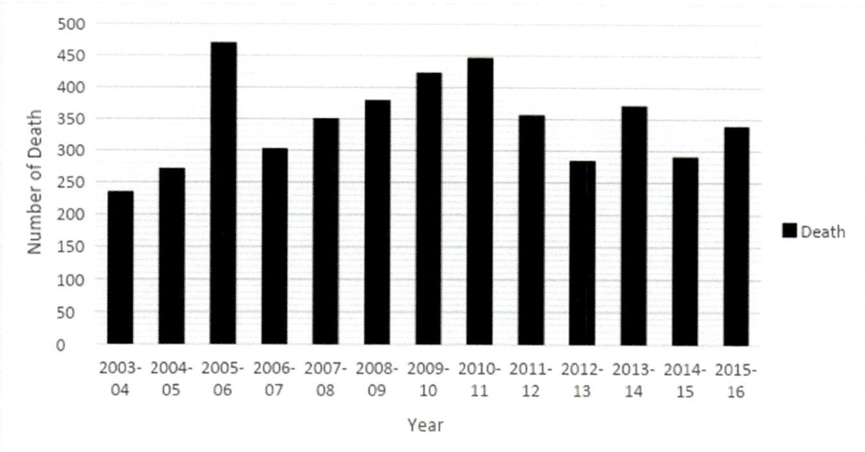

Fig. 9.2 Number of deaths. *Source* Delhi Fire Service, 2015–16

Density of Population in Delhi

As per 2011 Census, the density of population in Delhi worked out at 11,297 persons per square kilometer as against the national level of 382 persons per square kilometer. Density of population in Delhi was the highest among all states and union territories during the year 2011. District-wise density of population in Delhi in 2011 is presented in Table 9.4. In this table, the highest rank given to North East district which

Table 9.2 Showing district-wise population density

S. No.	District name	Population density (persons per sq. km)
1	South	11,060
2	South West	5446
3	West	19,563
4	New Delhi	4057
5	Central	27,730
6	East	27,132
7	North	14,557
8	North East	36,155
9	North West	8254

Source Census of India 2011

is showing population density is the highest there, whereas the lowest population density is recorded in New Delhi. Further information is shown in Table 9.2.

Built-Up Expansion in Delhi

Built-up expansion has been carved out which shows the actual picture of the analysis. In 2003 out of the 440 km^2 built-up area maximum is found in South West district (109.31 km^2) and minimum was existed in Central district (9.21 km^2) of Delhi (2003). In 2016 the total area of district was 727 km^2, and the highest percentage of built-up area was in North West district (183.68 km^2) and lowest was in Central district (13.62 km^2) among the various regions of district shown in Table 9.3.

Analysis of population density, built-up expansion (Fig. 9.3) and data of fire calls of the study area highlight that in the last few years South, North West, East and South

Table 9.3 District-wise built-up area 2003–2016

S. No.	District name	Area (km^2)	Built-up area (km^2) 2003	Built-up area (km^2) 2016
1	South	285.717	80.336231	1,28.5938
2	South West	428.527	109.3105	1,72,3689
3	West	138.864	62.646873	96.2686
4	New Delhi	37.7448	12.670907	1,77,9367
5	Central	20.8198	9.213265	13.62619
6	East	74.8596	36.924102	5,78,621
7	North	66.3171	19.40,8096	3,79,7411
8	North East	59.7952	23.677215	3,86,5411
9	North West	472.794	86.579571	183.687

West zone received the highest number of calls to report the fire incidents, and it has been seen that South, South West and West zones also had the built-up expansion very high though these areas have low population density. Only the North West zone despite the low level of built-up expansion had received many calls for fire incidents. North East zone has high population density but low built-up expansion and also received the significant number of calls to report the fire incidents. The Central and West zones of Delhi also received significant number of calls though lesser from the South, North West, East and South West zones, and these two zones have moderate population density but very high built-up expansion. This shows that though the population density and built-up expansion are directly linked to fire accidents but there are number of others reasons too which are responsible for making the places vulnerable for fire hazards.

Case Study of Malviya Nagar, New Delhi

Delhi, the capital of India, is witnessing rapid urbanization and in-migration in consequence facing the challenges of crowding, congested settlements and insufficient basic amenities for the residents. At the time of any disaster in these congested areas finding the space to rescue the people and property is very difficult. High population density, crowded streets, mixed occupancies, inadequate water supply, poor electrical services, unplanned siting of fire stations and encroachment are few examples of ineffective planning which adversely affect the disaster response time. Rapid urbanization and high density (11320 person per sq. km) can lead the city to different kinds of serious disaster. The above statistics regarding the fire accidents between the years of 2003 and 2015 advocates the trend that fire accidents with high intensity are going up in the future. Recent example of massive urban fire in Delhi is Khirki Extension, Malviya Nagar. Malviya Nagar is a locality in South Delhi. It is located in-between Saket and HauzKhas and near to IIT Delhi. It is named after the well-known freedom fighter Madan Mohan Malviya who was an educator too.

Initially in 1950 after the partition of India refugees from Pakistan came and settled in Malviya Nagar. Later a mixed population of Rajasthani, U.P., Bihari, Punjabis, Haryanvi, Bengalis and Sindhis started living. Malviya Nagar is also inhabited by people came from Africa and Afghanistan.

Now Malviya Nagar is encircled by Panchsheel Enclave in the North, Sheikh Sarai in the East, Saket in the South and Sarvapriya Vihar in the West. Enclaves that are part of Malviya Nagar include Geetanjali Enclave, Bhavishya Nidhi Enclave, Khirki Extension, Shivalik Colony and Sarvodaya Enclave.

Khirki Extension is located directly across from famous Shopping Mall Select City. Entrance of Khirki consists broken pavement, rubble and dirt road which are rarely constructed. Across the mall there is Khirki village which is officially demarcated urban village and then there is "Khirki Extension", which is unauthorized and exactly adjacent to the village. The Khirki Extension belongs to "Lal Dora" area.

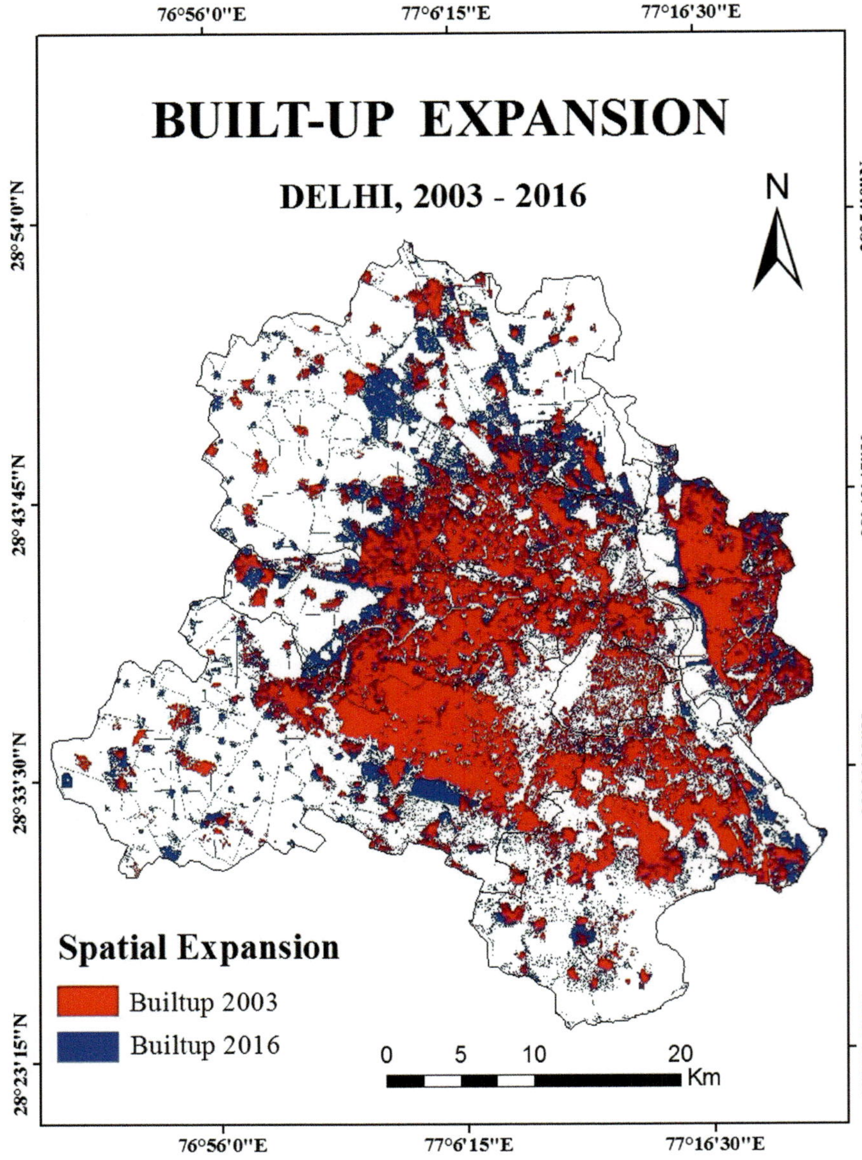

Fig. 9.3 Built-up expansion of Delhi. *Source* Satellite Imageries, 2003 and 2016

The word "Lal Dora" was used for the first time in the year 1908. It is a name classification given to that part of the village land which is part of the village "Abadi" (habitation). It was supposed to be used for non-agricultural purpose only. It is that part of the land which was supposed to have been an extension of the village habitation, wherein the villagers used to have their support systems, livestock, etc. In the past these areas were marked by the land revenue department by tying a Red Thread (Lal Dora in Hindi language) around it, to make a boundary and to distinguish it from the agricultural land. The Lal Dora also means that the jurisdiction of the municipal authorities or the urban development is not applicable, in fully. In 1957, the Delhi municipal corporation issued a notification and the government listed the lands under the Lal Dora classification, within and on the outskirts of Delhi (http://laldora.com).

Lal Dora was exempted from the building bye-laws and strict construction norms and regulations, as regulated under the Delhi Municipal Act. There is no need to apply for the building sanction plans, etc. In 1963, the MCD passed a notification which did not make it mandatory for building sanction plans to be passed for the urban Lal Dora lands too (http://laldora.com).

Though the term Lal Dora applies to both rural and urban villages, the thin dividing line has vanished over the period of time, and prime areas of Delhi today (though still classified as Lal Dora) operate commercial and high-end residential areas.

There are more than 1300 unauthorized localities such as Khirki Extension in the city, usually comprised of migrant population who came to search employment. These areas are not legally demarcated as residential or commercial development and therefore not getting the proper municipal services and amenities. But private utility companies are providing the services. Since they are not under the control of municipal laws these colonies are solving the dwell purpose of residing population by providing the residential and commercial place to them at cheap cost. Most of the gullies have shops or commercial establishment on the ground and houses on the upper floors. The narrow crowded streets, hanging wires, mixed occupancies, fragile structures of buildings and encroachments make the place vulnerable for any disaster.

However residents of Khirki Extension want the city to recognize their problems and provide services such as constructing the road, proper water and electricity supply, proper sewage system and paving the road. Legalization of unauthorized areas is not a simple issue in New Delhi because it sets examples and anticipations about how the city required to treat thousands of illegal colonies by using its limited resources. However the recent fire incident puts a big question mark on these unsettled issues of urban planning.

The recent fire started around 5 pm on May 29, 2018, Tuesday, in the warehouse at Khirki Extension, a crowded residential part of Malviya Nagar. As many as 80 fire engines were sent by the fire department and firemen were at work all night to stop the blaze from spreading to neighboring buildings, including a school. Thick black smoke rising (Photo 9.4) from the fire was visible from much of south Delhi. The police believe a truck loaded with rubber material parked near the godown caught fire and spreads in no time because of strong winds flowing on that night. Plastic and raw material stored in the factory fueled the fire. Thirteen buildings in the vicinity, including the school and a gym, have been evacuated for safety. It was tough for fire engines to access the narrow and congested lanes; therefore one firefighter has been injured.

Fire remain continues more than 19 hours. More than 50 fire tenders failed to tackle the plight, and Indian Air force took the situation in its hand. The fire detoriate

Photo 9.4 Fume spread after the fire, 2018. *Source* Authors

the quality of surrounding air, the nearest air quality monitoring station (Siri Fort) indicate PM10 314 (100 μg/m^3—24 Hourly average limit prescribed as per NAAQS 2009). Due to open burning tire and rubber the create atmosphere more toxic, which expose different kinds of pollutants like carbon monoxide (CO), sulfur oxides (SO_2), oxides of nitrogen (NOx) and volatile organic compounds (VOCs). It had created significantly short- and long-term health problems for surrounding peoples of that area. Burning such amount of rubber martial became cause of irritation in the eyes and mucous membranes, central nervous system depression and cancer. It is an alarm for the government and public for upcoming disasters inborn in the capital of India. Fresh instance of fire in the city forces to think how safe people are in the capital where such type of chemical factory (Photo 9.5) is working in the densely populated residential area. It was the first time when Air Force chopper was used to control the fire in the city. If fire spread around the buildings, it would have been unstoppable.

Urban Fire Management

Outinen (2012) suggested that fire-resistant building material, automatic fire detection, sprinkler system and smoke exhaust system should be used to develop the fire resistance of buildings in the city. He also suggested that with the help of advance

Photo 9.5 Chemical factory and warehouse, Malviya Nagar, 2018. *Source* Authors

technology detecting of fire and to control the fire automatic water extinguishing system can be installed at commercial place. Zhang (2013) recommended that fire accidents can be controlled in the cities by building of fire stations in every locality, undisrupted water supply to control fire and well-functioning of communication system. According to him lanes for firefighters, fire equipment fire, fire training and distance monitoring of urban fire are key to develop urban fire capacity. Ishiyaq and Sunil Kumar (2010) describe that slum are active hazard zone, the Central zone has the lowest number of slum, and South zone consists maximum number of slums in the city. South zone has 398 slums contributing 30.49% of the total slum in national capital territory, Delhi. Total area covered by slums is 1483 km^2. According to 2011 Census total population of slums is 1678, 7941 and density is about 11,320 population per sq. km. The city has history of brutal fire when hundreds of people died. The poor population living in slums is deprived of all the safety measures. Shanty houses build with plastics sheets, newspapers and cloths, with hanging wires for providing the illegal electricity supply, are highly prone for spreading the fire. Unplanned and overpopulated colonies with poor quality of housing enhance the vulnerability of the city. Prompt and timely response to any fire incident is required for controlling and managing the disaster as delay in the arrival of fire brigade at the spot can have significant consequences in terms of death, injury and damage of property. Kiran et al. (2015) suggested that there is an urgent need to identify the hotspots for manmade disasters in the city, and plan should be made to increase the capability of the city to prevent and control the fire. The city needs more fire stations, uninterrupted water supply, lanes for moving the fire-fighting brigade, installing of fire alarms and fire extinguishers and training of peoples to use them at the time of emergencies, etc. Despite the inappropriate safety measures in an unplanned colony Malviya Nagar fire incident is a good example of quick response and cooperation of public to control the massive fire.

Conclusion

Delhi is a populated metropolitan city in India, which has a long history of fire accidents. Historically Delhi always remains a center of attraction for outside population because of employment opportunities, education, medical facilities and in consequence facing the challenge of increasing population each year. The city is struggling to fulfill the needs of ever-growing population with its limited resources. Growth of population with limited physical space construed a complex structure of urbanization which is lacking in safety measures. Urban expansion analyses in this study show that North East has major urban expiation as well as South and South West are just after this. Mean high urbanization and congested built-up expansion are reasons for making the areas vulnerable for urban fire. High concentration of unauthorized colonies makes South and West zones of Delhi even more vulnerable.

The study also analyzed that for marking the vulnerable areas in cities Geographical Information System (GIS) can be used in the future, and for this one can study the history of fire accidents in past and locate the data of number of fire calls and analyze this pattern to find out the best location for the fire station in a particulate area. Fire hazards are big problem in front of developing and developed nations both; to tackle this problem the best option is awareness among society. Awareness about fire hazard can be generated by the organizing fire drills and mock drills in school, collages, multiplex and movie halls so that people should become capable to tackle this type of situation. As most of the time when people take wrong decision situation becomes even worst.

References

Delhi Fire service.

High Powered Committee Report on Disaster Management. (2001). *Department of agriculture and co-operation, ministry of agriculture, government of India.*

http://laldora.com-know and share.

Kiran, K. C., et al. (2015). Spatial and temporal patterns of fire incident response time: a case study of residential fires in Brisbane. In *Proceedings of the state of Australian Cities, Gold Coast, Queensland.*

Provisional Data of population, Delhi. Census of India, 2011.

Yao, J., & Zhang, X. (2016). Spatial-temporal dynamics of urban fire incidents: a case study of Nanjing, china. *The International Archives of the Photogrammetry, Remote Sensing and Spatial Information Sciences.*

Zhang, Y. (2013). Analysis on comprehensive risk assessment for urban fire: The case of Haikou City.

Chapter 10
Forest Fire Severity Mapping Using Geospatial Techniques: A Case Study of a Part of Bandipur Reserve Forest, India

Shashwati Singh, Prabuddh Kumar Mishra, and Varun Narayan Mishra

Abstract The changes in climate directly affect fire frequency and severity, which are likely to critically have an effect on the attribute of forest ecosystem. The present study aims to investigate the severity mapping of forest fire occurred in the year 2019 in Bandipur Reserve forest, India, using geospatial techniques. In the present study, distinct band ratios like Normalized Burn Ratio (NBR), Normalized difference Vegetation Index (NDVI) were obtained using Landsat 8 OLI images of bi-temporal data (pre-fire/post-fire). Using bi-temporal data, Difference of Normalized Burn Ratio (dNBR) and Normalized Difference Vegetation Index was obtained by subtracting the Pre/post-fire data to get the burn severity. Initially, the displacement of pixels in the burned and unburned area in the pre/post-fire NIR-SWIR and NIR-R was studied, the capacity of the two indices pre- and post-fire (bi-temporal) and post-fire (uni-temporal) was studied to analyze which one is more sensitive for severity levels discrimination. Based on the outcome, it was taken into consideration that the most appropriate way to study forest fire severity by index classification, and it was to differentiate between the pixels with unburned and burned with respect to NBR pre/post-fire difference value (dNBR). Further, both the image indices were reclassified to distinguish the pixels with high, moderate, low and very low severity. The two different sets of raster usually contain some numerical value. These two raster are overlaid and mathematically merged together to give new single output layer. The burned area were analyzed using MODIS fire points, helped in the locating hotspot area in Bandipur Reserve. This kind of study can avail valuable knowledge to prevent and monitor forest fires, and to understand the response of forest ecosystem.

S. Singh
Computational Biology Division, CSIR-Central Institute of Medicinal and Aromatic Plants, Lucknow, Uttar Pradesh, India

P. K. Mishra (✉)
Department of Geography, Shivaji College, University of Delhi, New Delhi, India
e-mail: prab19@gmail.com

V. N. Mishra
Amity Institute of Geoinformatics and Remote Sensing (AIGIRS), Amity University, Sector 125, Noida, Uttar Pradesh, India

V. R. Sharma and Chandrakanta (eds.), *Making India Disaster Resilient*,
https://doi.org/10.1007/978-3-031-50113-5_10

Keywords Forest fire · Satellite images: NDVI · NBR

Introduction

Forest is an enormous area predominantly covered with trees. It is part of human ecosystem since the time of immemorial (Satendra & Kaushik, 2014). Disaster is natural Catastrophe that give rise to huge destruction or loss of life, whereas hazard is factor which is likely to cause distress to a prone target. Hazard can be induced by human or can be natural. Similarly forest fire can be also human induced or naturally borne. Forest fire is most sensitive discussion because it is major threat to man and wildlife (Kulkarni et al., 2018). In India disaster like forest fire hit the large hectares of forest cover, mostly during 3rd week of February to 1st week of June. India covers around 7, 08273 km^2 of forest cover. Rising Interference of human in forest ecosystem it also increases in the forest fire incidences tremendously (Satendra & Kaushik, 2014).

The forest ecosystem is under major risk due to fire phenology, which is attributed for forest degradation, soil erosion, reduction in productivity (Brigitte, 2005). As per survey of India report, on an average 1.2% of total forest are prone to heavy fire, 6.28% to moderate and 45.27% to mild fires. Forest fire is majorly dependent on 3 elements, i.e., oxygen, heat and fuel. Vulnerability of the forest toward fire varies from one place to another place depending upon type of vegetation, the climate, topography, etc. (Satendra & Kaushik, 2014). It is difficult to map severity level in roughed and inaccessible areas using traditional methods. Remote sensing technologies play a significant role in providing real time and reliable data for monitoring and management of the forest fire (Xiao-rui et al. 2005). After fire a spectrum of alteration take place due to fire consuming the forest, destroying the chlorophyll, the soil losing fertility and mostly soil moisture are altered. In post-fire event the deficiency of chlorophyll results in high reflectance in visible region and SWIR (shortwave infrared) region of spectrum and low reflectance in the NIR region (near-infrared). Moreover in pre-fire event healthy forest has very high reflectance in NIR region and curtailing reflectance in the SWIR region and visible region of spectrum (Epting et al., 2005; Tran et al., 2018). To be more specific using remote sensing data, Landsat OLI/TIR images have been analyzed for fire severity evaluation. It provides data with adequate characteristics of spatial and spectral resolution. The approach used to derive indices from Landsat and the method to evaluate the level of severity in forest stands out.

In this work, the capacity of Landsat satellite image derived NDVI and NBR have been studied for fire severity evaluation and mapping (Ryu et al., 2018). For this purpose, fire occurring in Bandipur Reserve Forest in 2019 was studied. Initially, the displacement of pixels in the burned and unburned area in the pre/post-fire NIR-SWIR and NIR-R was studied, the capacity of the two indices pre- and post-fire (bi-temporal) and post-fire (uni-temporal) was studied to analyze which one is more sensitive for severity levels discrimination (Escuin et al., 2008).

Based on the above discussion, it was taken into consideration that the most appropriate way to study forest fire severity by index classification. It was to differentiate between pixel with unburned and burned with respect to NBR and NDVI pre/post-fire difference value (dNBR and dNDVI). Further, dNDVI was reclassified to null and moderate to low classes for severity mapping and high dNBR for burnt area and null for unburnt area (Escuin et al., 2008). The two different sets of raster usually contain some numerical value. These two rasters are overlaid and mathematically merged together to give new single output layer. Raster overlay is performed to create risk surfaces, using both NDVI and NBR image was used for creating raster overlay risk surfaces (Chang, 2008). And then reclassified image was overlaid and assigned percent weightage according to its importance and output image index are segmented according to pixel distribution on the basis of weightage assigned. The burned area were analyzed using MODIS fire points, helped in the locating hotspot area in Bandipur Reserve. Availability of daily, weekly and monthly MODIS fire points, determine which area is more prone to fire and therefore risk zone were evaluated accordingly. The entire map was prepared on scale of 1:400000

Geospatial techniques provide the facilities and the tools required to develop a forest fire vulnerability map in order to spot, categorize and map fire hazard area. The reliable and accurate data is vital during and after catastrophe. Making the records accessible via the worldwide net, people can percentage the facts to assess the state of affair and make choices. The present study aims to investigate the severity mapping of forest fire occurred in the year 2019 in Bandipur Reserve forest, India, using remote sensing derived indices.

Study Area

The Bandipur National park and Tiger Reserve covers part of Chamarajanagar and Mysore district of Karnataka, India. The extension of Bandipur National Park lies between the latitude 11° 35′ 34″–11° 55′ 02″ N and between the longitude 76° 12′ 17″–76°51′ 32″ E. The Reserve National Park covers an area of 868.63 km^2, and it was situated at altitude of 780 m to 1454.5 m above mean sea level (Fig. 10.1). This park is part of Nilgiri biosphere. The major irrigation project Kabini River is borderline linking the Bandipur Park and the Nagarhole Park. The Bandipur National Park shares borderline with Wayanad Sanctuary to its S-W (Kerala), to its south Mudumalai Sanctuary (Tamil Nadu) and to its N-W Nagarhole Park (Kerala) (Somashekar et al., 2009).

The park in 1973 was notified as Tiger Reserve National Park. The park holds up infinite number of wildlife comprising of elephant, tiger, and many other predators. This park is home to infinite number of bird species. The average annual rainfall varies from 914 to 1270 mm (Kulkarni et al., 2018). The reserve generally has moderate temperature with annual mean temperature of 24.16 °C and max temperature of 30 °C and min. temperature is 19 °C. The Bandipura has dry subtropical climate with distinct cold in November to mid-February, dry in March to June and wet in

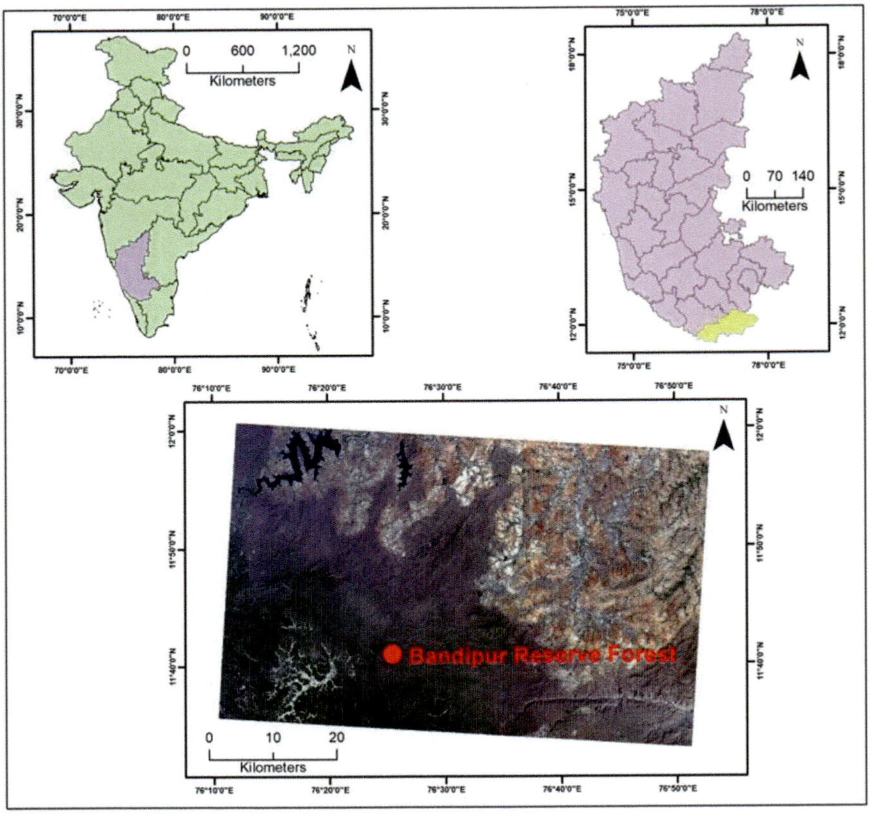

Fig. 10.1 Location map of Bandipur reserve forest materials and methods used: satellite data

mid-June to September. The park experiences good monsoon from southwest and less rain from northeast. Summer is basically hot and dry and December and January are coldest month. The forest has blend of different vegetation types; the park mainly consists of dry and moist deciduous forest, scrubs, grass lands, semi-evergreen forest; however, deciduous forest mainly dominates the park.

Landsat 8 was launched in the year 2013, 11 February; it has 11 numbers of bands, which measures the spectrum of frequency at different wavelength of electromagnetic spectrum (ranges from 0.433 to 12.51 μm). It comprises two types of sensors operational land imager and thermal infrared sensors, which has spatial of 15 meter for panchromatic band and 30 meter resolution for multispectral sensors. The satellite uses push broom scanner having ground swath of 185 km wide. To provide best possible result, it has receptivity or revisiting of satellite to same place is 16 days. Intensity resolution is 12bit/pixel (Landsat Report). The thermal infrared sensors having two wavelength bands, which use to measure or retrieve the thermal energy of Earth surface, it uses Infrared Photo detectors to measures the surface temperature. The details of the Satellite used is mentioned in Table 10.1.

Table 10.1 Details of satellite data used

Satellite-sensor	Data set	Acquisition date	Resolution
Landsat 8- OLI & TIRS	Pre-Fire data Post-Fire Data	22 February 2018 13 March 2019	PAN:15 m Multi-spectral: 30 m

MODIS (or Moderate Resolution Imaging Spectroradiometer) Terra is near real-time active fire data provided by The Fire Information for Resource Management System (FIRMS) of NASA. The spatial resolution of 1 km better for identifying hotspot area on ground, the projection is set to Geographic WGS84, with global coverage of 2 days. These hotspot data are fire point data can be downloaded in .shp file, TXT. file, KML file. The positional accuracy is about < 100 km. It was collected over night and day (NASA FIRMS MODIS). All the pre-processing tasks were performed using ENVI software. The radiometric and atmospheric corrections were done using FLAASH atmospheric tool. The Band math tool was used to calculation and produce NDVI and NBR images (HARRIS Geospatial Report). The image reclassification and operations like weighted overlay were performed in ArcGIS software to analyze the image burn severity of study area.

Image Pre-processing

In present study, bi-temporal data was processed (two different years but same season). To remove the flaws, present in the image a process called image pre-processing is done to remove errors. To process the data, data was imported to ENVI and applied radiometric and atmospheric correction to the image before applying any further mathematical calculation. The type of electromagnetic energy received by sensors are reflected energy, emitted energy, scattered energy, and these energies get recorded in every pixel of image as DN values. Sometimes the radiated energy is attenuated due to atmospheric complexities; therefore, exact signals of target are not recorded by sensors. The energy recorded by sensor can differ from what actual energy reflected or emitted from ground surface. This energy can be due to sun elevation and azimuth angle and atmospheric condition (water vapor, aerosols). Hence, it becomes necessary to remove the flaws and get actual irradiance. In order get real meaningful information DN values are directly converted to radiance, reflectance and brightness temperature (ENVI Atmospheric correction Module User's Guide).

It is not required to manual conversion. As in ENVI metadata file are imported because the entire conversion is automated system based so it reads datasets as well as Metadata. Rather doing long process of conversion radiometric calibration and reflectance, ENVI provides most desirable tool for all conversion in one tool "ENVI FLAASH" tool. But prior to that using radiometric calibration tool, applied FLAASH setting to multispectral image because FLAASH uses band interleaved image format which is scaled at 0.10 scale factor with float as output data type. FLAASH uses

MODTRAN atmospheric radiation transfer model. This provides true surface value without going to field, but before that in FLAASH tool every given parameters are required to be set like Latitude-Longitudes, sensor altitude, ground elevation, atmospheric and aerosol model, water and aerosol retrieval and many other.

Spectral Indices

It is mathematical blend or modification of bands that highlights the spectral attribute of features. The indices highlight the variation in feature so they seem clear from different image feature. Spectral enhancement methods are commonly used to draw out information this extra illustratable to the eye and extra suitable for evaluation. Band ratio is actually the divide one band by using another band to create a spectral index. The band ratio image intensifies the spectral variations between bands and may be beneficial while trying to discriminate between land cover types. With the aid of default spectral indices are displayed in grayscale, although to enhance interpretability a coloration scale may be implemented.

Normalized Difference Vegetation Index (NDVI) Calculation

NDVI is used to highlight the health and density of vegetation (Illera et al., 2006; Ryu et al., 2018). The infrared band of electromagnetic spectrum provides more significant reflectance of healthy vegetation in comparison to that of red band. Chlorophyll strongly absorbs in visible portion whereas cell structure strongly reflects infrared. The NDVI results in number ranging from -1 to $+1$; these values have some significance for interpretation of images. It can be calculated using formula given as:

$$NDVI = \frac{NIR - RED}{NIR + RED} \tag{10.1}$$

Normalized Burn Ratio (NBR) Calculation

NBR was developed to spot the burn area and evaluate fire severity; in healthy vegetation shortwave infrared (SWIR) portion have low reflectance, whereas near-infrared (NIR) portion have high reflectance in spectrum (Ryu et al., 2018). Similarly, burned areas have high reflectance in shortwave-infrared region and low reflectance in near-infrared region of spectrum. A recently burned area and bare ground have low values and healthy vegetation have high value index. It can be calculated as:

Table 10.2 Burn severity index

dNBR	Burn severity
< -0.25	High post-fire regrowth
– 0.25 to – 0.1	Low post-fire regrowth
– 0.1 to 0.1	Unburned
0.1 to 0.27	Low severity burn
0.27 to 0.44	Moderate-low severity burn
0.44 to 0.66	Moderate-high severity burn
> 0.66	High severity burn

$$NBR = \frac{NIR - SWIR}{NIR + SWIR} \tag{10.2}$$

Burn Severity

NBR is used to estimate frequent burn areas. Pre-fire image will have very low index in mid-infrared region and high index in near-infrared region, whereas imagery which is sensed after fire or post-fire will have very high index in mid-infrared region and low index value in near-infrared region. Higher the differential normalized burn ratio (dNBR) specify more severe damage. Area with increased vegetation productivity will have negative dNBR values. Basically, dNBR and NBR images are calculated in a very short period of time right after fire event just to support the initial evaluation of burn severity and fieldwork. The following table (Table 10.2) will be helpful in understanding dNBR image index value.

$$dNBR = \text{Pre - fire NBR} - \text{Post} - \text{fire NBR} \tag{10.3}$$

Differential Normalized Difference Vegetation Index (dNDVI)

It can be obtained by subtracting the pre-fire value from post-fire value. The vegetation index in burn area will have higher index value in pre-fire data than in post-fire data, so the difference of pre-post-fire will be positive.

$$dNDVI = \text{Pre - fire NDVI} - \text{Post - fire NDVI} \tag{10.4}$$

The all spectral indices including NDVI and NBR and burn severity including dNDVI and dNBR are calculated using Band math tool from ENVI tool box, with above expression like NDVI and NBR are mentioned float as output data types in Band math calculation expression. Float data types gives output image in decimal value otherwise by default it will take as integer which will have roundup figure.

Reclassification of the Maps

The images are imported to ArcGIS to reclassify spectral data images before performing weighted overlay operation. Reclassification tool enables you to easily reclassify your image. This process is mandatory before weighted overlay because it does not process float data types; image has to be in rounded figure. In reclassification tool the old values are assigned new value, and the new values are ranked according to old values e.g. the image having highest possible value are assigned highest rank and lowest value are given lowest rank.

Weighted Overlay Analysis

The images were classified into five categories using reclassification process. In the overlay analysis, every raster is given percentage influence. The raster cell value is multiplied by their percentage influence; the raster overlay output is always in integers. Every raster input is weighted on the basis of its importance. The sum of percent influence weight should be equal to 100. The following image was given 70% weightage to dNBR and 30% weightage to dNDVI.

Validation of the Hotspot Area

MODIS Terra sensor data provide near real-time active fire data which has positional accuracy of < 100 m, it is facilitated by FIRMS NASA, the data is actually active fire point data and best suitable to mapping fire spot with some extent less positional accuracy. The point data was overlaid on dNBR image to validate the hotspot area. Both fire point and burned area got adjusted to similar location. Using archival data of fire points, fire-prone area can be analyzed (NASA FIRMS MODIS Fire/Hotspot data).

Results and Discussion

Figure 10.2a, b shows the pre- and post-NDVI maps of the study area using satellite images of February 22, 2018 and March 13, 2019 respectively. These figures show on the basis of NDVI values that during this period Bandipur reserve forest experienced intense forest fire episodes.

Figure 10.3a, b shows pre- and post-NBR maps. These maps are used to detect the burn area.

Figure 10.4 a, b shows the dNDVI and dNBR maps respectively. Figure 10.4a shows the vegetation index in burn area. Figure 10.4b is used to estimate frequent burn areas. The burn severity index was used to categorize burn area

Reclassification was performed before weighted overlay analysis. In reclassification process the old values are assigned new value, and the new values are ranked according to old values, e.g., the image having highest possible value are assigned

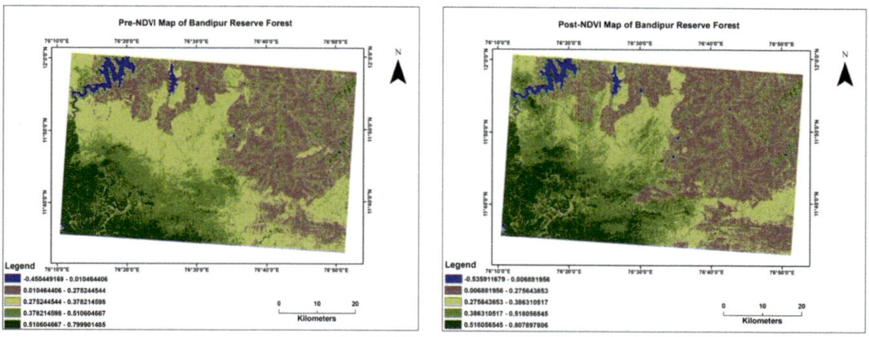

(a) (b)

Fig. 10.2 **a** Pre-NDVI, **b** Post-NDVI map

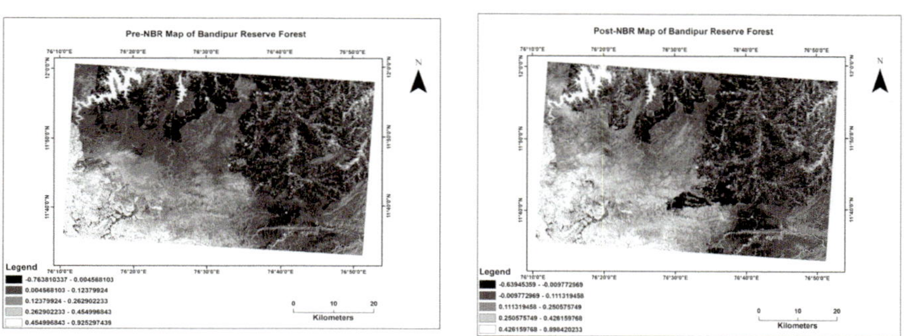

(a) (b)

Fig. 10.3 **a** Pre-NBR map **b** Post-NBR map

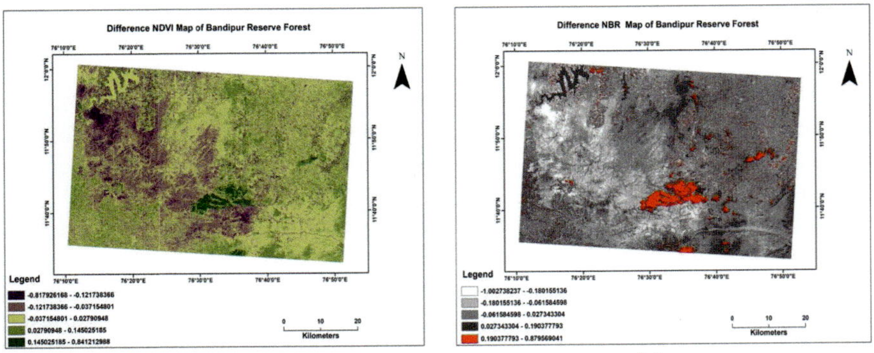

(a) **(b)**

Fig. 10.4 **a** NDVI map **b** dNBR map

highest rank and lowest value are given lowest rank. The reclassified maps are shown in Fig. 10.5. The weighted overlay maps of the study area are shown in Fig. 10.6.

The validation of hotspot area was performed using MODIS Terra sensor data and shown in Fig. 10.7.

The near-infrared and shortwave-infrared reflectance of Landsat 8 has a good signature to differentiate the before-fire and after-fire forest. The dNBR is more significant for mapping the fire severity location in Bandipur Reserve, even as the dNDVI is more likely to locate opening land from vegetation to non-vegetation without biomass burning. The NBR index showed itself to be more sensitive to the pre/post-fire displacements of the pixels affected by the fire in the space than NDVI in the R-NIR space. Both the indices are sensitive to the pre/post-fire spectral changes corresponding to the unaffected pixels attributable to "perturbing factors". The pre/post-fire difference indices dNBR and dNDVI are most perfect for studying

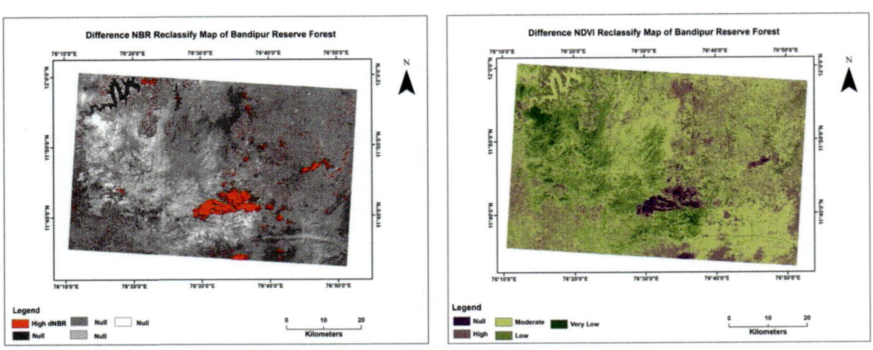

(a) **(b)**

Fig. 10.5 map for (**a**) dNDVI map reclassified dNBR (**b**)

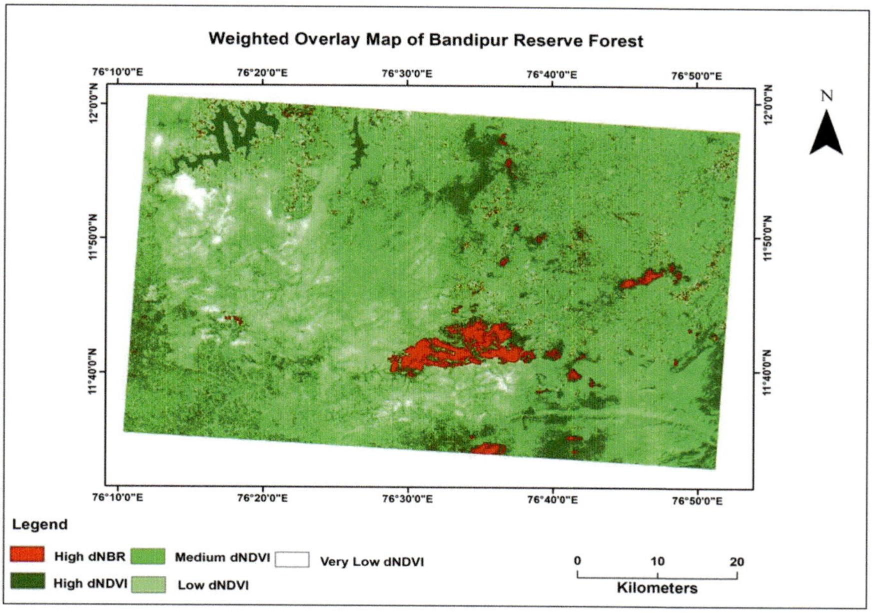

Fig. 10.6 Weighted overlay map

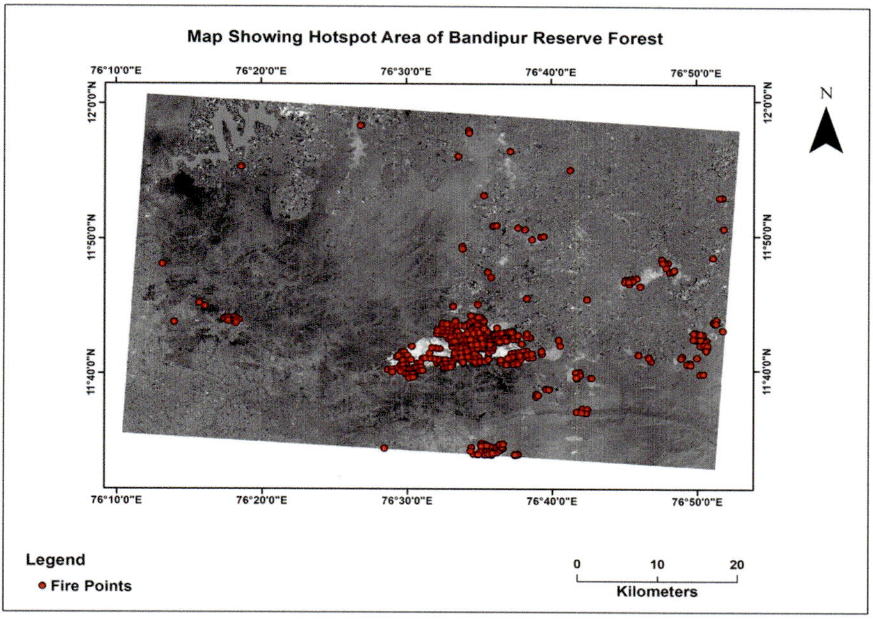

Fig. 10.7 Red points indicate hotspot area

out to differentiate between pixels not burned by fire (whose values are not significantly correlated with the previous values of the indices) and pixels affected by the fire. dNBR displayed moderately better output than dNDVI for separating the unburned moderate classes. However, the post-fire indices NBR-post and NDVI-post are better for discriminating between extreme severity pixels and pixels with a moderate severity. The NBR-post gave absolutely better outcome than NDVI-post for the severity between the moderate-extreme classes. The dNBR and NBR-post indices presented a clearly higher range than the dNDVI and NDVI-post indices, which suggests that they are more suitable for detecting different severity levels.

The best option for assessing fire severity by the segmentation of the indices studied is to do in two ways:

- Separating the unburned pixels from the rest on the basis of their dNBR value.
- Separating the extreme severity pixels from the moderate severity on based on their NBR-post values.

Prominently, the MODIS fire points helped in the locating hotspot area in Bandipur Reserve. Availability of daily, weekly and monthly MODIS fire points determines which area is more prone to fire, and therefore, risk zone can be evaluated accordingly.

Conclusion

The present study was carried out using Landsat 8 operational land imager (OLI). The major findings of this entire study revolves around to map burn severity, for this three different bands of Landsat (Red, NIR, SWIR) was studied. Further two major band ratios, NDVI and NBR, were calculated using earlier mentioned different bands of Landsat 8. It was seen that after calculating difference from pre/post of NBR and NDVI, dNBR gives more distinct result for severity mapping than dNDVI by employing the SWIR and NIR reflectance bands. These dNDVI was reclassified to null and moderate to low classes for severity mapping and high dNBR for burnt area and null for unburnt area, to validate the burnt area active fire points of MODIS was overlaid on dNBR image. Above generated maps are useful to forest department of India for evaluating the forest loss and restoring it.

References

Adagbasa, E., et al. (2018). Spatio-temporal assessment of fire severity in a protected and mountain ecosystem. In *IGARSS2018–2018 IEEE International Geoscience and Remote Sensing Symposium*.

Brigitte L. (2005). Monitoring forest fire danger with remote sensing. *Natural Hazards 35*, 343–359.

Chang, K., (2008). *Introduction to geographical information system*. McGraw Hill.

ENVI Atmospheric correction Module User's Guide, https://harrisgeospatial.com/portals/0/pdfs/envi/flaash_module.pdf.

Epting, J., Verbyla, D., & Sorbel, B. (2005). Evaluation of remotely sensed indices for assessing burn severity in interior Alaska using Landsat TM and ETM+. *Remote sensing Environment.*
Escuin, S., et al., (2008). Fire severity assessment by using NBR (Normalized Burn Ratio) and NDVI (Normalized Difference Vegetation Index) derived from LANDSAT TM/ETM images. *International Journal of Remote Sensing.*
HARRISGeospatial Report, https://harrisgeospatial.com/software-Technology/ENVI.
Keeley, J. E. (2009). Fire intensity, fire severity and burn severity: A brief review and suggested usage. *International Journal of Wildland Fire.*
Key, C. H. (2005). Remote sensing sensitivity in fire severity and fire recovery. In *Proceedings of the 5th international workshop on remote sensing and GIS application to forest fire management, Zaragoza (Spain).*
Kulkarni, M., Adarsh, & Manikiam, B., (2018). Fire risk zoning of Bandipur National Park using remote sensing and GIS techniques. In *19th ESRI India user conference.*
Landsat Report, https://landsat.gsfc.nasa.gov/landsat-data-continuity-mission/.
NASA FIRMS MODIS Fire/Hotspot data, MODIS collection 6 NRT hotspot/Active Fire Detection MCD14DL. Available on line, http://earthdata.nasa.gov/Firms.
Ryu, J. H., Kyung-Soo Han, K. S., Hong, S., Park, N. W., Lee, Y. W., & Jaeil Cho, J. (2018). Satellite-Based evaluation of the post-fire recovery process from the Worst Forest Fire Case in South Korea.10, 918
Satendra, & Kaushik, A. D. (2014). *Forest fire disaster management.* National Institute of disaster Management, Ministry of Home Affairs, New Delhi.
Somashekar, R. K., Ravikumar, P., Mohan Kumar, C. N., et al. (2009). Burnt area mapping of Bandipur National Park, India using IRS 1c/1D LISS III data. *Journal of the Indian Society of Remote Sens, 37*, 37–50.
Tran, Nguyen, Tran, et al. (2018). Evaluation of spectral indices for assessing fire severity in Australian temperate forest. *Remote Sensing.*

Chapter 11
Fire Hazards in Anaj Mandi (Grain Market), Old Delhi: Vulnerability and Resilience

Shubham Kumar Sanu

Abstract Fire is one of the crucial resources since its discovery and played a pivotal role in civilization to the modernization of Anthropos. Its misuse by human converts fire as a resource to resistance and when the resistance of fire becomes furious that creates fire disaster. India and particularly its capital city is highly vulnerable to fire, which is evidenced through the 31,000 and 85 calls yearly and daily respectively by Delhi population to fire department to douse it. To explain this vulnerability of Delhi, disastrous fire incidence of Anaj Mandi, dated 8th of December 2019, which deceased 45 people and impacted economically, socially and psychologically to thousands of population, so pronounced as "Disastrous Devil of December" has been taken into account. Why and how Anaj Mandi is vulnerable and what are the solutions to make it resilient are described in this research paper. For the study, both primary and secondary data sources are taken into account. The study found that poor design, haphazard construction, improper electric wiring, lack of public awareness, illegal industries, congested lanes, absence of fire safety measure, etc. makes Anaj Mandi vulnerable to fire. Disastrous Devil of December was developed and decorated by authorities, societal organizations, locals and individuals by their close coordination. Unawareness is at utmost as more than 75% of the population are unaware of fire safety measures. The best strategy for making Anaj Mandi and Delhi fire resilient is the three-tier model based on the A4 Phase of the fire disaster development cycle, a society-based and K3 principles or individual-based model. This research paper provides a detailed analysis of the fire incidence of Anaj Mandi in old Delhi.

Keywords Fire disaster · Vulnerability · Anaj Mandi · K3 principles · Disaster management

---

S. K. Sanu (✉)
Indraprastha College for Women, University of Delhi, New Delhi, India
e-mail: shubhamsanu@ip.du.ac.in

V. R. Sharma and Chandrakanta (eds.), *Making India Disaster Resilient*,
https://doi.org/10.1007/978-3-031-50113-5_11

Introduction and Literature Review

Anaj Mandi situated in the periphery of Delhi's walled city area near Filmistan, Sadar Bazar Old Delhi, witnessed one of the horrific fire accidents after the Uphaar Cinema Tragedy of 1997, in the early hours of Sunday 8 December 2019. The second floor of the five-story building caught fire and abruptly engulfed other floors in its blaze. This incident assassinated more than 45 people; more than 5 were minor in that due to inhalation of toxic gasses like Carbon mono oxides produced by burning of raw material like plastic to rexine which was used in various manufacturing units of the building. Various units of manufacturing were running illegally as told by the officer in the building, without any permission or no-objection certificates from concerned departments and authorities. As the building do not have a proper ventilation facility, the exit gate of the roof was locked and another exit gate was blocked with the extensive storage of raw material near that leading to complete blockage of workers inside the building as the main exit gate was engulfed by fire. This incidence was disastrous incidence as it caused loss of life to economic loss and social loss to psychological trauma among affected communities. As this incident occurred in the month of December and derived like as a devil which disrupted the lives of thousands of population so named it as "Disastrous Devil of December" or "Devil of December".

Fire as a disaster is a serious threat to human life and property. As per estimate, 70 thousand people between 2012 and 2015 lost life due to fire incidences in India. The difference in actual occupancy and defined land use are the main cause (Tomar et al., 2017). All across the country the fire department faces problem of funds to modernize and upgrade their equipment as well as manpower. According to the international standards fire station after every 3 km and standard response time is 3 minutes but India's fire service is not in such a situation that can fulfil these standards. So there is an urgent need to focus on fire preparedness immediately (Pal & Ghosh, 2014). Most of the fire-related deaths occurs in women in the age group between 15 and 34. This particular age group suggests that kitchen accidents, self-immolation and different form of domestic violence are the major cause of it. The total death figure was 5 times more than the figure reported by the police. This shows that death caused by a fire is a neglected public health issue that needs to give more attention to solve very soon (Sanghavi et al. 2019). Fire disaster is one of the serious disasters in Delhi. Out of the total fire, 99% are of small category fire. From residential to industrial every location is showing the trend of increasing fire incidences that causes huge loss of wealth and life. Delhi's semi-arid climate, electric short circuit, high level of pressure on the fire department, slum cluster with a huge population and less fire safety measures, industrial landscape of Delhi, etc. are some of the major causes of an increasing trend of fire indices (Kapur, 2005). Vulnerability analysis of Delhi and appropriate measures are required to make the city resilient in terms of fire disasters.

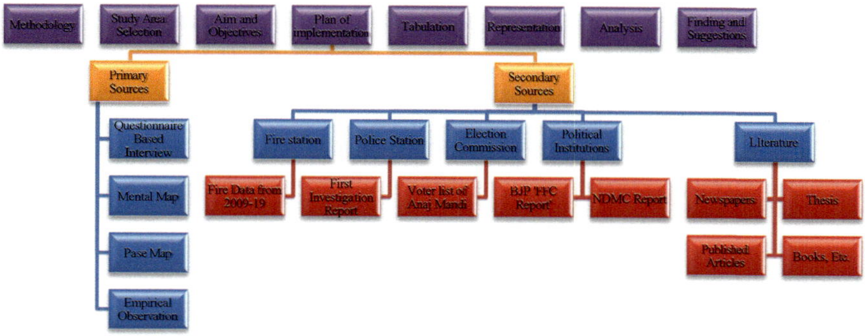

Fig. 11.1 Research Methodology. *Source* By Author

Data Sources and Methodology

Primary data for this research work were collected through questionnaire-based interviews and empirical observation. A total of 50 samples were collected from the study area that as per locals estimation have four thousand population. A stratified random sampling method was applied and strata of age, sex, job type has been considered to acquire samples from all the diverse group. Empirical observations were made during a personal visit to the area, and a field diary was prepared to record observations. Different land use and land cover in and around the area were observed and then correlated to the topic of research. Through this observation, photographs of the area were taken to substantiate the problems and conditions of the area. Secondary data were mainly assessed from newspapers, magazines, the Delhi Fire Service, reports of committees, etc. The collected data were filtered, tabulated, verified and analysed to achieve the objectives of this study. The methodology of the study is represented using Fig. 11.1.

Study Area

New Anaj Mandi (Map 11.1), The Mandi of grain in earlier days of the 1980s was developed to release some pressure from one of the biggest Grain Mandi of that time Ruhi Mandi, of Sadar Bazar. New Anaj Mandi as the name suggests were newly developed Grain Mandi, near Filmistan, Sadar Bazar Delhi 11,006 comes under ChowkNaiBasti surrounded by Bara Hindu Rao road in the north and Maharaja Agrasen Marg in South and Mohammad Ismail or Chamelian Marg in East and Rani Jhansi Road in West. This place has transformed itself from Mandi of Anaj where mostly gain selling wholesalers and grinding machines to the hub of household industries and illegal manufacturing industries ranging from cap manufacturing to bag manufacturing and other products according to the demand of the market. Its proximity to New Delhi railway station, Old Delhi railway station, Bus terminals, one of the biggest wholesale markets of India Sadar Bazar, make this site hub of secondary and tertiary activities or SECTER. The study area of New Anaj Mandi is such a location where the SECTOR factor has a pivotal presence, which makes it a lifeline for millions, but it also possesses vulnerability to fire hazard, due to its congestion, crowd, corruption and carelessness of people as well as the authorities responsible for the well-off and take care of the society.

Analysis

Profile of Fire in Delhi and Anaj Mandi and Why Is It Vulnerable?

Fire is one of the prominent hazards of the capital city which hit city each day with lesser to moderate and high to severe intensity. Each day Delhi fire department receives on average 85 calls to dose fire. Decadal growth is 32% as it was 21,314 calls in 2009–10 and increased to 31,157 in 2019–20 as per the official data of Delhi Fire Service (Fig. 11.2). Fire in general terminology is the furious disaster of the national capital as it can be evidenced through the fact that within an hour Delhi fire department receives on average more than 3 calls to douse a spreading blaze. Incidence of fire is increasing day by day in Delhi as the numbers of calls and severity of incidences like the Anaj Mandi fire indicates. These all-mentioned facts indicate that capital is highly vulnerable to fire disasters.

Anaj Mandi comes under the jurisdiction of Rani Jhansi Road Fire station which receives approximate 616 calls yearly which means approx. 2 calls on a daily basis in such a small location (Fig. 11.3). In the study area, also number of fires are increasing with a decadal increase of 28.7% from 2009–10 to 2018-19. The site of Anaj Mandi in 2019 as per the opinion of locals evidenced 3 major fires in the same lane (Chimliyan Road fire, Bara Hindu Road, Anaj Mandi Plastic Factory fire and Fire in the Building of PradipSrashanWals) which with property and economic losses also engulfed life of human beings. Anaj Mandi's population is highly vulnerable to fire in terms of physical, social, economic and environmental vulnerability. Physical vulnerability in terms of loss of life during past fire incidence and as the scenario is not changed very much yet so maybe causative in future also. In terms of economics, it causes loss of property, business, loss of livelihood, disruption of essential services, local national economic loss, etc. Social vulnerability as the inhabited population particularly the inability of labourers to prepare themselves from such incidence make than social vulnerable to fire. Labourers know that the area is very prone to fire but they are unable to do anything about this because all things lie in the hand of factory owners, and they cannot do any things. This shows the social vulnerability of labourers and the population of Anaj Mandi. Environmental vulnerability leads to pollution of air to impact on the natural component of nature human beings.

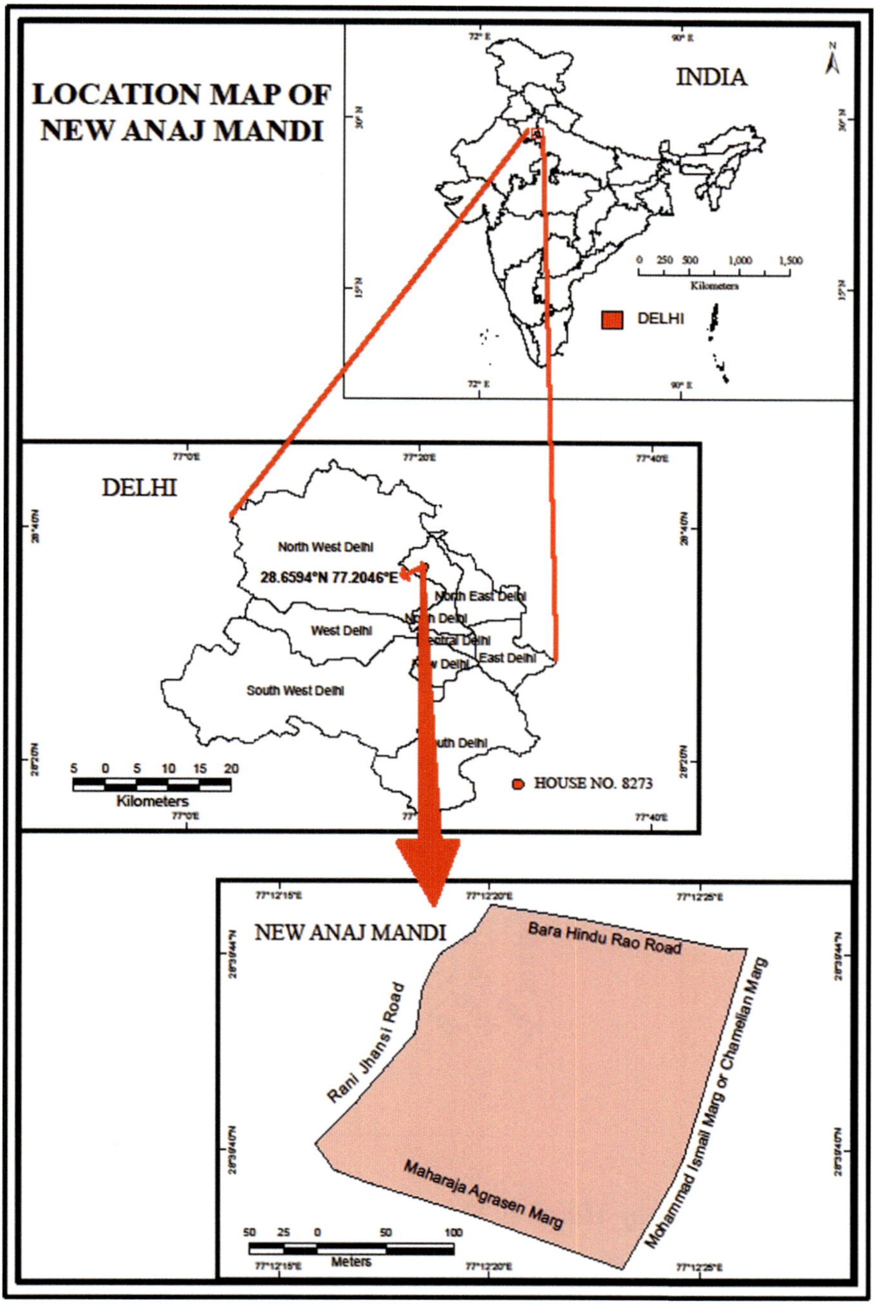

Map 11.1 Study area *Source* By Author

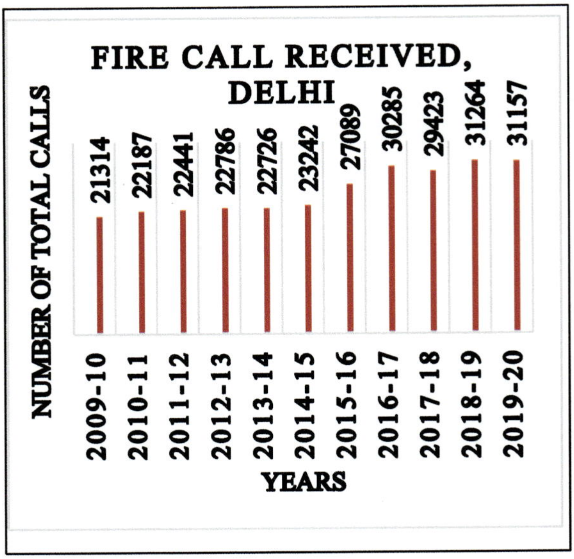

Fig. 11.2 Number of calls to douse fire in Delhi. *Source* Delhi Fire Department, 2020

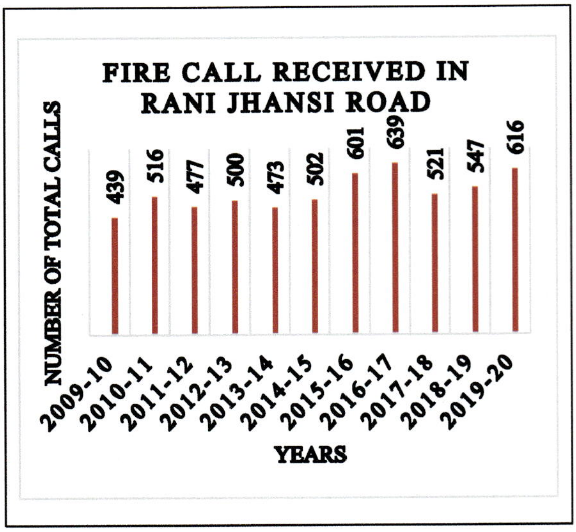

Fig. 11.3 Number of calls to douse Fire in Study area. *Source* Delhi Fire Department, 2020

Why Delhi and Anaj Mandi is Vulnerable to Fire?

Poor Design and Haphazard construction of buildings:

As the study area is a hub of secondary and tertiary activities that lead to more and more concentration of people in this small area from different parts of the country, ultimately increased value of land and space in locality. Builders and landowners of the site constructed buildings up to the heights of more than the prescribed limit of government by taking a consensus of local authorities as told by the locals. Tall buildings, with narrow lanes, absence of appropriate exit system, windows and ventilators makes this site prone to fire. The primary survey found that more than 50% of the houses have equal to or more than 5 floors in their buildings. Even in the buildings of more than 500 ft^2 of the area, there is only one exit. The primary survey shows that about 96% of buildings of Anaj Mandi has only one exit gate. Only 4% of buildings have 2 exit gates.

Lack of Public Information

This is the responsibility of various authorities particularly of the Delhi Fire Service to make the population aware of fire safety. Authorities run awareness campaigns that lead to awareness and an increase in precautionary measures among the population. During primary survey found that out of total surveys 98% told that authorities do not run any awareness campaign about fire safety (Fig. 11.4). Neither had they come to us to teach us what we should do to protect ourselves from fire when it goes out of control. This is the serious point that leads to the development of casual behaviour among the population about fire safety.

Illegal Industrial Activities in Residential Area:

There may be differences of opinions about legal or illegal aspects of industry in study area. One respondent during the survey told that "The legalness and illegalness of industries depend on individual perspective". FIR filed for The Devil of December, in that report it was clearly mentioned that illegal industrial activities were running

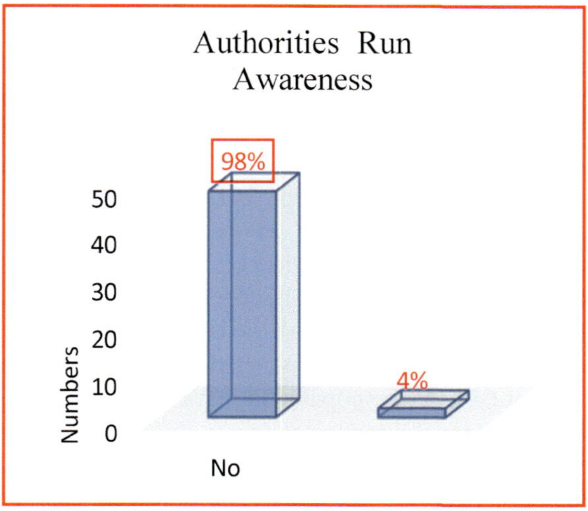

Fig. 11.4 Opinion of population about fire safety awareness campaign. *Source* Primary Survey, 2020

in a residential area. As per the second opinion of respondents during the survey, it was found that most of the industries are household in this area, only some are big factories. During the primary survey, about 54% of population told that we are not aware of illegal factories in this area and not any illegal factory run here. On the other hand, 44% of the population agrees that yes in this area illegal factory run and there is need to tackle it strictly so that this area can be saved from any such incidences in future. Overall it can be said that when illegal factories run in any residential area that increase the vulnerability of huge number of innocent population by neglecting various rules and regulation, which are made for the safety of mass.

Congested lanes and haphazard electricity wiring:

Lanes of the Anaj Mandi are highly congested from the entrance gate to the exit gate. Even a bike cannot pass through Anaj Mandi main gate to Katra Atma Ram Gate as near temple lane become so congested that even people cannot walk parallel there and the instalment of a stone pillar on the lane makes it impossible for bike to cross it. With such congested lanes, industrial activities of various categories thoroughly go through all floors of the buildings. On the other hand, the wiring system is also in haphazard manner. It can be said that there is a jungle of electricity wire in the lane.

Absence of Fire Safety Measure in House:

In most houses and industries, there is absence of fire safety measures like acid extinguishers, sand buckets, house pine, fire panic alarm system and other measures. After the disastrous incident, the concerned authority of Anaj Mandi decided to install fire safety measures in all households and factories of the study area. But despite this effort, the survey found that about 28% of factories or households don't have primary fire safety measures. This situation becomes worse when most of the workers of the factory say that we don't know how to use these fire safety measures.

Unawareness of Population:

Being unaware of any hazardous events are the biggest cause for turning any hazard into a disaster. As from local to national level, all the societies are made up of individuals, so awareness of individuals is highly crucial to fight any kind of disaster. In the case of Anaj Mandi, the survey found more than 76% of the population are unaware of the disastrous aspect of fire. Most of the respondents told that we even can't think that fire can be so disastrous. More than 56% of population don't know the helpline number of fire service. This is the condition after suffering from the disastrous event just 3 months before. Only 36% of population know about do and don'ts of fire safety. The respondent told that they are not highly educated and also they don't get too much time to do inquiries about all these things so they don't know do and don'ts of fire. As the population, those who work here are not highly educated so about 64% of the population don't know fire safety measures. About 70 of the population never participated in any fire mock drill programmes. When asked to a respondent why you don't know do and don't and why haven't participated in any mock drill? He responded that "I have just studied up to 5th… We were never told even in school… I have been living here for 15 years… But nobody ever came to tell us about the safety measures of fire... So you tell me how we will know." Despite all these facts, only 34% of the population has life insurance. On the other hand, 66% of the population don't have any type of life insurance to secure their family from any adverse condition of future. Why the Devil of December become so adverse can be visualized through the analysis of the above mentioned facts and findings. From personal casual behaviour towards disaster to the inappropriate behaviour of stakeholders are the main causative factors for the high level of vulnerability of Anaj Mandi to fire.

Results

Anaj Mandi Fire

Usually, the down of Anaj Mandi comes up with light, liking and leisure for the inhabitant as it activates the activities of the hub of SECTER and opens the way of earning for thousands and put smiles on the face of the most population connected to it. But the Down of Sunday, on winter's day of 8th December, was different as it comes up with the news of fire incidence in one of the houses of Anaj Mandi (Sanu and Sharma, 2022). For the site of Anaj Mandi, both the facets of fire as a resource and resistance is not a new thing, but the resistance of such a great extent was new and shocking. Down of Sunday Morning was the evilest Down for most of the occupiers of Anaj Mandi and particularly the tenants or residents of house number 8273 of Anaj Mandi where fire engulfed the five-story building and deceased 43 people and injured more than 20 victims seriously. Initially, 43 persons died on a disastrous day and 20 persons were injured but later on, two more injured men died at Lok Nayak Hospital between 15th and 22nd December which increased the death toll to 45 persons and injured 18 persons. It was the worst fire tragedy in Delhi after the incidence of 1999, Lal Kuan in the last twenty years (Sanu et al. 2023).

In this five-story building at least five different unlicensed units of manufacturing bags, plastic toys, jackets, etc. were running, by providing night stay and food to the workers at the same site of work. Only a few windows and just one narrow entrance, and a stock of combustible raw material like plastic, cardboard, rexine, etc. turned the building into an inferno around 5 am. Most of the workers were relaxed on Saturday night, and they were planning their Sunday, a day off from the gruelling routine of 12 to 15 hours of work. On usual Saturday workers got together, made phone calls to families and planned Sunday outings with friends. Most of them were asleep by 2 am, huddled in whatever space they could find on sheets spread on a cold floor, in congested spaces between machines and on top of packed goods in small rooms. These were the same spaces where they worked on machines, cooked and ate every day. The workers were used to waking up late on Sundays. But this time, for most occupants, Sunday never came. And for survivors, night gave way to a nightmare.

Most people said the fire broke out between 4 and 4:30 am. By the time, when the fire department received the first call at 5:22 am it was too late. Atul Garg, Director of DFS told after call four fire tenders from the nearest Rani Jhansi Road fire station reached to the spot within minutes but only one managed to enter into lane due to its narrowness and congestion. Another firefighter told "there were two exits in the buildings. One in the rear was locked from outside. We broke the lock but found the staircases blocked with packed goods. The people living on the second and third floors couldn't access the other staircase as it was on fire." The rescue operation of a firefighter to combat a disastrous event lasted for around four hours and thirty minutes. In rescue total, 150 firefighters were involved with 30 fire tenders in service and use of 1.5 lakh litres of water to sprinkle. The main agencies involved in the rescue were Delhi Police, Delhi Fire Service, Forensic Science Laboratory, CATS Ambulance, National Disaster Response Force and Delhi Civil Defence.

In the rescue process "The firefighters first entered the building's ground and the first floor and rescued six to seven occupants, many of whom couldn't walk without support. When we reached the spot, the sole exit point was open. Many women and children living on the ground and first floor had escaped early," said Gard. To rescue the people on the third and fourth floors, firefighters used a ladder between two buildings to cut open a window and rescue the engulfed population. But when they entered the building through the window they experienced the horrific scene. The fire officer told "There was hardly anyone seeking help by the time we gained entry. People were lying huddled in congested spaces. They were sleeping in spaces where one could not even sit. Their faces appeared tired but peaceful. Most of them were asphyxiated in their sleep; they probably never even woke up." In this furious fire, only those survived who deal the disaster with dare as Mohammad Afzal of 22 years old told "I dipped my muffler in water and wrapped it around my mouth to escape the fumes before rushing to a window to get some air. I was semi-conscious when I felt someone carry me to safety."

All the rescued victims were immediately sent to Lok Nayak, Lady Hading and Ram Manohar Lohia hospital which are situated in the proximity of within five kilometres, for further treatment. Most of the victims were sent to Lok Nayak hospital flowed by Lady Harding. During the rescue process, the emergency response system of the concerned department also looked inefficient as in the initial four hours there was a shortage of ambulances at the site of fire as the victims were sent to the hospitals by using PCR vans and auto-rickshaws. Most of the deaths in this furious fire occurred due to inhalation of toxic gases like carbon monoxides, etc. explaining this situation Dr Sudhir Gupta, head of Forensic Science, AIIMS told worker of the gas chamber of Anaj Mandi where all the windows were gilled and packed with cardboards that lead to accumulation of the huge amount of gases in rooms which caused suffocation, dizziness, unconsciousness and finally death of the victims. The fire was of serious category which took about 5 hours to douse it. A summarized timeline of the event is given in the Table 11.1.

In less than ten hours, police arrested the building owner, who had fled to Uttar Pradesh as well as his manager, who was hiding at a relative's house in Delhi. Monika Bhardwaj, Deputy Commissioner of Police (north) said: "Mohammad Rehan 43 years

Table 11.1 Timeline of Anaj Mandi fire

Time	Sequence of events
4:30 am	The fire started on the second floor in five-storeyed buildings
5:22 am	Fire department receives a call
5:27 am	Fire fighter and tenders reached to spot
5:35 am	The first occupant of the building, a man was rescued. He was rushed to a hospital
6:00 am	Delhi fire service receives a request for backup. More fire tenders deployed
6:30 am	Fire on the first and the ground floors brought under control
7:00 am	The senior officer from DFS including the director reaches the spot to supervise the rescue operation
7:35 am	11 people taken out, with firefighters making multiple rounds
8:03 am	22 people are rescued and sent to hospitals
9:58 am	The last charred body of a man was taken out of the building and sent to the hospital
10:08 am	A total of 63 people rescued, out of that 43 declared dead by hospitals
10:30 am	The cooling operation concludes
11:30 am	Rescue Operation declared over

Source Times of India and Hindustan Times, 9th December 2019

of age, the owner of the building and his manager Furkaan (39) were booked under the Indian Penal Code (IPC) sections 304 (culpable homicide) and 308 (attempt to commit culpable homicide)." After his arrest, Rehan told to police that "The building is owned by several people in Rehan's family. While Rehan owns around 60% of it, his brothers and in-law owns the other portions." The police officer told "It was also found that multiple small scale factories, managed by contractors, were operating out of the buildings. All of them will be questioned and if found negligent will be booked." By evening Delhi Police Commissioner transferred this probe of investigation to the crime branch. Rajesh Deo, Deputy Commissioner of police told "The first step will be to ascertain whether the fire was a result of some mischief, deliberately sparked or an accidental blaze. We will examine reports from the forensics department, electricity department, and the Delhi Fire Service to determine criminal liability." Order of Magisterial probe was given by Delhi Chief Minister, Fact-Finding committee was formed by Delhi BJP and NDMC also form a three-member committee to investigate the causes of fire. Overall, it can be said that the furious fire of the 8th of December was devastating that shook the capital city. In the proceeding section, there is an attempt to explain how and why the Fire of 8th December was a Disaster.

Disastrous Devil of December

Why this incidence of Fire was a disastrous devil to understand this, first of all, there is a need to simplify what I mean by a disaster. So for me and in my perspective disaster is an accident, event or situation that occurs due to natural or manmade causes in any area which results in loss of human life, ecosystem, economy and other aspects of human life, which is beyond the capacity of affected communities to recover itself from it. So "bad star" or Dis Astro or disaster is an adverse event that affects human beings and a large section of society adversely. Adverse, abrupt, abnormal, acute and all-encompassing or "5A" are the terminologies that decorate and characterized any event as a disaster. In this way, this incidence of fire caused the loss of life, adverse impact on the economy as thousands of people lost their livelihood, huge economic loss for the inhabitant of Anaj Mandi. This incidence also socially and psychologically impacted the life of the affected population.

As 45 persons were died and caused injuries to more than 20 people, Indirectly it also caused the death of a relative of a deceased person after getting news of the death of his nephew. Economically this incidence leads to the loss of jobs for thousands of population, crores of money were lost during the fire and damaged the business of Anaj Mandi. Explaining this situation a resident told "There is a lot of loss for all the people here... most of the people here have to close their factories... many labourers lose their jobs...even a tea seller suffered a loss of Lakh of rupees... So you can understand the loss of a big businessman." Psychologically, this incident has shaken the affected communities as many families lost their breadwinners and the main pillar of their family. Many women became widows, hundreds of children get orphans, and so many parents lost their old age reliance. All these traumatic incidents traumatized victims and kin of them psychologically. This incident socially shook the owner and as well as the creators of the disaster. They lost their reputation among their communities and society. These all facets impact of furious fire makes this a disaster. In Anaj Mandi fire, there is the presence of A5 Characteristics of a disaster which are Adverse, Abrupt, Abnormal, Acute and All-encompassing. The damage caused to life and property makes it adverse, its sudden appearance in early winter morning when most of the workers were sleeping explains the characteristics of abruptness. Thirdly in normal conditions, fire is doused easily without any huge loss, but it caused a huge loss on all facets making it abnormal. The fourth facet of acuteness can be visualized through the difficulties and suffering faced by the engulfed population, who lived the disaster. And finally all-encompassing means coverage at a large scale, in this case, it impacted the lives of thousands of population directly or indirectly, which makes it all-encompassing. A detailed analysis of all the disastrous facets of this fire incidence explains it as a disaster that impacted large numbers of population. So decided to call this incidence of fire "Disastrous Devil of December or Devil of December".

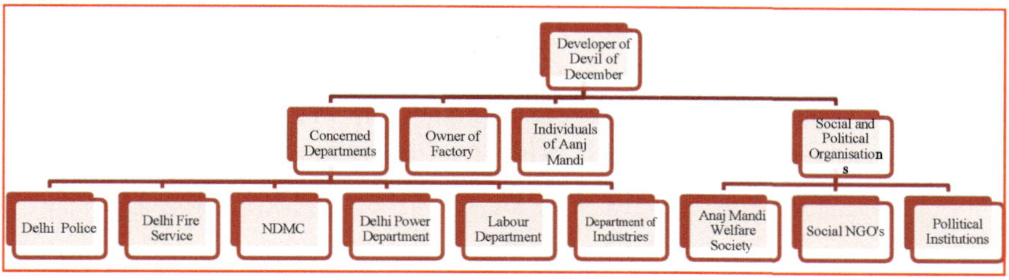

Fig. 11.5 Developer of Devil of December. *Source* Primary Survey, 2020

Discussion

Developer and Decorator of Anaj Mandi Fire

The death of maximum labours during sleep in Anaj Mandi was due to inhalation of dangerous gases and burn injuries. This was the immediate cause for the death of innocent people, but the real causative factor that developed the Anaj Mandi fire was concerned authorities who were neglecting their role and responsibilities. Under their nose, illegal industries in the residential areas were running without proper emergency management and violation of all rules and regulations. In Delhi, there are number of laws, rules and regulations to prevent any disaster and for proper management of the city but due to corruption, political patronage and relaxed behaviour of concerned authorities, all rules and regulations are just a written document at the grassroots level. After any incidence, various concerned authorities reached the site, give an order of investigation, come up with the report and recommendation and with the passage of time forget about it. Anaj Mandi fire incidence was created with the cumulative effort of the various departments (Delhi Police, Delhi Fire Service, Labour Department, North Delhi Municipal Corporation, Electricity Department, etc), owner of factory, unaware individuals and different social organizations like Anaj Mandi Welfare Society. All the concerned stakeholders responsible for the development of disastrous devil of December have been highlighted in Fig. 11.5.

Models to Make Anaj Mandi and Delhi Fire Resilient

The area of Anaj Mandi is highly vulnerable to fire disasters, as we observed through the discussion of the earlier section of this work. So here I tried to suggest a three-tier model for Anaj Mandi make it fire resilient. The first general model to fight fire is based on the A4 phase of the fire disaster development cycle. Second is the Community-Based Model and finally the Individual-Based Model (IBM) or K3 principle.

Development Cycle (A4 Phase)-Based Model

This model's strategy has been suggested to deal with fire in a different phase of the fire disaster Development Cycle. This tries to explain and give a simple and comprehensive method to deal with fire and check the resistance aspect of fire.

Ante Phase

Disaster Preparedness:

It means being or keeping ourselves prepared to cope with the disastrous aspect of fire. It includes the study of the vulnerability of Anaj Mandi to fire disaster, preparation of vulnerability map, and understanding the nature, severity and mode of occurrence of it. After that, it talks about the creation of awareness among the population about the adverse impact a fire can cause. It empowers people to deal with a fire disaster.

Disaster Management:

It works to minimize the disruptive forces of fire disaster and suggest strategies to reduce the magnitude or intensity of fire disaster. There should be a proper land-use plan for Anaj Mandi, government authorities and locals should stop all activities of this area which can cause a fire in the future. There should be proper management of raw materials that are used in factories. To manage fire disasters in the Ante Phase, occupants of Anaj Mandi Should be abode by all lawful activities.

Amid Phase

Robust Rescue and Emergency System:

In the Amid Phase, majorly response team work to save people and property from the fire. To cope with this phase effectively, there must be a well-equipped team with modern tools and techniques a robust rescue and emergency team. As in the case of Anaj Mandi fire, there was observed that initially there was an absence of Ambulances, because of that response team used auto and PCR vans to send victims to hospitals, and in this process, many people also lost their lives during transportation. This indicated the loopholes of the emergency system. So there is a need for an efficient rescue team to tackle the Amid Phase of fire disaster.

Annexation Phase

Relief:

This phase requires an appropriate relief measure that should be taken to assist the victims of the fire. Victims should be facilitated with the necessary resources to live their rest lives properly. Under this, there should be an adequate arrangement of relief materials like food to shelter and monetary funds for the victims.

Awakening Phase

Disaster recovery and rehabilitation:

Both these processes are long-term process, which comes under the Awakening phase. It involves adaptation and adjustment of disaster victims to the condition created after a fire disaster. This process helps people to forget about the past incidences and make way for another fire disaster. This strategy should be implemented to recover and rehabilitate the victims of fire, but also there is a need to keep people aware of such incidences so that they will be active and not allow any fire incidence to convert into a disaster.

Community-Based Model to Make Anaj Mandi Fire Resilient

Anaj Mandi with its unique characteristics in terms of physical, economic, social and cultural entity makes it a special area. Society of Anaj Mandi which is composed of a businessman to labourers and from residents to shop keepers needs to develop a societal or community-based model to protect themselves from fire and make Anaj Mandi fire resilient. As it is well-proven fact that the participation of communities to deal with disasters is the best way to cope with disasters and manage them properly. As when any fire incident occurs in any hose of Anaj Mandi that directly or indirectly impacts the life of the whole community, so community-based approach to deal with fire is the best way to combat fire disasters in Anaj Mandi. This approach for Anaj Mandi is based on a decentralized bottom-up approach to deal with their problem of fire, which will make them more aware, informative, cooperative and confident to deal with fire disasters at the grass route level. This approach is economical, heterogeneous by considering all section of society, gives equal participation to all and also develops a sense of accountability amount participants.

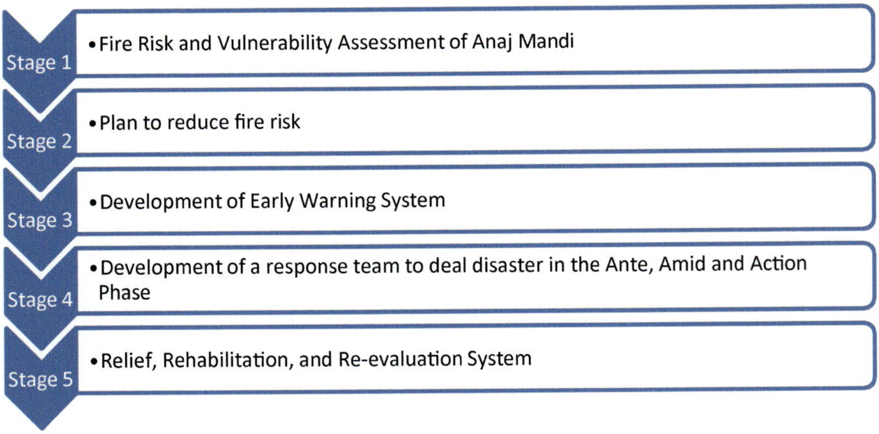

Fig. 11.6 Stages of community-based model. *Source* By Author

Stages of Community-Based Model to Deal with Fire Disaster in Anaj Mandi (Fig. 11.6) Fire Risk and Vulnerability Assessment of Anaj Mandi

Under this person of Aanj Mandi should do the assessment of fire risk-prone areas and vulnerable sections of society based on the past incidences of fire. They should make a detailed map of vulnerable, most vulnerable and non-vulnerable areas of Anaj Mandi. In this process, they should collect detailed data about the past fire incidences by including the date of occurrence, the timing of occurrence, losses caused by then on physical to economic aspects and frequency of occurrence, etc. Based on this data, community should identify the most vulnerable sections like workers, minors, or any specific locality, house, or spot and they should locate them on the base map of the community. After making a detailed report of the fire vulnerable section, location and a spot of an area the next step comes into existence.

Plan to Reduce Fire Risk

This is the second step to making the community fireproof. Under this, after getting the scenario of fire vulnerable spot of Anaj Mandi there is a need to make a detailed plan to reduce the risk of fire. Under these different steps comes are:

- Increase awareness of community population about fire safety
- Organizing community meetings from time to time
- Make a detailed road map for disaster response.

Development of Early Warning System

Early warning is a crucial aspect to deal with a disaster, but in the case of fire, people get a very less early indication. So in every factory of Anaj Mandi, there should be an instalment of Emergency fire panic alarm, smoke detector and other warning systems to make aware people in the early stage of fire occurrence. To get such equipment in every working unit community, representatives should do attempts to get it from the fire department or any other department.

Development of a Response Team to Deal Disaster in the Ante, Amid and Action Phase

Under this, Anaj Mandi society should develop a different team of locals to deal with different aspects of fire disasters like rescue and response, early warning and communication, relief coordination and other help.

- Rescue and Search Team
- Relief Coordination Team
- Early Warning and Communication Team
- Water and Sanitation Team
- First Aid, health and trauma counselling team.

Relief, Rehabilitation and Re-evaluation System

In Anaj Mandi, also there is a need to establish an R3 system of Relief, Rehabilitation and Re-evaluation. Under the relief, the community authority should take appropriate measures to provide relief to the affected population, by developing coordination with authorities. After the occurrence of any disastrous incidence, ex-gratia is announced by the government and various organizations, but the kin of victims, most of the time face lots of problems to get this money so community representatives should attempt to help them. The re-evaluation system should take into account the occurrence of any fire incidence in society, and they should take appropriate measures to control such mistakes in the future. When the community of Anaj Mandi will follow these steps automatically, they will develop Anaj Mandi as a Fire Resilient unit of Capital city and can become a role model for another part of the city too.

Individual-Based Model

Individuals are the smallest unit of any society or community, and they are the main causative factor for the creation of fire disaster at various levels as well as the main sufferer of it also. It is well-established fact that if all individuals start doing their duty honestly then they can easily handle the adversity of any disaster and particularly fire, a manmade disaster. So considering the smallest unit as the main cause of fire incidence, I tried to develop a model or principle based on individuals. This model tries to explain that if an individual will start following certain strategies, with continuation in it, then they can easily protect themselves from the fire disaster. There are many laws, rules and regulations to deal with fire disasters but still day-by-day incidence of fire is increasing in the capital city as well as in Anaj Mandi. The basic reason is that people are not following these rules and regulations. So we have to shift towards the individual-based model of K3 principle (Fig. 11.7). K3 which stands for three terminologies of Hindi words called Kartavya Baudhayata (literally meaning knowing about disaster), Kartavya Nisthata (means implementing the knowledge in real life) and finally Kartavya Anupalana (means long-term follow-up of disaster preventive measures). If an individual honestly starts following these measures in their life, it can be said that they can easily fight with fire disaster. In the following paragraph, there is a detailed description of these three principles.

Step I: Kartavya Baudhayata

The literal meaning of this word Kartavya Baudhyata is knowing your duty. This step is all about gathering knowledge about fire and fire safety at the individual level. This step says that every individual should be aware of their duty. In the case of fire disaster and protection from it, this explains that first of all individuals have to know

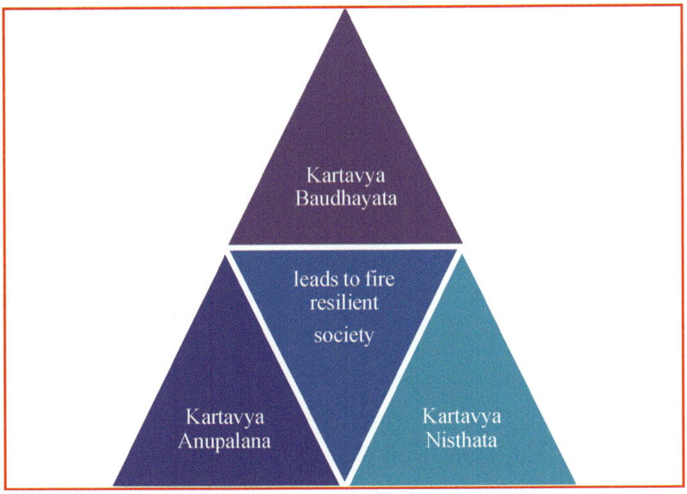

Fig. 11.7 K3 principles of fire disaster management. *Source* By Author

about fire and fire as a disaster. Fire is one of the crucial resources of our life, we cannot live life without fire, so it is a great resource of our life, but the problem is why fire converts into resistance, so in this process, we should know why fire is a resistance, what factors make fire furious, etc. As during survey found the maximum population of Anaj Mandi was unaware of the disastrous aspects of fire, so there is a need to increase knowledge about it. After knowing why fire is a disaster and what factors make it furious, now individuals should focus on what are the preventive measure to check or restrict fire from becoming resistant. Under this, individuals should increase their knowledge about the do and don't of fire, different fire safety measures to deal with fire in different stages and locations. An individual should also make aware of their family members who are unaware of it. Under this step, individuals with the development of knowledge about fire also collect and make fire safety equipment and detailed road map to deal at their own level form house to working place.

Step II: Kartavya Nisthata

Kartavya Nisthata stands for the implementation of knowledge in real life. As in Step I of Kartavya Baudhayata individuals, first of all, get the knowledge about the disaster and how to protect themself, so this step says individuals to implement that knowledge in real life. An individual should arrange fire safety equipment, and other necessary things like escape routes during a fire, etc. in their house. They should develop house specific plans and implement them in day-to-day life. It should not just be a word written on paper but should be will implement in real life.

Step III: Kartavya Anupalana

The simple meaning of this is to implement all the duties for fire safety for a longer period without break in it. As in most of the cases, it is observed that initially after the occurrence of any fire incidence all individual to stakeholders become very active, they start doing their duty properly, but after a certain time they leave, it in between and start neglecting the safety measure of fire, which become the main cause of any fire. This step is the most important step to restrict fire from becoming resistant. As we know that fire disaster occurs mainly due to the negligence and fading up of the past event. So if an individual wants to protect themselves from the fire, they should follow the fire safety measure for a longer period without break in its continuation.

These principles to fight fire at an individual level is the best way to protect from fire disaster. If the people who died in the Anaj Mandi fire and the owner of that factory had followed K3 Principle, then Anaj Mandi would not have suffered such a huge fire disaster. They had suffered such a big devastating disaster because they were not following it. Just imagine if workers and factory owners were aware of fire safety and were following it, without breaking the continuation, so the economic loss and loss of life they faced does not happen. Overall it can be said that once everyone starts following these K3 principles at an individual level with a hundred per cent honesty, they can easily restrict the resistance of fire because an individual is also a factory owner, fire officer, labourers, an officer at various posts and the decision-makers. So everyone should deal with this disaster at their own individual level.

Conclusion

The site of Anaj Mandi due to congestion, corruption and carelessness is highly vulnerable to fire disaster. Anaj Mandi fire incidence of Delhi was one of the horrific incidences in the recent history of the capital city. It has impacted the lives of thousands of population. Lack of awareness and activeness of responsible authorities and individuals were the main cause for it, so they should work properly by following the model, measures and methods suggested. The K3 principle is the best way to deal with fire in Anaj Mandi as well as in Delhi as this focuses on the role and responsibility of every individual. By following these suggestive measures Anaj Mandi and Delhi can develop themselves fire resilient sites.

Acknowledgements Will like to thank Prof. Anu Kapur, Dr Subhash Anand, Dr.Vishwa Raj Sharma, Manoj Tiwari Ji, Advocate Sanket Gupta Ji, Anaj Mandi Fire Department officer Somvir sir and the locals of Anaj Mandi.

References

Kapur, A. (Ed.). (2005). *Disasters in India: Studies of grim reality*. Rawat Publications.

Pal, I., & Ghosh, T., (2014, January). Fire incident At AMRI Hospital, Kolkata (India): A real time assessment for urban fire. *Journal of Business Management & Social Sciences Research (JBM&SSR), 3*(1). ISSN No: 2319–5614. www.borjournals.com (Blue Ocean Research Journals 9).

Tomar, S. K., Kaur, A., Dangi, H. K., & Saram, K. (2017, August). Fire risk analysis using geospatial approach and mitigation measures for South-West Delhi. *International Journal of Emerging Research in Management &Technology, 6*(8).

Sanghavi, P., Bhalla, K., & Das, V., (2009). Fire-related deaths in India in 2001: A retrospective analysis of data. Retrieved from www.thelancet.com

Sanu, S. K., & Sharma, V. R. (2022). Fire Disaster Development Cycle: A Case Study of Anaj Mandi Fire, Delhi, India. *Research Journal (Arts)*, XLVIII (July-December 2021), 130–155.

Sanu, S. K., Rai, S., & Sharma, V. R. (2023). Description of Fire as a Disaster with a Case Study of Delhi's Mundka Fire. *Disaster & Development Journal, 11*(2).

Newspapers

Adak, B. (2019, December, 11). MCD have 20 days to shut 9000 illegal factories, but only 15 inspectors to do it.*Hindustan Times,* p. D3.

Correspondent, H. T. (2019, December, 9). 43 dead in fire at illegal factory.*Hindustan Times,* p. D1.

Correspondent, H. T. (2019, December, 9). Exits locked, they were trapped by blaze.*Hindustan Times,* p. D3.

Correspondent, H. T. (2019, December, 9). Building owner, manager held; crime branch will probe how fire started.*Hindustan Times,* p. D3.

Correspondent, H. T. (2019, December, 9). A body pulled out every 5 mins.*Hindustan Times,* p. D3.

Correspondent, H. T. (2019, December, 9). Not fire but smoke killed men: Docs.*Hindustan Times,*. p. D4.

Correspondent, H. T. (2019, December, 9). At hospitals, rush for news of beloved.*Hindustan Times,* p. D4.

Correspondent, H. T. (2019, December, 9). 'There is no escape...take care of my family'.*Hindustan Times,* p. D4.

Correspondent, H. T. (2019, December, 9). Illegal building spotted, wasn't sealed.*Hindustan Times,* p. D5.

Correspondent, H. T. (2019, December, 9). Centre faults fire dept, AAP govt says bid to shield MCDs. *Hindustan Times,* p. D5.

Correspondent, H. T. (2019, December, 9). Cramped, unliveable: Hub of illegal factoreis. *Hindustan Times,* p. D5.

Correspondent, H. T. (2019, December, 9). Faceless victims: Aspirations for a better life bring you migrants to jobs in 'tinderboxes'.*Hindustan Times,* p. D5.

Correspondent, H. T. (2019, December, 9). Delhi learns no lessons from past fire tragedies.*Hindustan Times,* p. D7.

Chapter 12
Assessment of Fire Disaster Risk Reduction in Higher Educational Institutions in Delhi

Abhay Shankar Prasad, Bindhy Wasini Pandey, Rajesh Kumar Abhay, and Priyanka

Abstract Disaster risk reduction (DRR) in education is a key issue in all countries to build resilience to disasters. In the post-2015 framework for DRR, the important of education and awareness-raising program has been agreed as the top priorities in the policy of disaster. Awareness program of education and efforts is everyone's responsibility rather than governments and the media implementing and promoting agencies only. The aim of the study is assess the extent of understanding about disasters knowledge among college students and staffs. This research used a cross-sectional study design for disaster risk reduction planning wherein it attempts to assess the disaster-related knowledge, preparedness, adaptation, awareness, and risk perception among the students. This paper has been based on the primary survey. The questionnaire was prepared with open- and close-ended questions. This questionnaire was used to collect basic data and respondent's perception about the disaster management. A survey was conducted in one of the Higher Educational Institute of Delhi University using questionnaire, field observation, interview techniques, informal interviews, interaction and discussion with students, teacher and non-teaching staffs. A sample of 200 respondents was taken consisting of students, teachers and non-teaching staff in Dyal Singh College. The data has been analyzed using statistical and cartographic techniques. This mitigation measure of disaster risk reduction is important in enhancing resilience capacity and ensuring sustainable development pathways and provides various man-made and natural disaster relief processes for disaster management. These research papers suggest a policy to improve the transformation of scientific knowledge into policy and to increase mutual understanding, partnership and collaboration for better policy outcomes in the goal of sustainable development and disaster management.

A. S. Prasad (✉) · R. K. Abhay
Department of Geography, Dyal Singh College, University of Delhi, New Delhi, India
e-mail: abhaydse@hotmail.com

B. W. Pandey · Priyanka
Department of Geography, Delhi School of Economics, University of Delhi, New Delhi, India

V. R. Sharma and Chandrakanta (eds.), *Making India Disaster Resilient*,
https://doi.org/10.1007/978-3-031-50113-5_12

Keywords Higher institutions · Fire hazard · Vulnerable area · Preparedness plan and disaster risk reduction

Introduction

In general, disaster can be defined as a hazatter of natural, man-made or technological mistakes by human, which will be causing serious physical damage or destruction, loss of life, health impacts, property damage and also may destroy the entire area/ environmental system such as the economic, cultural and social life of people (Castro, 2013). Thus, disaster risk can be explained most simply as the function of a specific hazard, physical exposure of elements at risk and human vulnerability (Asia-Pacific Disaster Report, 2012). Education has been identified under the Hyogo Framework for Action (HFA) as a key element in mitigating the impact of natural disasters and disaster risk reduction (United Nations Hyogo Framework for Action, 2015). The United Nations Children's Fund (UNICEF) is working hard to secure the school building and facilities in the disaster area. Disaster Risk Reduction (DRR) activities are also included in the school education system to increase awareness and readiness among students and teachers (UNISDR, 2000; UNISDR, 2015).

Disaster management is a systematic process that consists of several components based on the key management principles of disaster management planning, organizing and leading, including coordination and control policy of disaster risk reduction (UNISDR, 2007). A more proactive approach is beyond disaster management response and mitigation and aims to minimize the negative impact or consequences of adverse events. Fire hazard is a potential fire accident that can happen at any workplace. The main responsible causes of fire hazards behind a major fire accident is negligence and carelessness, which can lead to a wide variety of accidents, such as death, injury, environmental damage, business damage and so on (Sharpe & Kelman, 2011). Fire hazard includes all types of live flames, cause of sparks, hot objects, chemical that has potential for ignition or that can aggravate a fire to become large and uncontrolled. A fire hazard is a threat to life, property and livelihood. Therefore identifying, eliminating or minimizing the risk of fire caused by potential hazards is a key objective of security systems. An emergency response plan is prepared to address such occupational hazards and prevent further damage (Joshi, 2014).

Management systems for fire hazard emergency situations include communication, procedures, manpower, materials, transportation, recovery, evacuation plans and shelter (Bosher & Dainty, 2011). Various scientific measures can be taken pre, post and during fire emergency situations with the use of GPS and GIS, especially for assessing vulnerabilities, risks of various groups. Children and students are the most vulnerable group in society. They spend maximum time in school or engaged in extracurricular activities in the colleges and schools. Disaster education therefore plays an important role in Disaster preparedness and mitigation plan among the students and teaching-non-teaching staffs (Mitchell et al., 2008; Joshi,

2009). The New Delhi Municipal Corporation (NDMC) has taken the main focus on to community engagement and preparedness plan for DRR through introducing Community Disaster Management Programs (CBDRM) since 2014. The official website and social media accounts, as well as conducting awareness sessions for schools, colleges and other preparedness projects are going on from 2016 onward. Natural and man-made disasters can have devastating consequences worldwide and locally. To accelerate the process of disaster-related knowledge dissemination the research paper makes an attempt to discuss how academic institutions can play a larger role in influencing young minds to develop appropriate values and attitudes necessary to create a culture of disaster preparedness and mitigation through developing appropriate knowledge modules, demonstrating activities related to developing institutional disaster management plans, carrying out mock drills and conducting capacity building programs for Disaster Risk Reduction (DRR). In addition to developing a better knowledge base, schools, colleges, universities and higher education institutions play a leading role in conducting relevant research in the field of disaster management, reviewing and revising disaster management policies and policies followed by the country (Johannes, 2020; Gangalal et al., 2014). Degree courses in Disaster Management specifically provide students with a basic knowledge of all the dangers, hazard, vulnerable and possible consequences. Institutional arrangements for the Disaster Management / Disaster Risk Reduction (DRR) are established or strengthened with new or updated legal directives and often form inter-ministerial policies (systems, councils, commissions, etc.) directly related to the government or especially the compulsory ministries. National Disaster Management Agencies (NDMAs) that act as coordinators of inter-ministerial arrangements have executive responsibility for disaster and emergency response and DRR coordination in the country (Kanyasan et al., 2018; NDMA, 2006). Higher education institutions often provide technical assistance and guidance in civil service training or in specialized DRR training institutes at the national and local levels (Petal, 2009; Rajesh et al., 2011).

Geographic information systems and databases need to be developed at the organizational level to focus on the development of methods and decision support tools using GIS to integrate, manage and display risk-related information for Disaster Risk Reduction. Such systems may include technologies to assess building damage; influence of infrastructure and public risk to assist local authorities in decision making in emergencies and it will be help in design the policies for the disaster management and preparedness plan for DRR (Roy & Pandey, 2016).

Study Area

The study area of this research work is one of the prestigious institutes of Delhi University, i.e., Dyal Singh College which is centrally located at Lodhi Road in South Delhi near Lodhi Garden, Jawaharlal Nehru Stadium, and Indian Habitat Center. The college is very much linked with road transport and by Delhi Metro (nearby station is Jawaharlal Nehru Stadium on violet line). Its latitudinal extent is from 28′ 24″N to 28′35″ N and longitudinal extent from 77′13′48″E to 77′14′00″E (Figure 12.1).

Presently, the college has the following infrastructure (Based on the Primary Survey, 2020):

- 59 classrooms including 7 eco-friendly bamboo classrooms and tutorial rooms;
- Approximately 22 labs including computer labs which are linked to Delhi university;
- A seminar hall with a seating capacity of 120 people (Approximately);

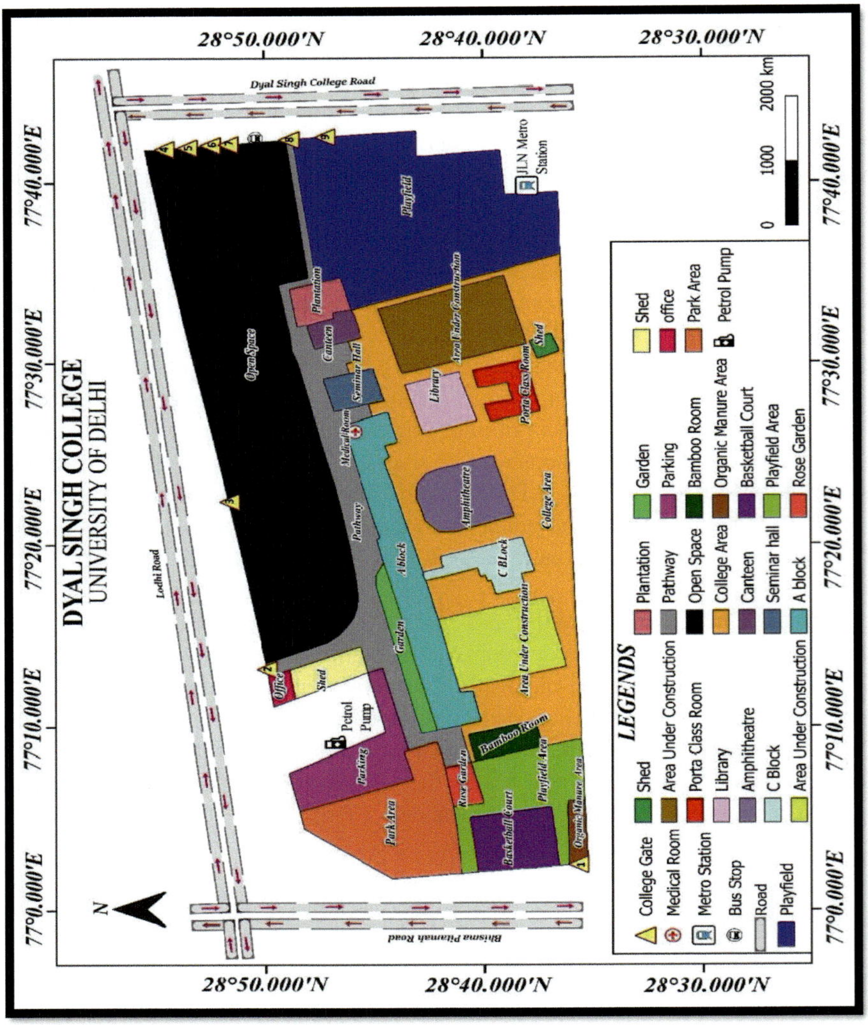

Fig. 12.1 Location of the study area: Dyal Singh College, Delhi University. *Source* Based on the College Administrative Map and Primary Survey, 2020

Table 12.1 Block-wise division of college campus

A	B	C	D
It includes Principle Office, Administrative Office, Medical Room And Staff Room	It includes Seminar Hall, Canteen, Lift, Auditorium and Computer Labs	It includes New Science Block	It comprise of the Library

- An open air amphi theater and auditorium with a seating capacity of 500 people (Approximately);
- The college campus is divided into 4 blocks as described in Table 12.1 (Based on Primary Survey, 2019)

The college campus is over 11 acres is centrally located on metro map at Lodhi Road in South Delhi and imparting education to over 5000 students and more than 300 teaching and non-teaching staffs (Source: College website, 2020)

Database and Research Methodology

Both qualitative and quantitative methods have been used in the study. Primary data has been collected from field observation, interview techniques, informal interviews, interaction and discussion with students, teacher and non-teaching staffs. The main technique of primary data collection was done through structural and non-structural questionnaire. A questionnaire survey was conducted to collect data and information from the students, teacher and non-teaching staffs of the college through online and offline platform such as Google Form and WhatsApp. Literature reviews have also been done in this project to gain and understand more about Disaster risk reduction and its preparedness plan. The questionnaire was divided into five parts: the fire risk assessment, education opinion in DRR, culture and adaption practice in DRR, experiences in disasters and the method of exploration on DRR awareness-related program. For field survey in the college various methods have been used such as Random sampling, interviews, questionnaire etc. A *field survey* was conducted via a questionnaire which was prepared in the context of the topic, in which both the types of questions, i.e., *open-ended* and *close-ended* questions were included among the multidisciplinary students in the college. After the *questionnaire* was prepared on the specific type, the area of study was visited for the survey and asked the students, teachers and non-staff members, the formed questions. The respondents were selected on the basis of *random sampling*. Some of them were interviewed, and a little information was gained by observing the area.

The secondary data were obtained from various newspapers, college website and articles related to the topic of the study and then have been used in the report to make it more reliable. Help from Internet to review other published reports and articles has been done to procure very useful data.

In this present study, 200 respondents were asked to have their opinion and fill the questionnaire.

Few were interviewed to have a diverse and good quality of information. The areas were observed, and then photographs were clicked of those areas which are more prone to fire hazard. Proper blocks and other specific areas have been shown through maps, so that analyzed information can be easily understood and would have better representation.

Results and Discussion

Adopted Fire Safety Measures

By regulations, schools, colleges and universities operating in buildings must have fire safety measures arrangements that include fire engines, fire extinguishers, fire buckets, terrace tanks, fire pumps, down comer and hose reel etc. It is mandatory for every educational institution to obtain a No-Objection Certificate (NOC) from Government organizations and University, which ensures that management, has adequate fire safety arrangements, but unfortunately most schools and college are unaware of such an important problem (UNISDR, 2015).

With reference to this point, it was found that the respondents of the college are highly aware regarding the disaster management. Disaster management is a core subject in the syllabus of Graduation. This has been deduced from the primary survey that there are fire safety measures in college (Fig. 12.2). Out of the total number of respondents, 24% of them thinks that the college do not have fire safety measures, whereas 76% are aware of the fire safety measures in the college campus.

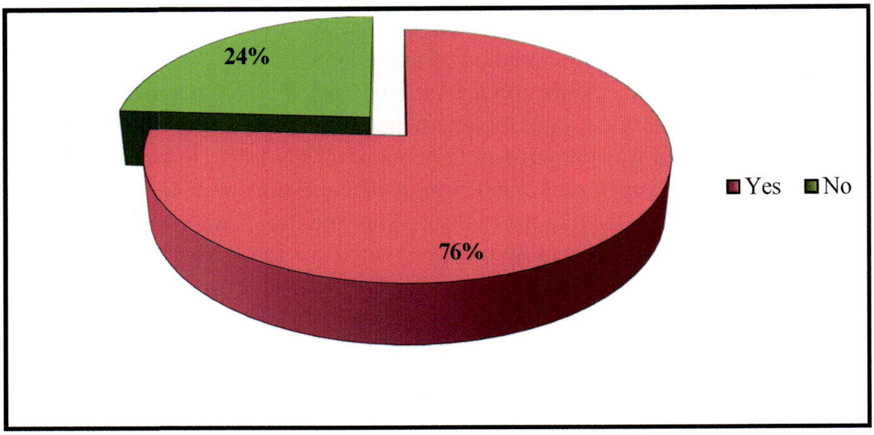

Fig. 12.2 Fire safety measures in college

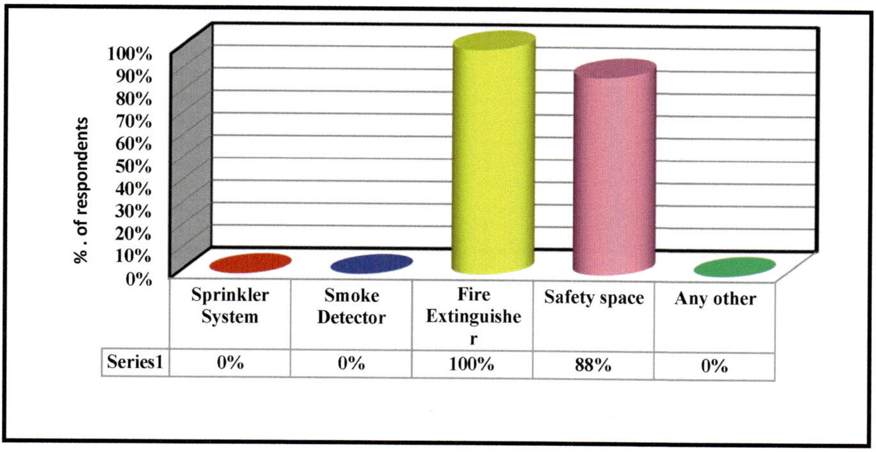

Fig. 12.3 Types of fire safety measures. *Source* Primary survey, 2020

Disaster cell of the college with the support of CISF wing and Government and non-government agencies is conducting evacuation drills to make both the management and students aware of what to do in case of an unexpected fire breakout.

Types of Safety Measures Available in the in College

Disaster Management for Disaster Risk Reduction (DRR) is a multidisciplinary subject thus there is necessity to aware every sector of the society including Educational, Institutional, Industrial, Agricultural, Defense Services, Government agencies, NGOs, etc. Safety measures of disaster management are empowered and indicator of capacity building of disaster risk reduction as an emergency planning, risk assessment, community development, and Humanitarian aid (UNESCO, 2013).

It seems that the respondents are aware of the fire extinguishers located in the college as each and every person of them has given this answer, i.e., total no of percentage of respondents agree that the college has fire extinguishers (Fig. 12.3).

Also, it can be concluded from the respondent's point of view, that the college has much safety space in case of a fire outbreak whereas according to them, the college does not have sprinkler system and smoke detector at all.

Potential Causes of Fire Outbreak

All students must keep in mind certain safety precautions and measures. Some responsible causes of fire hazards in the campus are canteen fire while cooking, faulty electrical equipment, during laboratory work, sparks during arid conditions,

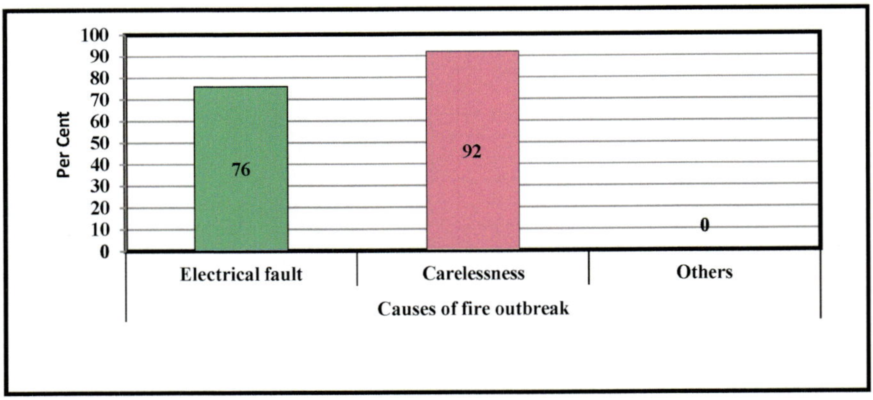

Fig. 12.4 Causes of fire outbreak. *Source* Primary Source, 2020

electrical distribution systems, etc. (Pandey et al., 2017). It is situated to close to the petrol pump, which is nearest to the college parking and bamboo classes. Both areas (Parking area nearby Chemistry lab and Bamboo Classes) are high vulnerable area of the college.

It can be observed from Fig. 12.4 that carelessness can be the main cause of fire outbreak. So, it can be deduced that either there is lack of staff that keeps a continuous watch on such issues or may be the staffs has no training.

Electrical fault can be the second major cause of fire outbreak, according to the respondents. From this, it is clear that there is lack of proper wiring or might be the case of mismanagement.

Fastest Way to Evacuate Students and Staff in a High Rise

According to respondents, fastest way to evacuate students and staff in a high rise is fire exit doors. It means that there are fire exit doors within the campus which can be used during a fire outbreak or in case of a high rise (Fig. 12.5).

None of the respondents thinks that elevator could be the fastest way to evacuate, whereas a few of them (20%) thinks that fire escape staircase and Windows can also be used to evacuate.

Time Taken to Escape to a Place of Safety During Fire

The principle on which escape rules are based is that the time available to escape (estimated between the time the fire starts and the ways in which it escapes from the office is unsafe) is longer than the time required to escape (it takes time to evacuate everyone after the fire is detected and alerted).

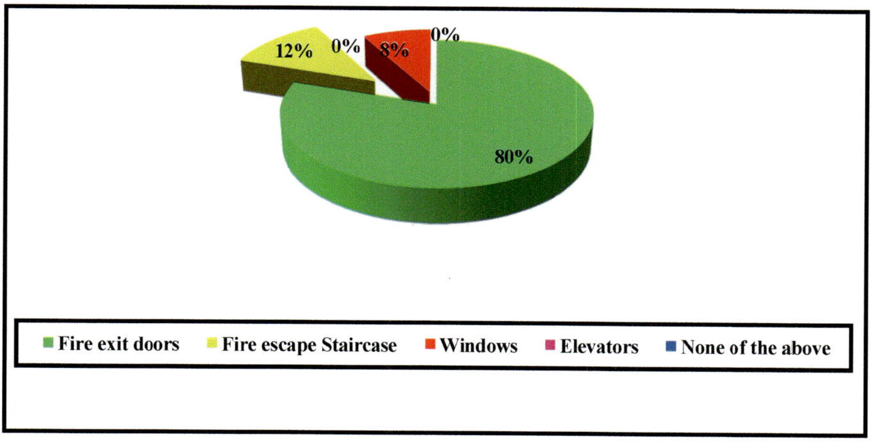

Fig. 12.5 Ways of evacuation in a high rise. *Source* Primary Source, 2020

According to respondents, it will take two hours (40%) or more than two hours (52%) for all the occupants to evacuate to a place of safety once a fire has been detected (Fig. 12.6).

If there are any students or staff with disabilities on your premises, their needs should be taken into consideration when planning an evacuation strategy.

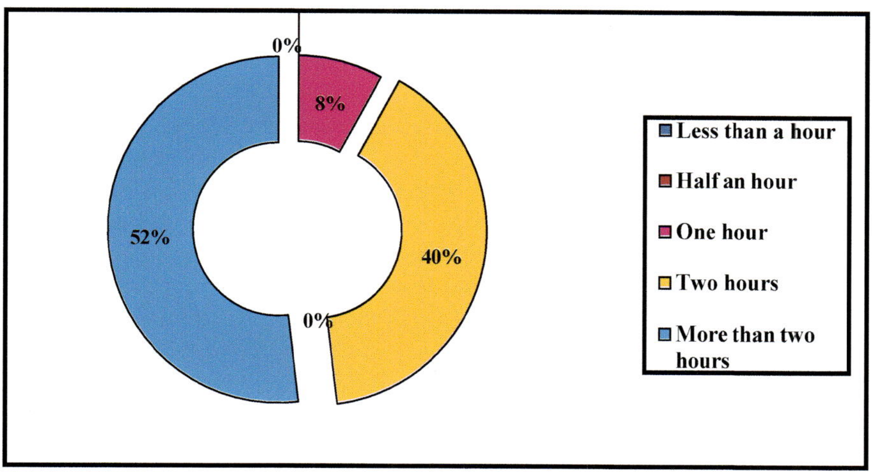

Fig. 12.6 Time taken to escape to a place of safety once to a place of safety once. *Source* Primary Survey, 2020

Plate 12.1 Highly vulnerable narrow exit gates. *Source* Primary Survey, 2020

Are All Escape Routes Easily Identifiable, Free from Any Obstruction?

All offices, classrooms, staff rooms of the studied institution must clearly identify escape routes in the event of a fire. These escape routes should be kept clear at all times to ensure that everyone can leave the college campus in the event of a fire or other emergency. Take care if placing notice boards in escape corridors/ routes as any paper on the board could be fuel in the event of a fire. Arrangements must be conveyed to all those occupying the workplace and particularly to personnel such as fire wardens who will be assisting in overseeing any emergency evacuation. To understand what types of emergency evacuation routes are required, the relevant building regulations must be considered.

Majority of the respondents think that all escape routes are not easily identifiable and not free from obstruction (Plate 12.1). It means that either there is a dearth of escape routes or they are at such a place where there is obstruction lies in between.

It can be inferred from this that, either the building is not well designed or the staff and students is not trained enough to make all occupants to escape to safety place. It is also possible that might be the place of safety is not near to the campus or no place of safety within the campus. Few of them (8%) also thinks that it might take minimum one hour for all occupants to escape to a place of safety, however, it might be possible that they know more safety places nearer to campus or aware of fire safety equipment and fire exit doors.

Sufficient Extinguishers Sited Throughout College Campus

It is imperative for any institution that fire safety measures and equipment must be kept in working order. This includes all fixtures and fittings such as fire doors, stairs, corridors, fire detection and alarm systems, fire-fighting equipment, notices

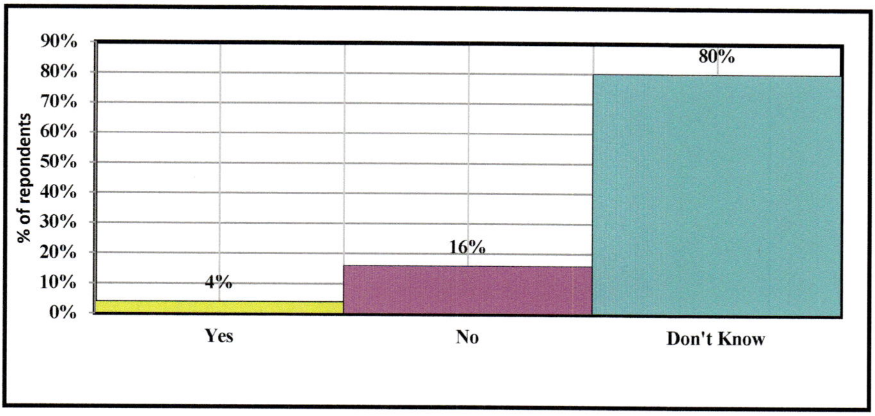

Fig. 12.7 Sufficient extinguishers sited throughout college. *Source* Primary Survey, 2020

and emergency lighting. Regular inspections, periodic servicing and maintenance must be carried out no matter how large the workplace. Any errors should be put as soon as possible.

The employer or nominated employee or disaster cell committee of college may perform inspections and routine maintenance tasks. However, it is important to ensure the reliability and safe operation of installed equipment such as fire extinguishers and fire alarms and emergency lighting. This is best done by periodically servicing and using a competent person to make the necessary repairs. The record of work undertaken on such equipment and systems helps to show that it is in accordance with the law.

Fewer of the respondents (4%) are aware of the sited extinguishers in College Campus. Majority of them, i.e., 80% don't know whether there are sufficient extinguishers are or not (Fig. 12.7).

But it has been found that many fire extinguishers have been installed in the college campus (Plate 12.2).

16% of the total respondents said that there are not sufficient extinguishers in college.

Fire Hazard Prone Area/Vulnerable Area

The college campus is located near the petrol pump, and there is also college parking area. There are always many cars of the college staffs in the college parking area. College's lab and Bamboo rooms are situated nearby the college parking area. In this way parking area, labs and petrol pump are highly vulnerable area. Employees should be made aware of all escape routes from the high-risk zone or vulnerable area

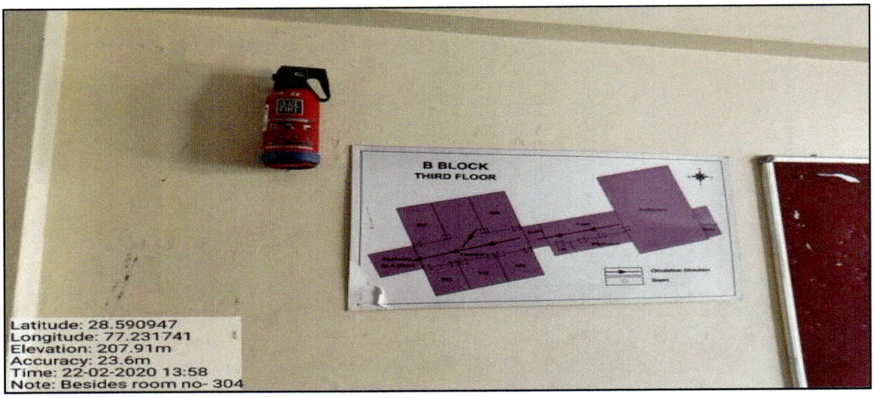

Plate 12.2 Installed fire extinguisher in the corridor. *Source* Primary Survey, 2020

and use emergency drills regularly to practice among the students and staffs using them as part of their emergency routine for the disaster risk reduction.

From the respondent's point of view (40%), canteen is the most vulnerable to fire hazard, whereas parking area, sciences labs are equally prone to fire hazard (Fig. 12.8).

None of the respondents think that library and geography department are prone to fire hazard (Fig. 12.9).

Hence, it can be inferred that more care is needed in these areas which are prone to fire hazard (Plate 12.3).

Careful handling of chemicals and electrical equipment in labs is necessary to avoid such hazard.

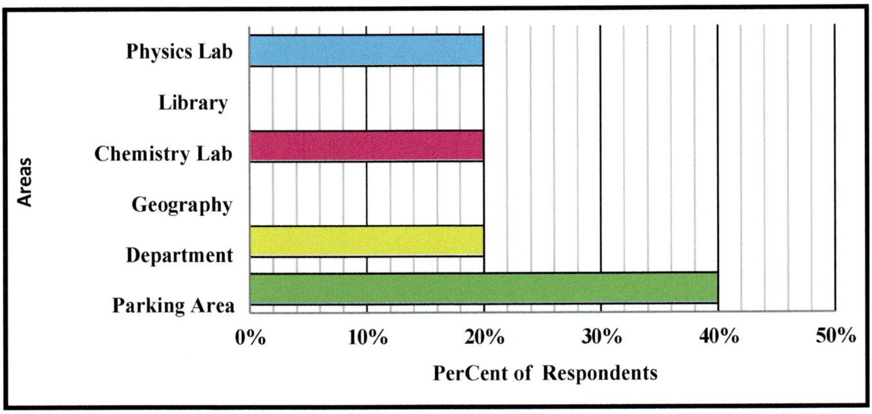

Fig. 12.8 Areas vulnerable to fire. *Source* Primary Source, 2020

Fig. 12.9 Fire hazard vulnerable area in the college. *Source* Based on Primary Survey and Respondent's Perceptions, 2020

Are the Right Type of Extinguishers Located Close to the Fire Hazards and Can Users Gain Access to Them Without Exposing Themselves to Risk?

As about 80% of the respondents think that there is a lack of right type of extinguishers close to fire vulnerable areas, and even if they are located, then users cannot gain access to them without exposing themselves to risk (Figs. 12.10 and Plate 12.4).

If we talk of those 20% of the total respondents who think that there are right type of extinguishers and they can get access to them, might be they are more aware of the sited extinguishers or they know how to get access to them.

Plate 12.3 Fire hazard vulnerable area in college campus. *Source* Primary Survey, 2020

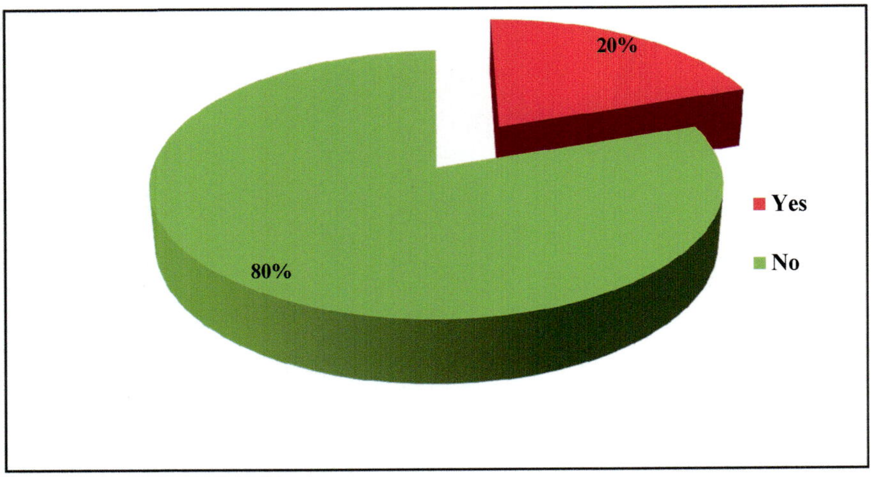

Fig. 12.10 Right type of extinguishers. *Source* Primary Survey, 2020

Emergency Plan or Disaster Cell in College

10% of the total respondents said that there is a Disaster cell in college to take quick actions in case of fire outbreak, means that they are aware of it and know how it take actions, whereas a large percentage of respondents, i.e., 90$ don't know anything about such disaster cell or emergency plan within college campus (Fig. 12.11).

This means that there is a need to make them aware of such cell if it is there in the college. It can be deduced from the respondents answers that the disaster cell do not take account of all reasonably foreseeable circumstances, or it means that there is a lack of proper management and necessary actions need to be taken.

Plate 12.4 Fire extinguisher, bamboo room. *Source* Primary Survey, 2020

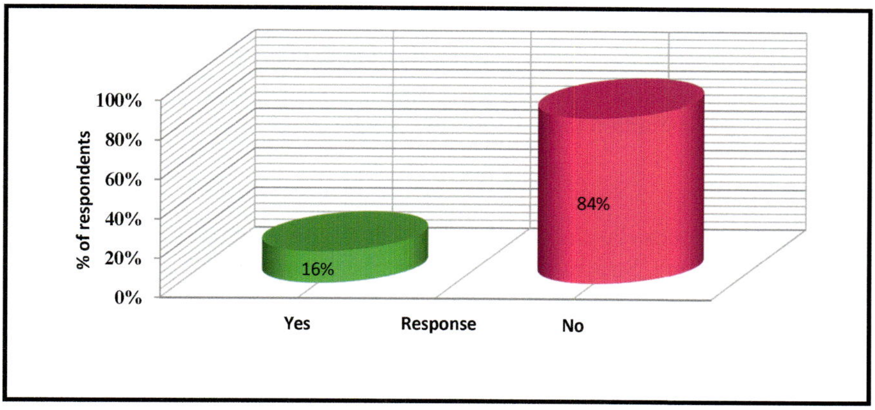

Fig. 12.11 Emergency plan in college. *Source* Primary Survey, 2020

Emergency Contact Numbers

Almost 88% of respondents (Fig. 12.12) know about emergency contact numbers which are most common, i.e., 101 (Fire Brigade), 100 (Police Station) and 102 (Ambulance).

Also, there are about 12% of respondents, those don't know about any emergency number in case of a fire hazard, it means there is need to make them aware and necessary campaigns need to be carrying out to reach basic and important information to the farthest one.

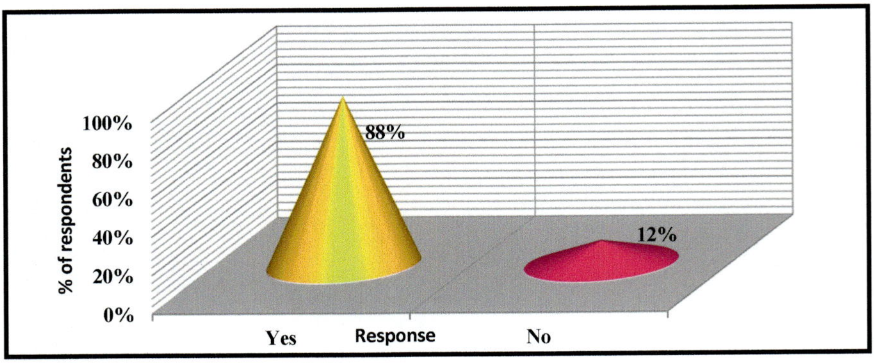

Fig. 12.12 Emergency contact numbers. *Source* Primary Survey, 2020

Preparedness Plan

Disaster preparedness refers to measures taken to mitigate and manage the effects of disasters for Disaster Risk Reduction (DRR). The main purpose of this research paper is to secure institution through education and preparedness plan.

Organized source

Organized resources can be classified into both structural and non-structural disaster management planning. Structural measures such as engineering technology to help in achieve the risk resistance and resilience in any physical structure or system, and reduce the possible impacts from disaster (Pathak, 2009). Common construction measures for fire risk reduction include fire extinguishers, emergency exit doors, building codes and safe and open ground. Non-structural activities include research and assessment, information resources, campaigns and planning awareness plans (Pandey et al, 2019).

Capacity analysis

It is important to emphasize people's capacity to anticipate, cope with, resist and recover from disasters, rather than simply focusing on the vulnerability that limits them. It is an integral part of disaster preparedness and contributes to the creation of community-based disaster preparedness programs at the grass roots level (Victor and Rachel, 2016). Preparedness measures including establishing early warning systems are seen anywhere in neither the college area, nor any program rising of risk is being initiated in the college or if such program exists, majority of students and staff members are not aware of it.

Communication system

Emergency communication plans ease the stress and helps to keep the recovery process running smoothly. But in the college, there is hardly any internal emergency-response team in case of a fire outbreak and if such exists, students and staff don't know about it. There is neither emergency mass notification system, nor any alarm to raise emergency.

Medical facilities

It is very essential to have a proper medical room and necessary facilities to provide immediate help. There is a medical room in college, but in case of a fire outbreak or in case of any other disaster it can't serve its purpose to a large extent. There should be a good communication system and better transportation facilities so that necessary actions can be taken as quick as possible.

Hazard Identification and risk

Hazard identification and risk assessment helps in determining the level of risk by combining the likelihood of a hazard occurring with its severity using the Risk matrix. It consists of a number of sequential steps such as hazard identification, consequence & frequency assessment, risk estimation based on the existing controls and recommendations to reduce those risks which are not under acceptable limits (Singh, 2005). In this survey, there has never been an incidence of fire hazard and no other damage has estimated yet. But, of course, there is risk of fire outbreak, as many areas are vulnerable to fire.

Developing mitigation plan

Some important mitigation measures for disaster risk reduction as follows in college campus:

- Design the improvements of adaptation infrastructure for the mitigation or prevention of disaster.
- Land use planning and design decisions that prevent development and community infrastructure in hazardous areas.
- Community participation in decision of policy maker or awareness campaigns to increase the indigenous knowledge of how to prepare for disaster events and its management planning (United Nations International Strategies for Disaster Reduction, 2007)
- Community education programs and participation of stockholder to build knowledge of the appropriate actions to prepare for and respond to a disaster management
- Resilience activities including partnership building and engagement between sectors
- Annual programs (e.g., vegetation management around essential services and essential infrastructure such as power lines).
- Mapping My Neighborhood for fire risk should be taken up by NCC, NSS, etc.

Retrofitting of old buildings

It is difficult and expensive to build all new structures in old buildings, Therefore, it is important to include security measures through this currently connecting pole and beam support buildings by providing fire safety measures and appropriate protective fittings for expensive items.

Adaptation and implementation

Disaster-inspired development initiatives are influenced in a number of ways, but two aspects are especially important. First, disasters can highlight particular areas of vulnerability, for example where serious loss of life has occurred, or where the economic damage is disproportionate to the strength of the impact. The outcome of this is usually to highlight the general level of underdevelopment (UNESCO, 2013; CBSC, 2006). Second, for a few weeks or months, the political environment may favor a much higher rate of economic and social changes than before, in areas such as land reform, new job training, housing improvements and restructuring of the economic base (Castro, 2013). The value of direct international assistance given after disasters may partially compensate for economic losses, although the amounts are usually rather small in relation to the total loss.

Conclusions

This can be inferred from the survey and analyzed information that there is lack of awareness among the respondents regarding fire hazard and are also not efficiently trained how to act in such circumstances. Majority of the respondents were not aware about the sited maps in college, location of fire extinguishers and emergency exit doors. Poor awareness was also found that whether there exits any disaster cell or emergency plan in college for disaster risk reduction.

According to the respondents, there has never been an incident of fire hazard in college, but majority of the respondents think that carelessness can be a reason of fire outbreak in future. If proper measures will not be taken then there can be big loss of property and lives as majority of people are untrained as well as staff of the college do not know the use of fire extinguishers and also about the functioning status. So, this can be concluded that the college building as well as people within the campus is vulnerable to potential fire hazard. This is the need of the hour that necessary actions should be taken and an effective disaster management plan need to be set up, and if exists, it must work effectively and efficiently to mitigate risks of all kind of hazards.

Disaster risk reduction should start at home, in communities and in colleges and schools. The educational institutions play an important role in this field of Sustainable Development Goals (SDGs) and disaster risk reduction is to shape the mind of the child / students to become a responsible citizen in the future. Culture of Prevention and indigenous knowledge for disaster management is important not only for youth, adults or seniority but for everyone. In the hazard-prone areas, policymakers have the responsibility to ensure and protect children from unwanted incidents no matter inside school and on the way to school. In order to improve the study on awareness of DRR among students and raise the awareness, further research needs to be conducted to increase the disaster awareness and preparedness.

References

Asia Pacific Disaster Report (2012). *Reducing Vulnerability and Exposure to Disasters*. https://www.unescap.org/publications/asia-pacific-disaster-report-2012-reducing-vulnerability-and-exposure-disasters

Bosher, L., & Dainty, A. (2011). Disaster risk reduction and 'built-in 'resilience: towards overarching principles for construction practice. *Disasters, 35*(1), 1–18. https://doi.org/10.1111/j.1467-7717.2010.01189.x

Castro, R. (2013). *Disaster Risk Reduction and Management*. https://www.slideshare.net/irpex/disaster-risk-reduction-and-management-28415360

Central Board of Secondary Education (CBSE). (2006). *Natural Hazards and Disaster Management. Delhi.*

Gangalal, T., Ryuichi, Y., Ranjan, K. D., & Netra, P. B. (2014). Knowledge of disaster risk reduction among school students in Nepal. *Geomatics, Natural Hazards and Risk, 5*(3), 190–207. https://doi.org/10.1080/19475705.2013.809556

Johannes, M. L. (2020). Disaster-resistant schools for disaster-resilient education. *Quality Education*, 1–17.

Joshi, P. C. (2009) *NIDM. Status of Disaster-related Courses in Indian Universities*. http://nidm.gov.in/idmc2/session_education.asp

Joshi, S. (2014). Knowledge and practices of school teacher regarding disaster management. *International Journal of Health System Disaster Management., 2*, 98–102.

Kanyasan, K., Nonaka, D., & Chatouphonexay, A. (2018). Implementation of disaster risk reduction and management policies in a school setting in Lao PDR: A case study. *Tropical Medical Health, 46*, 42. https://doi.org/10.1186/s41182-018-0124-7

Mitchell, T., Haynes, K., Hall, N., Choong, W., & Oven, K. (2008). The roles of children and youth in communicating disaster risk. *Child Youth Environment, 18*(1), 254–279.

National Disaster Management Authority (NDMA). (2006). *Draft National Disaster Management Framework*, pp. 14–17.

Pandey, B. W., Prasad, A. S., Mishra, H., & Godara, S. (2017). Urban dynamics and resource consumption: a case study of NCT of Delhi. In: Poonam, S., & Swati, R. (Eds.), *Sustainable Smart Cities in India, Challenges and Future Perspectives*. pp. 333–352 Springer International Publishing. ISBN 978–3–319–47145–7 (E-Book) ISBN 978–3–319–47144- 0

Pandey, B. W. Mishra, H., & Pathak, U. K. (2019). Planning for healthy and sustainable urbanization: a case study of national capital territory, Delhi. In: V. R. Sharma & Chandrakanta (Eds.), *Making Cities Resilient, The Urban Book Series*. https://doi.org/10.1007/978-3-319-94932-1_5

Pathak, K. (2009). *Role of Engineering Education in Development of Human Resources for Preparedness and Management of Natural and Man-Made Disasters*. http://nidm.gov.in/idmc2/session_education.asp

Petal, M. (2009). Education in disaster risk reduction. In R. Shaw & R. R. Krishnamurthy (Eds.), *Disaster Management: Global Challenges and Local Solutions* (pp. 285–301). Universities Press.

Rajesh, G., Chhabra, K. G., Shetty, P. J., Prasad, K. V., & Javali, S. B. (2011). A survey on disaster management among postgraduate students in a private dental institution in India. *Am Journal Disaster Medicine, 6*(5), 309–18.

Roy, N., & Pandey, B. W. (2016). *Concepts and Practices of Disaster Management, World Focus, Disaster Management and Mitigation*, pp. 122–126.

Sharpe, J., & Kelman, I. (2011). Improving the disaster-related component of secondary school. *Geography Education International Research in Geographical and Environmental Education, 20*(4), 327–43. https://doi.org/10.1080/10382046.2011.619810

Singh, R. B. (2005). *Risk Assessment and VulnerabilityAnalysis, IGNOU PG Diploma in Disaster Management- MPA-003, New Delhi*, pp. 355.

UNISDR. (2015). *Sendai Framework for Disaster Risk Reduction 2015–2030*. Retrieved 7 November 2017.

United Nations International Strategies for Disaster Reduction. (2007). *Towards a Culture of Prevention: Disaster Risk Reduction begins at School.*

United Nations Educational Scientific and Cultural Organization, UNESCO. (2013). *Disaster Preparedness: Education for Disaster Risk Reduction at UNESCO.*

United Nations Hyogo Framework for Action. (2015). *Building the resilience of Nation and Communities to Disaster; 2000–2015.*

Part IV
Earthquake and Other Related Hazards

Chapter 13
Earthquake Awareness and Preparedness Survey of Yamuna River and Surrounding Region of Delhi

Vishwa Raj Sharma, Neha Arora, Swarnima Singh, and Kshetrimayum Krishnadas

Abstract Earthquakes are caused by sudden shaking of the earth, and its occurrence in ever-expanding city like Delhi poses a serious problem in the city. Delhi, falling in the seismic zone IV, is highly vulnerable to such a disaster. If an earthquake of great intensity strikes the capital, the extent of damage to life and property is unimaginable. In the coming years the frequency of occurrence of such a disaster has increased. In such alarming situation there becomes a need to identify vulnerable areas of the city which will be most affected by an earthquake of moderate to high intensity. The present study thus attempts to identify vulnerable areas likely to be affected by an earthquake. Primary survey forms major part of the study. A sample of 350 respondents from five selected areas of the city was conducted with the help of questionnaire, and results were obtained. Different maps and diagrams have been prepared with the help of GIS technique and other software. Also some suggestions were also given to reduce the impact of such a disaster that will be helpful for an individual, government department, policymakers and planners.

Keywords Earthquake vulnerability · Disaster risk reduction · Soil liquefaction · Seismic zonation · Earthquake preparedness · Risk assessment

---

V. R. Sharma (✉)
Department of Geography, University of Delhi, New Delhi, India
e-mail: vrsharma2002@gmail.com

N. Arora
Department of Geography, Shaheed Bhagat Singh College, University of Delhi, New Delhi, India

S. Singh
Department of Geography, Deen Dayaal Upadhyay University, Gorakhpur, Utter Pradesh, India

K. Krishnadas
Department of Mathematics, Shaheed Bhagat Singh College, University of Delhi, New Delhi, India
e-mail: kshetrimaym.krishnadas@sbs.du.ac.in

V. R. Sharma and Chandrakanta (eds.), *Making India Disaster Resilient*,
https://doi.org/10.1007/978-3-031-50113-5_13

Introduction

An earthquake is a sudden shaking of the earth caused by the breaking and shifting of rock beneath the earth's surface and followed by a series of vibrations. Earthquake is one of the most devastating disasters that cause loss of property and innocent lives. Delhi, capital of our country with an ever-increasing population well above 16 million, has a majority living in urban parts of the city. The city being the socio-political and economic face focus of India commands much more responsiveness in case of disaster preparedness especially from decision-makers, planners and engineers, where the city is positioned in highly earthquake-prone belt Zone IV of IS Code (IS 1893:2002). The heavily populated city like Delhi with an ever-increasing population above 16 million, large numbers of man-made structures have been created is more prone to severe damage due to moderate earthquake, where micro-zonation should be done to prepare the city with any impounding disasters. The word means to divide the capital city in micro- or very small zones are the subdivision of a seismic zone into smaller one. This may vary with definite measure to simplify the enactment of seismicity. Based on extensive and site-specific studies, no micro-zonation studies have been taken up so far for any Indian cities. Protection alongside the earthquake hazard has two broad facets: the structural protection beside possibly destructive energetic forces and the protection of a place itself due to amplification, liquefaction, land sliding and subsidence. Right after the disturbing and distressing 2001 Bhuj earthquake the Delhi NCR attracted attention of several technical and methodical studies, because this NCR has had undergone many earthquakes in the recent past and faces the danger of falling in severe seismic zone of occurrence due to threat on the central Himalayan seismic gap. Most recently in 2005 and 2014 Delhi has shaken for more than 20 seconds thrice a day. The seismic zonation map of India classifies Delhi in the category of moderate to high earthquake-prone zone (IV), with intensity of VIII on modified Mercalli scale, where the magnitude up to 6.2 and more has been reported in NCR. In the present report detailed geological, geophysical, geotechnical and seismological studies have been carried out to prepare micro-zonation map of the Delhi region based on seismicity (Methodological Flow Chart 13.1).

Site and Situation of the City

Delhi is circumscribed by the Ganga-Yamuna alluvial plains in the North and East, by Thar Desert in the West and by Aravalli hill ranges in the south, where the topography is flat in general overall barring the NE-SW trending with low Delhi-Haridwar ridge that is an extension of the Aravalli hill ranges of Rajasthan. The region around the city is deliberated for seismic hazard assessments in corporate numerous active tectonic features including the Main Central Thrust, the Himalayan Main Boundary Thrust and the Delhi-Haridwar ridge, the Aravalli-Delhi fold, the

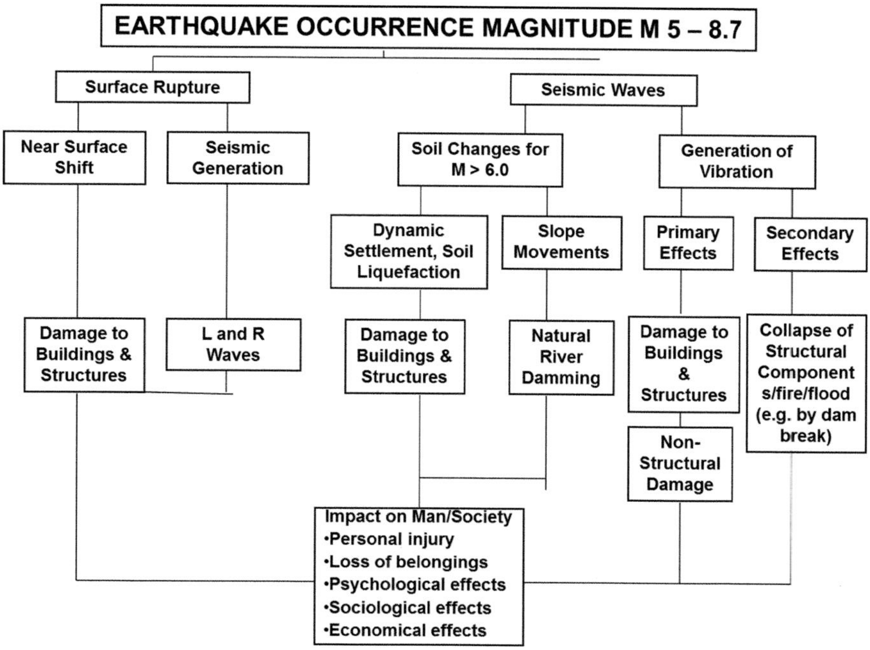

Chart. 13.1 Methodological flowchart. *Source* Singh and Singh, 2011

Delhi-Lahore ridge, the Sohna fault, the Mathura fault, the Rajasthan Great Boundary Fault and the Moradabad, etc. to name a few. Tectonic activities on these landscapes have been shaking the city with minor and major earthquakes since chronology. The city has different physiographic divisions; the ridge area, older alluvium (west of ridge), newer alluvium (east of ridge) and the Cenozoic Yamuna flood plains, and the structurally different strata of these layers encounter different ground signals during seismic occurrence. The bed rock in Delhi is overlaid by Aeolian deposits as depicted by Central Ground Water Board (CGWB) that is further overlaid by alluvial deposits. The Delhi-Haridwar ridge in its bed rock topography is characterized by many anticlines and synclines with varying depth of bedrock from position to position. The buried Delhi-Haridwar ridge passes through Chanakya Puri to Connaught Place where bed rock is very thin and varies from 5 to 20 m, and the thickness of soil cover in the eastern part of ridge area in North Delhi varies from 0 to 30 m, with a continuing east ward slope toward the Yamuna River.

Fragile river bed, unauthorized colonies, high-rise buildings, flyovers, slums and several other unplanned settlements dominate the city. Every year reports of several incidences of occurrence of earthquake come out (Table 13.1; Fig. 13.2), and its number is also increasing day by day in the city. If such trend continues it will have serious impacts on the lives of residents and uncountable loss to the property as well.

Table 13.1 Recent earthquakes in Delhi

Date	Latitude 0 N	Longitude 0 E	Magnitude
June 6, 1992	28.65	76.69	2.8
Mar 27, 1993	28.63	77.20	3.6
Dec 03, 1993	28.60	77.40	3.5
July 28, 1994	28.51	77.25	2.8
Nov 16, 1994	28.50	76.95	2.9
May 25 1999	28.62	77.80	4.1
April 25 2015	28.52	77.60	5.3

Source National disaster management authority

Photo 13.1 Delhi earthquake, 2015

On October 26, 2015, at 5:15 pm, New Delhi, earthquake measuring 7.5 on the Richter scale had its epicenter at a place about 50 km south west of the city of Jarm near Afghanistan's border with Tajikistan?

Delhi falls in the seismic zone IV which will have a fairly high seismicity where the general occurrence of earthquake is of 5–6 magnitude on Richter scales and occasionally of 7–8 magnitude. It thus lies among the high-risk areas. Hence, there becomes need to assess the likely affected areas so that at the time of occurrence of large-intensity earthquake the loss in these vulnerable areas can be reduced. As we know that it is difficult to stop the occurrence of such natural hazard, such type of study can surely help in reducing the impact of an earthquake. Risk assessment involves hazard and vulnerability analysis. The probability of occurrence of an earthquake varies from location to location, and local site conditions play a vital role in determining the intensity of earthquake. This study aims at identifying vulnerable areas likely to be affected along with level of awareness, management techniques and practices involved in reducing such a disaster.

Research Problem/Hypothesis/Objectives

Aims and Objectives

- Create awareness among students, government functionaries, technical institutions, NGOs and communities about earthquake vulnerability and possible preventive actions
- Development and institutionalizing of earthquake preparedness and response plans and practice these through mock drills
- Development of regulatory framework to promote safe construction and systems to ensure compliance
- Networking knowledge on best practices and tools for effective earthquake risk management
- Community perception regarding awareness through primary survey.

Data Sources and Research Methodology

The research methodology adopted includes identification of critical steps in the understanding of the problem, analysis and interpretation of data and other aspects of team work. Purposive stratified random sampling techniques have been used for conducting survey.

- For the above-stated objectives mainly primary data is used with the help of well-structured questionnaire (Appendix I), and secondary data is also considered to substantiate the study.
- Primary survey includes selected localities of Delhi, namely Chirag Delhi/Khirki, Khanpur/Madangir, Chawri Bazar/Chandni Chowk, Cannaught Place and Trans-Yamuna (Patparganj, Wazirabad and Laxmi Nagar). Totally 350 respondents were undertaken to judge the levels of awareness and perception of the people of city regarding earthquake disaster vulnerability assessment and management.

Secondary data has been collected from different sources, namely:

- IMD—Indian Metrological Department
- NDMA—National Disaster Management Authority
- NIDM—National Institute of Disaster Management
- DDMA—Delhi Disaster Management Authority
- NRSA—National Remote Sensing Agency
- Census of Delhi, 2011

Methodology—Techniques/Sampling/Tools/Materials

The methodology of the study includes many steps to achieve its various objectives. The objectives have been achieved by going through various literatures available and then the secondary data was obtained. Several computational analyses have been done for the following:

- Liquefaction hazard assessment and mapping
- Seismicity of the area
- Standard penetration test
- Iwasaki method.

In the present study, empirical equations have been used for estimation of amplification factors in terms of fundamental period based on equivalent soil layer properties. To recommended an empirical formula for computation of amplification factor (μ) by combining the results of actual measurements with theoretical responses, for a single surface layer system of predominant period (Tg) for shear wave velocity (Vs) and equivalent soil density mapping for assessing vulnerability of city due to earthquake (Singh and Singh, 2011).

Different maps were prepared using ArcGIS 10.1 to delineate areas prone to earthquake that were ascertained. Finally from the selected areas, with the help of purposive stratified sampling five areas of Delhi were identified and mapped. These include Chirag Delhi/Khirki, Khanpur/Madangir, Chawri Bazar/Chandni Chowk, Cannaught Place and Trans-Yamuna (Patparganj, Wazirabad and Laxmi Nagar). A total 350 respondents (approximately 70 from each) were interviewed with the help of well-structured questionnaire. It deals with the household characteristics of the respondents related to their level of awareness in case of a disaster like earthquake. Management issues were also raised and also how they are equipped to deal with the disaster. After completing the filling of questionnaires the data was compiled and tabulated. It was then represented through various figures, charts and diagrams.

Result and Discussion

Geology of Delhi

Geology of NCR is fascinating due to exposed ancient Aravalli mountain ranges extending NE in this area and its adjoin region is surrounded in the north and east by Indo-Gangetic plains, in the west by the extension of the great Indian Thar desert and in the south by the Aravalli ranges. The rocks have undergone multiple folding and different phases of metamorphism. The quartzite is bedded and highly jointed with intrusive, the post Delhi intrusive is enclosed by the quaternary sediments the form of Aeolian and alluvial deposits. The alluvial sediments belong to the Pleistocene period, i.e., archaic alluvial sediments and Cenozoic era. Several soil profiles have been created masking the entire Delhi region based on the collected geotechnical borehole data to define and analyze the sub-soil heterogeneity. Silt is found predominant in the Trans-Yamuna region, where the data has been collected for grain size allotment and distribution (GSD); therefore, the GSD curves are illustrated at 4 different depths (3.0, 4.5, 6.0 and 9 m) for north, south, east, west and central blocks across the city. These graphs showing varying soil depth grain size have been used during the primary estimation of liquefaction potential to bring out ground response analysis (GRA) in seismic micro-zonation studies across the city. The bedrock information of the city has been divided into zone-wise soil distribution as follows (Singh and Singh, 2011):

1. The bedrock depth is less than or equal to 30 m in south and central part of Delhi. In north and western part of the area the sedimentary thickness goes up to 150 m. But the soils in these locations are comparatively dense silty sands with clay seams.
2. In Trans-Yamuna region the bedrock depth is at around 200 m, and soils are loose sandy silts and silty sands. Also, the groundwater contour map is prepared using this data. The water table is high in Trans-Yamuna region and very low in south and central Delhi.

The microscopic studies in Delhi alluvium based on wave diffraction and scanning system have been done, where quartz is a chief mineral with lesser quantities of chlorite, mica, feldspar, kaolinite and calcite; it has been found that southern part of the city is having predominant quartz percentage in soil sample locations as compared to others. On the contrary to it kaolinite percentage is moderately high in Trans-Yamuna and north-western side of city. The detailed site with latitude and longitude locations across city is also measured using the GPS. The produced GSD-induced seismic trend data has been processed using SEIS manager for refraction (deviation in GSD path waves) for acquiring 2D, primary (P) and secondary (S) wave velocity for the given model. For 18 GPS locations the 2D P and S wave velocity models have been run to generate 2D seismo-contour map at every 10 m interval. Soil augmentation and amplification factor has also been estimated from these given wave velocities using ArcGIS software and the micro-zonation map to recognize the soil amplification factor (SAF) that is generated with a correlation matrix for seismic vulnerability overlay analysis. The comprehensive GPS site characterization has been constructed or the velocity of primary (Vp) and secondary wave (Vs), the secondary wave velocity at 30 m (Vs30) is done by dividing the area into four zones ZA, ZB and ZC, where these zones are closely corresponding with the geology and soil attributes across region. That is the A zone which has been depicted as ZA (Vs30>350 m/s) is delimited in the southern and central part with quartzite across Delhi, and the dense sands silts and silty sands with clay in the zone B are having velocity varying (Vs30=250 to 350 m/s) among the Pleistocene soil. The zone ZC is calibrated in the Trans-Yamuna region where soils are very loose sandy silts with low N value (Holocene) (Table 13.2).

The equation has been derived for Delhi region both near- and far-field earthquake based on attenuation law sources as follows:

Table 13.2 Correlation matrix of three zones (ZA, ZB and ZC) across Delhi under seismic hazard vulnerability and micro-zonation modeling

Zone A: ZA (South and south-central Delhi)	Zone B: ZB (West and north-western Delhi)	Zone C: ZC1, ZC2 (Trans-Yamuna)
$V_{s30} > 350$ m/s	$V_{s30} = 250$ to 350 m/s	$V_{s30} < 250$ m/s
$V_s = 280.6\ D^{0.08}$	$V_s = 216.7\ D^{0.13}$	$V_s = 140\ D^{0.24}$
$V_s = 66\ N^{0.48}$	$V_s = 48.02\ N^{0.54}$	$V_s = 39.2\ N^{0.61}$
$V_p = 1.6\ V_s + 310.0$	$V_p = 1.8\ V_s + 66.5$	$V_p = 0.99\ V_s + 208.5$

Source Based on Primary Calibrations

$\log(PGA) = a \log R + bMw$

where $a = -1.94$; $b = 0.15$ and $R < 100$ km (for near-field sources) $a = -1.23$; $b = 0.14$ and $R>150$ km (for far-field sources)

This observed attenuation law is very beneficial as well as convenient when any other crucial information is missing or not available for analyzing liquefaction potential. That means the resonance frequency of soil is dependent inversely on the soil thickness. It can be conclude that locations where susceptibility for the liquefaction is high having high vulnerability index which can be used to identify possible vulnerable or susceptibility areas to shown on-structural and structural damages probability propounded by future earthquakes.

Seismicity in Delhi

Delhi is bounded by the Indo-Gangetic alluvial plains in the north and east, by Thar Desert in the west and by Aravalli ranges in the south. The terrain of Delhi is flat in general except for the ridge extending to southern portion of the state which is the extension of Aravalli of Rajasthan. The city's settlement pattern is much skewed in relation to location and geological characteristics. Seismicity around Delhi is associated with Delhi-Haridwar ridge. The city falls under zone IV of the high seismicity with magnitude of 5 to 8 on the Richter scale (Figs. 13.1 and 13.2).

According to Delhi Disaster Management Plan, seismicity is defined as an engineering technology determined by the depth of the bedrock below ground level. It further states that the buildings on the alluvial soils are more vulnerable to long-distance earthquake. Geological Survey of India (GSI) reports that the bedrock depth is 60 m in the Patel road area, 15 m in Cannaught Place central park, 40–50 m near Rajghat and 150 m and beyond in the Yamuna river bed. Similarly, the depth is reported to be 80–100 m in the Aurobindomarg-Hauz Khas area. The process of liquefaction (i.e., during ground shaking cracks are developed or filled and the traveling seismic waves transfer material or soils resulting in loosening of foundations and collapse of structures), physical location and hydrogeology also contributes to the extent of damage caused by an earthquake. It was noticed from the liquefaction

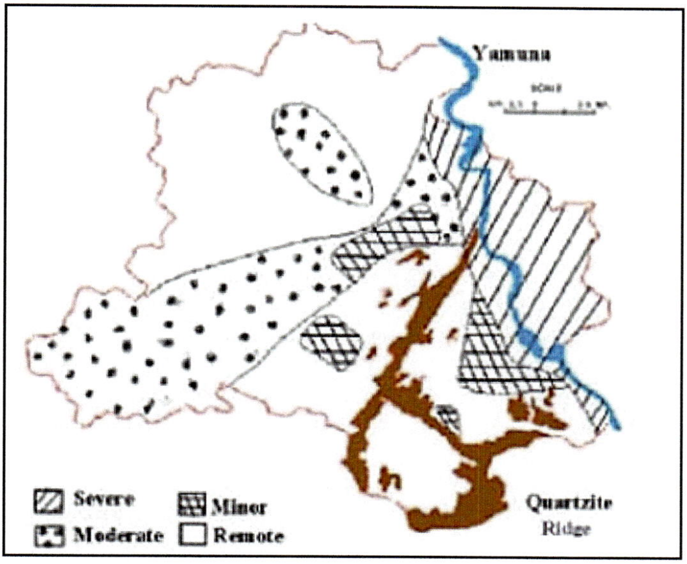

Fig. 13.1 Liquefaction potential map of Delhi region. *Source* Prepared by students based on calibrations

zonation map of Delhi which also indicates that the areas along Yamuna river bed extending till Noida and Faridabad and Hauz Khas area, due to soft alluvial soils, will face a grave problem of soil liquefaction during an earthquake. The land use map of Delhi (Fig. 13.2) clearly shows the city dominated by built-up area with very sparse vegetation. Pockets with high-rise buildings exist without specific consideration of earthquake resistance. Similarly, unplanned settlements with sub-standard structures are also prone to heavy damage even in moderate earthquake. The frequency of occurrence of earthquakes in Delhi is increasing over the years. The Central Business District namely Connaught Place and sprouting high-rise group housing schemes are high-risk areas due to the vertical configurations. The walled city area, the Trans-Yamuna area and scattered pockets of unplanned settlements are also figured as high-risk zones due to their sub-standard structures and high population densities. Therefore Delhi is highly prone to earthquakes (Figs. 13.3 and 13.4).

As per Vulnerability Atlas of India (1997), for shaking intensity VIII, 6.5% houses in Delhi have high damage risk, and 85.5% houses have moderate damage risk. These estimates are based on very simplistic assumptions. Kapur confirms not only the earthquakes that are located as close as Gurgaon and Mathura but ones with their epicenter in Shimla, Kangra, Kumaon and even in neighboring countries like Nepal, Pakistan and Afghanistan have an impact on Delhi. Recent tremors in Afghanistan during February 2015 were widely felt in different parts of the city. Experiences of past earthquakes both in India and abroad have clearly outlined the vulnerability of multi-story reinforced concrete buildings if not designed and constructed correctly. Huge number of multi-story reinforced concrete buildings in Delhi, particularly those

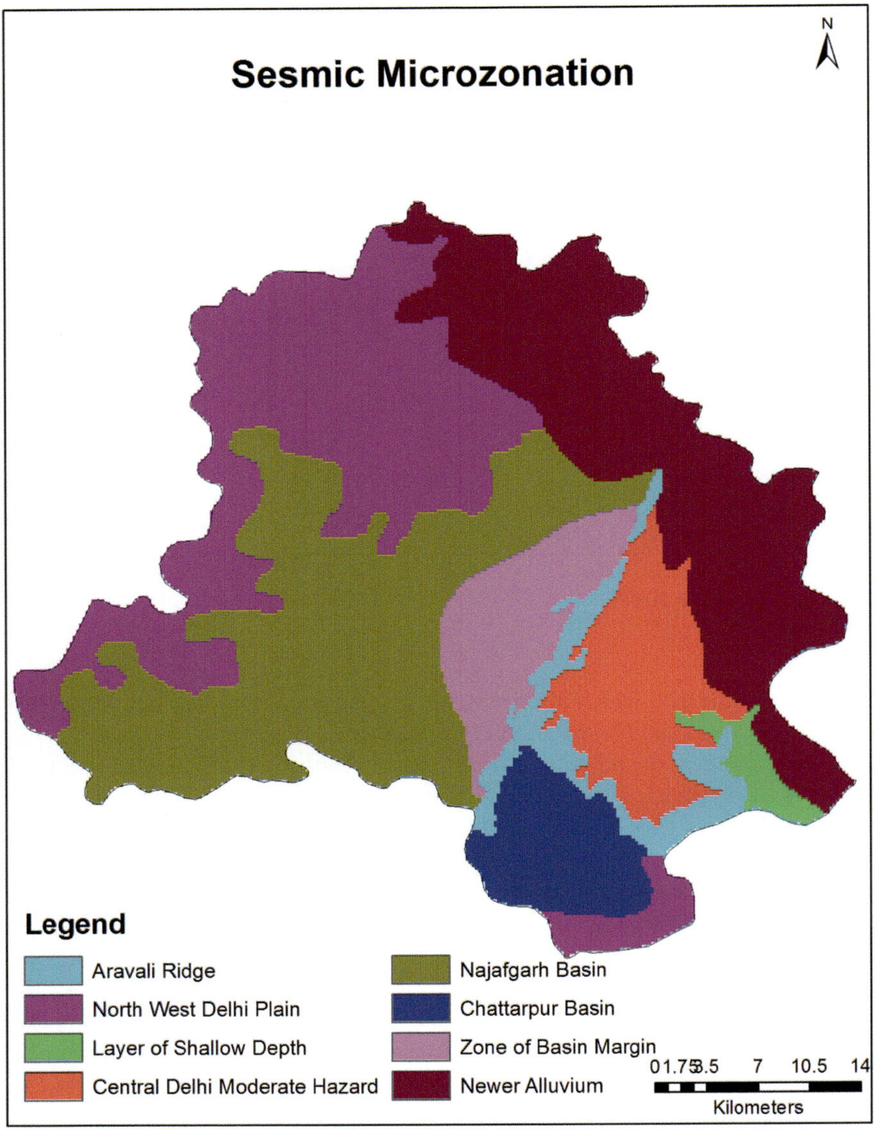

Fig. 13.2 Micro-zonation of Delhi seismic wave record under GSD students based on calibrations and National Center for Seismology data

with open ground story to accommodate vehicle parking, could also pose a major challenge in the event of a strong earthquake (Fig. 13.4a–d).

Delhi is currently passing through a major infrastructure development phase with a large number of bridges, flyovers and the metro project under construction. Indian Seismic Code (IS: 1893-1984) is not applicable for major projects which require

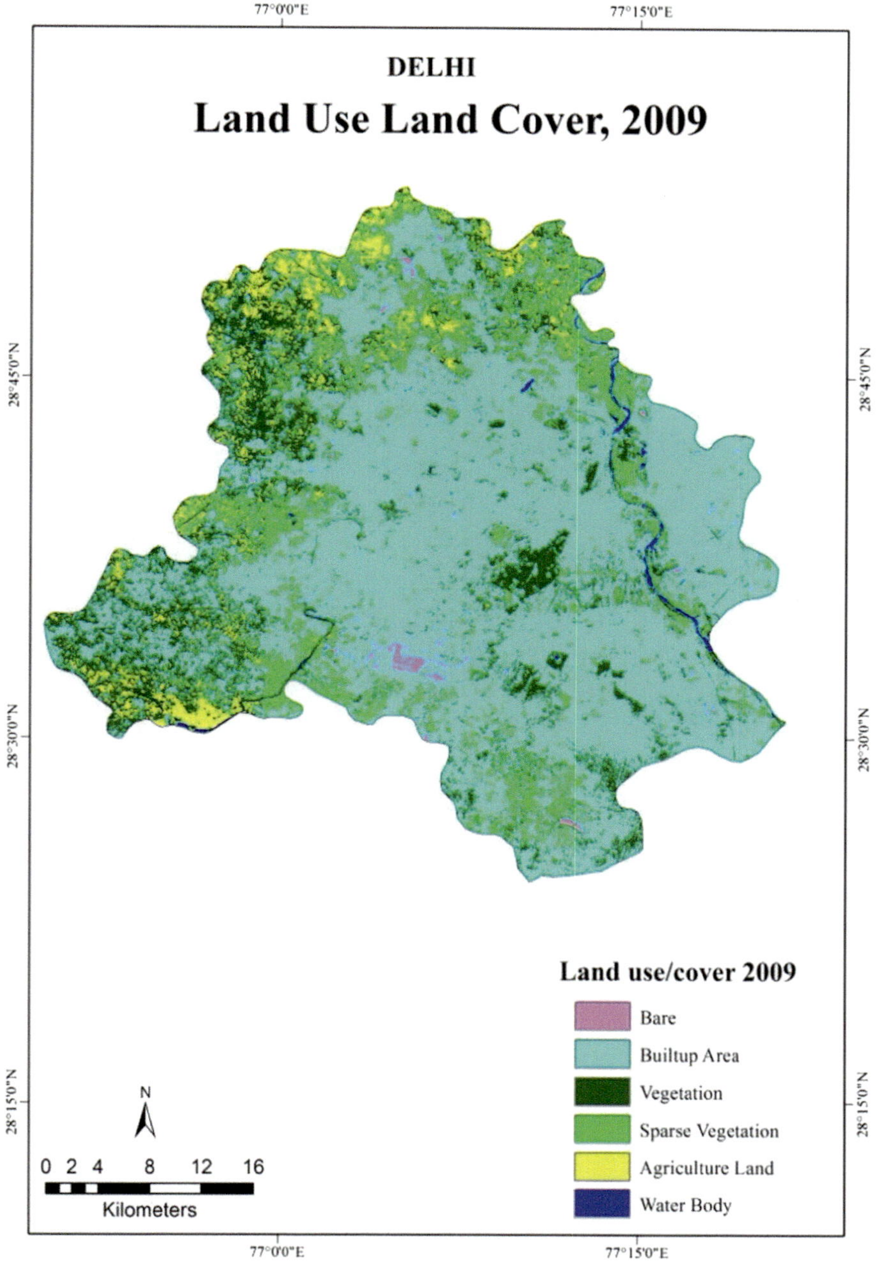

Fig. 13.3 Land use/land cover map of Delhi. *Source* Bhuvan Portal, NRSC

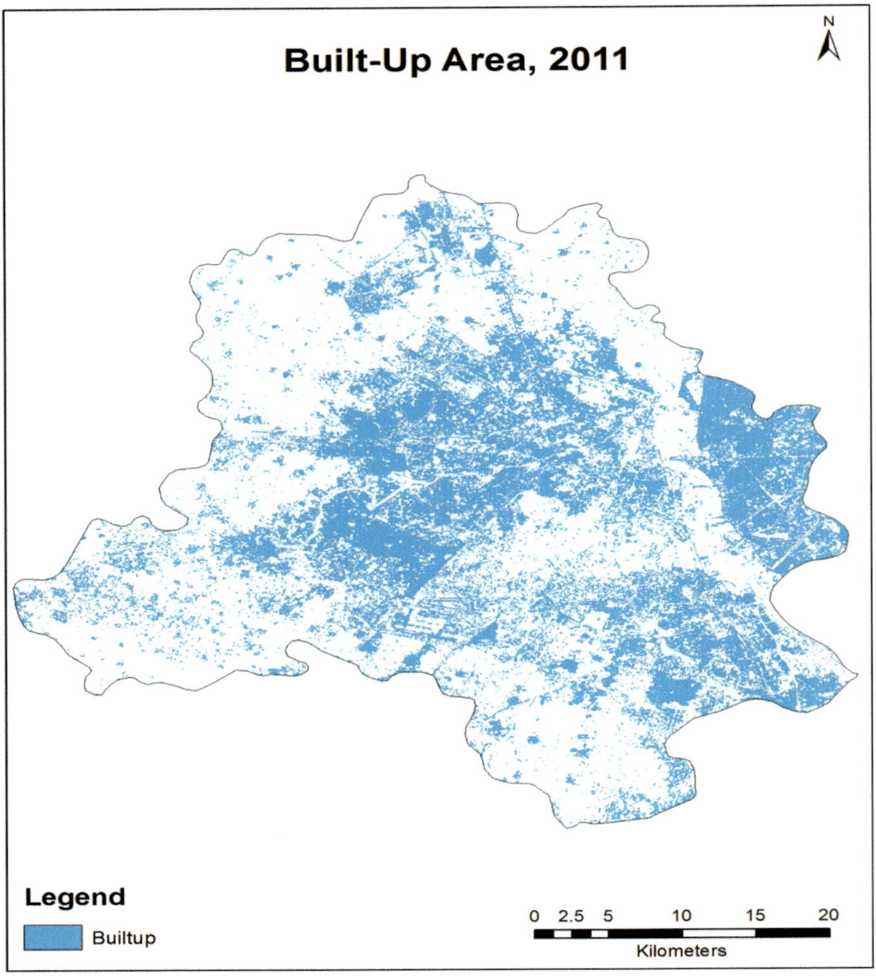

Fig. 13.4 Built-up area in Delhi. *Source* Bhuvan Portal, NRSC

special studies on seismic design criteria. Moreover, the Indian Seismic Code provisions on bridges as these exist today are obsolete and inadequate (Jain and Murty, 1998).

Finally a composite earthquake vulnerability map of Delhi is prepared using selected indicators with the help of ArcGIS 10.1 software (Fig. 13.8). The indicators are

- Built-up area of Delhi, 2009
- Land use/land cover map of Delhi, 2009
- Micro-zonation map of Delhi, 2011.

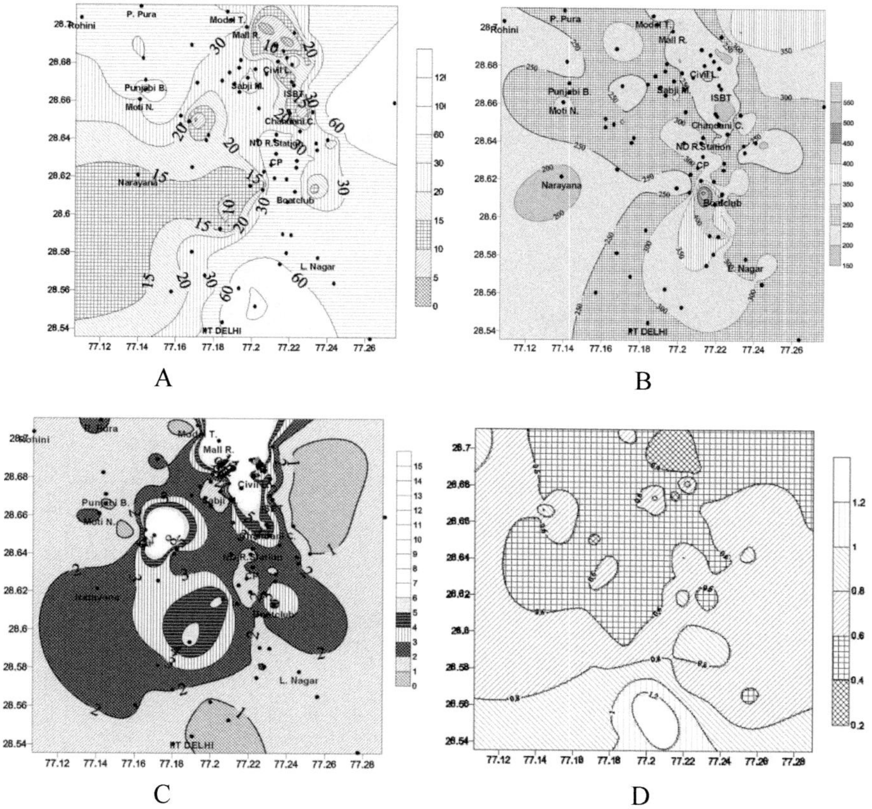

Fig. 13.5 **a** Thickness of the soil layer in the area under study. **b** Equivalent shear wave velocity for further estimation of amplification factors. **c** Map showing predominant frequency of soil. **d** PGA levels at the ground surface calibrated. *Source* Calibrations based on Singh and Singh, 2011 and primary data prepared by students

The map clearly indicates the fact that the areas along Yamuna river bed come under very high category of vulnerability to earthquake disaster in Delhi. In the upcoming times it has the potential to go well beyond the statistics of deaths and injuries. Such a disaster in the country's capital, which also happens to be a major commercial and industrial center, will have huge economic and political implications which will affect the entire country and not just the population of Delhi.

Hence, an attempt has been made thereafter with the help of primary survey to assess the level of awareness and the need for management of disaster like an earthquake in the selected areas of Delhi (Fig. 13.6).

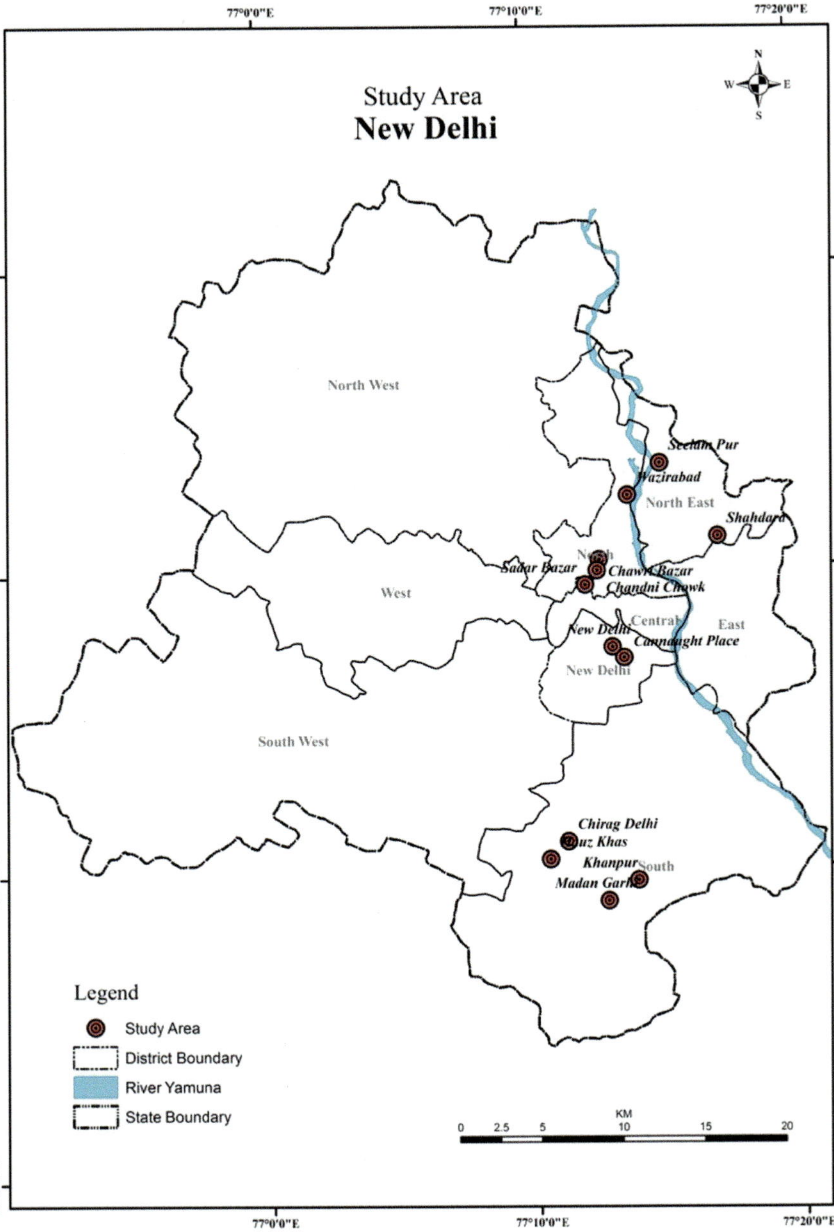

Fig. 13.6 Primary survey sample sites

Results and Analysis of Field Survey

With the help of stratified random sampling method five selected areas of Delhi, namely Chirag Delhi/Khirki, Khanpur/Madangir, Chawri Bazar/Chandni Chowk, Cannaught Place and Trans-Yamuna (Patparganj, Wazirabad and Laxmi Nagar) have been surveyed by a group of students. In such an exercise 350 respondents were interviewed (70 from each) with the help of well-structured questionnaire. The questionnaire is divided into several sections; the first part deals with the household data of the respondent like education, male/female status, income level, etc. The second part deals with the level of awareness and preparedness among the respondents with regards to earthquake like building material used, year of construction, number of lifts/exits, fire safety measures, etc. Several questions were raised with regards to their experience during an earthquake, first step when earthquake strikes and government help. The final section is devoted to the management issue. It deals with the assessment of role of government, media or NGOs and government policy in handling the future occurrence of earthquake in the city. The result of the survey is presented in the preceding paragraphs.

Of the 350 respondents, male residents dominate in all the five selected areas of Delhi. People with less than 15 years of age dominate in Khanpur/Madangir area, while 15–30 years of residents dominate in Cannaught Place and Trans-Yamuna area. In matters of educational status, Cannaught Place leads with the maximum number (90%) of educated people with education till senior secondary level. The Trans-Yamuna area has maximum number of illiterate respondents. The rest surveyed areas have respondents with education till secondary level.

In matters of occupation, mostly respondents in the Trans-Yamuna area are engaged in agricultural activities (Fig. 13.7), whereas people engaged in private jobs are noticeable in Chawri Bazar/Chandni Chowk and Khanpur/ Madangir.
Occupational Status of the Respondent

Cemented material is used for the construction as revealed by the respondents of the building in nearly every area except for the Trans-Yamuna area where brick and mud are still in use. Buildings are mostly recently constructed not older than year 2000. In question related to the number of stories in the building of the residence, it was found that respondents in Chirag Delhi/Khirki/Hauz Khas area have more than five-storied buildings, whereas respondents in the Chandni Chowk/Chawri Bazar area are residing in three-storied buildings.
Number of Stories in Buildings of the Respondent

It is clearly visible from Photos 13.2, 13.3 and 13.4 that people are residing in multistoried buildings with lack of open space and are largely illegal and unauthorized constructions.

In all the photographs, there is only one exit in the building and is also proved from the primary survey that 80% of the respondents in the Trans-Yamuna/Chandni chowk area have only one exit. Chandni Chowk and Chirag Delhi are the areas of commercial business and private centers largely dominated by shops of local/private

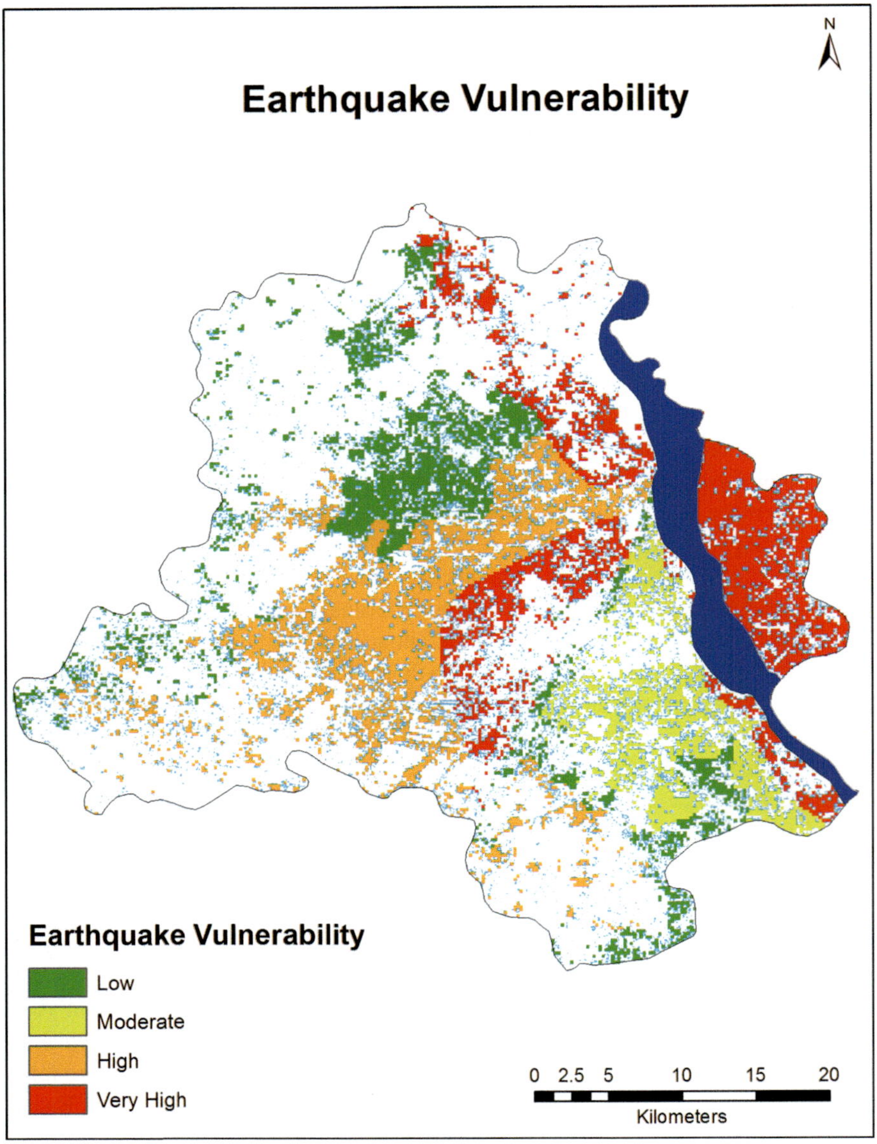

Fig. 13.7 Composite earthquake vulnerability in Delhi. *Source* Administrative Atlas of Delhi, Census of India, 2011

businessman. Nearly 50% of the respondents have no access to fire safety measures like extinguisher, etc. which might be used if fire broke out during an earthquake. Respondents were even unaware of such measures in Chandni Chowk and Trans-Yamuna area. Photo 13.5 clearly shows that if fire broke out in this area and a strong tremor occurs, people will find no space to escape.

Photo 13.2 Students in Chirag Delhi. *Source* Field survey

Photo 13.3 Chandni Chowk. *Source* Field survey

During an earthquake, there might be a requirement of different civic amenities like hospitals, water tanks, etc. Therefore, question was asked from the respondents regarding their accessibility and nearness to water tanks, hospitals, fire station, police station and disaster management center.

Photo 13.4 Trans-Yamuna. *Source* Field survey

Photo 13.5 Electric wires hanging in Chandni Chowk. *Source* Field survey

Nearness to Civic Amenities

In matters of nearness to amenities like water tanks and police stations, more than 150 respondents reside close to a distance of one kilometer, whereas hospitals are located within 1–3 km. People are unaware of disaster management centers located in their residences, and also no one knows about the helpline number.

Level of Awareness Among Respondents

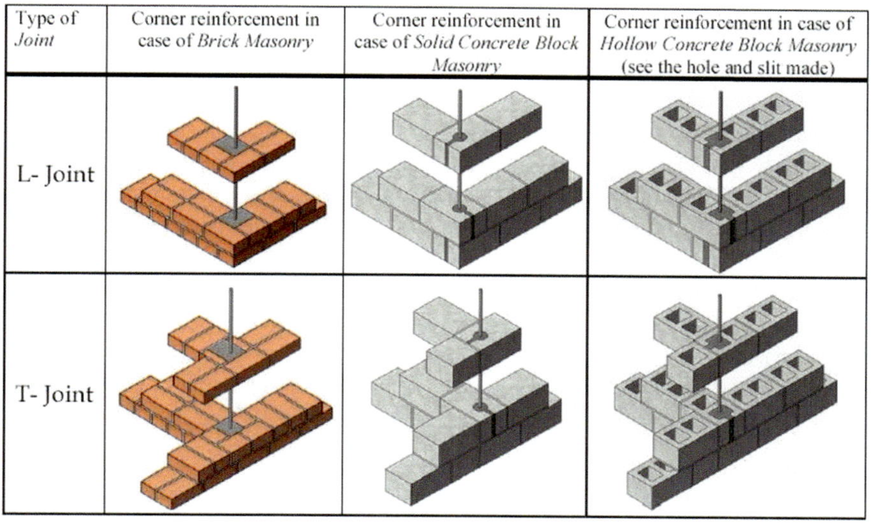

Fig. 13.8 Recommended joint details with vertical R/F at corners. *Source* DDMA

Though every respondent has experienced earthquakes in the recent past but they are still not insured against the calamity. Except Connaught Place, none of the selected sites have earthquake resistant structure in their buildings/homes.

First Step When Earthquake Occurs

In order to know the level of awareness among respondents, questions were asked related to their first step when earthquake occurs. Most of them preferred to move out in open area. More than 40% of the respondents are unaware of mock drills and surprisingly more than 65% in Connaught Place.

Participation in Mock Drills by the Respondents

In the role of NGOs and government in any combating situation resulting due to earthquake, respondents were highly unsatisfied. The reason being slackness in application of disaster prevention techniques and measures.

Question Arises What Is the Solution?

Here comes the need for disaster risk reduction (DRR).

DRR means the formulation of a disaster management plan which will help to reduce the risk at pre- and post-disaster. It aims at providing guidance and help to the residents by involving them in a disaster management plan. It includes government, public and private partnerships and networking in the city so as to ensure quick mobilization of resources on the aftermath of any disaster. This process involves certain steps:

- Awareness generation
- Development of preparedness and response plans at local level
- Development of technological and innovative measures

- Capacity building
- Knowledge networking.

The first step involves creating mass public awareness. It is generated through campaigns, literature and posters, emergency and evacuation drills, observing safety day or safety week (local and national level), newspaper and other source of media. Hence, the role of media is vital in public awareness. In schools, children can prepare the mock drills by getting under the desks. Such drills should be done on a regular basis in schools, colleges, universities, offices, resident welfare organizations, etc. Early warning signs like abrupt behavior of animals in case of an earthquake should not be ignored. Fast communication networks should be developed in remote areas of the cities.

The next step involves measures to develop disaster preparedness plan by understanding and identifying vulnerable parts of the city like the exercise done in this research.

- Analyzing the soil type before construction and do not build structures on soft soil
- Follow Indian Standard Code for construction of buildings
- Land use control and restriction on density and heights of buildings
- Preparation of alternative sites for evacuation
- Providing efficient and essential services like ambulance, water tanks, fire brigades, rescue teams, etc. at the time of earthquake.

This also involves preparing post-disaster plan for recovery and rebuilding like

- Repair and reconstruction of infrastructure like water, sewer, electrical services and roads
- Financial assistance for repair and reconstruction of houses and public buildings
- Providing insurance by government and private agencies.

Innovative ideas need to be evolved; different projects can be supervised in different parts of the city. A large number of local masons and engineers can be consulted and trained in disaster resistant techniques. Retrofitting of buildings can be done as it is undergoing for five places in Delhi like Delhi Secretariat, Police Headquarters, Ludlow Castle School, Guru Teg Bahadur Hospital and Divisional Commissioner's Office complex. Such innovative work is more required in a sensitive city like Delhi. This involves public as well as private investments in such projects. Mobile applications (app) can be created which could give instructions to people whenever required regarding pre- or post-disaster.

- Fourth step involves capacity building. It means strengthening resources to combat disaster like an earthquake in the city. The Government of Delhi has established a nodal agency, Delhi Disaster Management Authority, to facilitate, coordinate and monitor disaster management and mitigation practices in the city. Besides this the local authorities like NDMC, DDA and MCD have to constitute hazard safety teams. The role of local government is to carry out the reconstruction and rehabilitation activities in accordance with the policies and guidelines specified by DDMA. These authorities must be strengthened and given training together with financial assistance to tackle any situation arising out of the earthquake.

- DDA and MCD must follow the land use pattern as given in the development plan.
- Land use change of vulnerable areas like Delhi ridge, river banks, etc. must not be allowed by law.
- Legal framework and building byelaws should be enforced strictly to mandate that all constructions in Delhi must implement seismic code provisions (Fig. 13.8).
- Unauthorized colonies in the city must be undertaken for redevelopment on compulsory basis or must be improved through land pooling techniques. Thus, upgrading them to a level where certain earthquake-resisting techniques can be introduced or if not then at least relief can reach the areas in event of earthquake. This may vary from case to case and area to area (Fig. 13.9).
- Public–private partnerships must be improved for disaster mitigation.

The last but not the least step is networking which includes a portal based on knowledge sharing between local and nodal authority. National database can be prepared on earthquake risk management and earthquake preparedness. GIS and remote sensing techniques should be used to create maps of different parts of the city which are more vulnerable and likely to be affected by occurrence of an earthquake. This will certainly help to reduce the damage to property and life on a large scale in the city. No project or a plan is successful without people participation; hence their involvement at every level of planning is the best way to reduce the risk and vulnerability of earthquake as a disaster in Delhi.

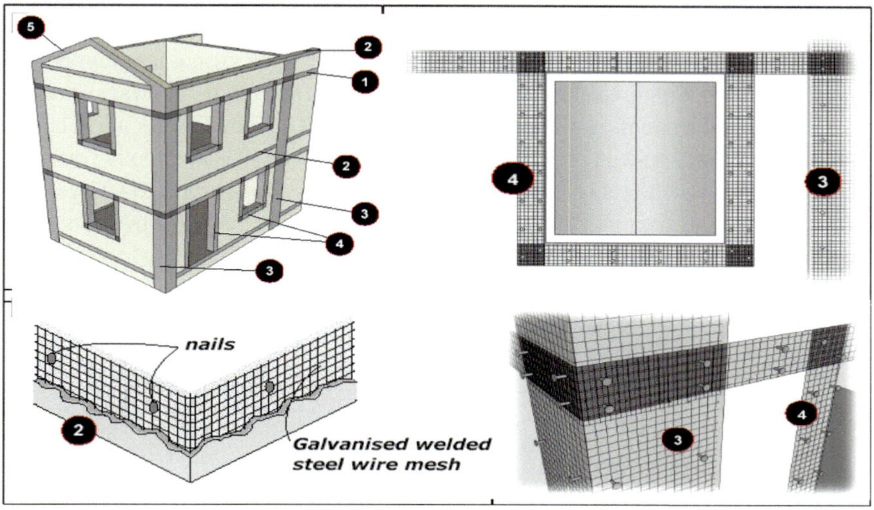

Fig. 13.9 Seismic retrofitting of masonry buildings. *Source* DDMA

Innovation Shown by the Project

This study is basically based on primary data which was collected from field survey. Secondary data was also collected on the theme of the project, and maps were prepared. Based on these maps selected survey sites were identified, and about 350 respondents were interviewed. These respondents have been interviewed according to a well-structured questionnaire which includes the personal characteristics of the respondents, level of awareness in case of a disaster like earthquake, management issues were also raised, and also how people are equipped to deal with the disaster. After completing the filling up of questionnaires the data was compiled and tabulated. It was then represented through various figures, charts and diagrams.

Conclusion and Future Direction

At the end it can be concluded that people's participation is the key to make any plan or initiative successful. Delhi today is in a grim situation in terms of increasing incidences of tremors caused by an earthquake with its epicenter within or outside the country. Therefore, it becomes a necessary and innovative step to conduct such research that will highlight sensitive/vulnerable areas within the city. This project has also helped to highlight the level of awareness among residents of the city related to earthquake. It will also help the government and private agencies to have in-depth study of the issue raised in the project. Hence, this research holds importance in today's times. Areas of soft alluvial bed around River Yamuna, namely Trans-Yamuna and Chirag Delhi/Hauz Khas, are more vulnerable as people have more unauthorized constructions, more than four-storied buildings, less of open space, less number of exits, ill-equipped with fire safety and first-aid measures and totally unaware of mock drills. Similar is the situation among the respondents of Chandni Chowk and Chawri Bazar area. The local as well as central authorities need to evolve an effective disaster management plan for the city. Disaster risk reduction is the solution to every issue in this regard. Retrofitting of buildings and enforcement of building byelaws with people participation are the need of the hour. Simple steps to stay safe during an earthquake are represented in Table 13.3

Table 13.3 Simple steps to stay safe during an earthquake

Do's	Don'ts
• Drop to the ground and take cover by getting under a study table or any other furniture; hold on until shaking stops	• Do not stand close to buildings, trees, street lights and utility wires • Do not stop under buildings, trees, overpasses and utility wires if in a vehicle
• If there is no shelter near you, cover your face and head with your arms and crouch at the corner of the building	• Do not take roads, bridges or ramps damaged by quake
• Stay away from glass, windows, doors and walls and anything that could fall	• If trapped under debris, do not light a match
• Stay in bed if you are there when earthquake strikes. Protect your head with a pillow	• Try not to shout as you may inhale dust; knocking or tapping is better

Acknowledgements I am very thankful to the mentor of my innovation project Dr. R B Singh for his guidance and being the source of inspiration. I convey my special thanks to Dr. P K Khurana, Principal, Shaheed Bhagat Singh College, for financial and infrastructural support. I also acknowledge the help and support extended by co-investigators of the research project, Dr. Neha Arora and Mr. Kshetrimayum Krishnadas. I would also like to give my special thanks to Dr. Swarnima Singh, Assistant Professor, Department of Geography, Shaheed Bhagat Singh College, for her support in preparation of maps and data analysis. Last but not the least I am also thankful to the entire team of research students of the project, Kuldeep Shukla, Nishant Ketu, Aniket Chandra, Arun Singh Dhaliwal, Akanksha Singh, Rishikesh Jha, Harshdeep, Shubham Kumar Sanu, Shubham Kumar Sanu, Sreyashi Bhattacharjee and Nikita Malik.

References

Administrative Atlas of Delhi. (2011). Census of India.

Bhuvan Portal, National Remote Sensing Centre.

Delhi Draft Disaster Management Plan (http://delhi.gov.in)

India Meteorological Department (2014). A report on Seismic Hazard Micro zonation of NCT Delhi on 1:10,000 scale, Centre for Seismology, Ministry of Earth Sciences, Government of India

Iynegar, R. N. (2000). Seismic status of Delhi megacity. *Current Science, 78*(5), 568–574.

Iyengar, R. N., & Ghosh, S. (2004). Seismic hazard mapping of Delhi City. In *13th world conference on earthquake engineering Vancouver*, B.C., Canada, Paper No. 180.

Jain, S. K., & Nigam, N. C. (2000). Historical developments and current status of earthquake engineering in India. In *Proceedings of the 12th World Conference on Earthquake Engineering, Auckland, New Zealand, Paper No. 1792.*

Muneeza, N. (2016). "Is Delhi ready to withstand a major earthquake? (Source: http://indianexpress.com).

Prerna Vijay Kumar, M. (2007). Safe Cities An Assessment of Earthquakes in Delhi with Short case study of 1999 Earthquake. In: Project Submitted to World Bank Institute & NIDM in partial fulfilment of the requirements for the award of a certificate in online course on Natural Disaster Risk Management – safe cities

Ravi, S. Risk assessment for a postulated earthquake in Delhi. Earthquake risk analysis, Module 2, Case Study 2, India.

Rodgers, J. E., Tobin, L. T., & Kumar, H. (2008). Building capacity in Delhi: India to reduce earthquake risk from existing buildings. In Project Manager, Geo-Hazards International, Palo Alto, California, Geo-Hazards Society, Delhi presented in The 14th World Conference on Earthquake Engineering October 12–17, 2008, Beijing, China.

Shukla, K., Rajesh, P., Dal, S., Ravi Kant, S., Pandey, A. P., Mandal, H. S., & Nayal, B. M. S. Seismic micro zonation of NCT Delhi, Earthquake Risk Evaluation Centre, India Meteorological Department, New Delhi

Singh, R.B., & Singh, S. (2011). Rapid Urbanization and Induced Flood Risk in Noida, India, *Asian Geographer Journal*, Vol. 28, Issue 2, pp. 147–69. ISSN-1022–5706.

Srivastava, L. S., & Somayajulu, J. G. (1966). The seismicity of the area around Delhi. In *Proceeding of the third symposium of earthquake engineering, Roorke*, pp. 417–422.

Chapter 14
Mapping Fire, Earthquake and Bio-hazard in Delhi: A Micro-level Study

Vineka Sanoria and Chandrakanta

Abstract Delhi, being India's capital territory, is a massive metropolitan area that is extremely vulnerable to various types of disasters because of the widely spread built-up area that houses the population from all over the country. Delhi lies in Seismic Zone IV[14], which makes the area sensitive to disasters. Another major problem that Delhi is currently facing is of proper garbage disposal, since the density of the population is high, tons of waste is generated. A fair share of the waste generated also includes biomedical waste. Delhi generates more biomedical waste than it can process. The area chosen for the present study is Chirag Delhi and Sheikh Sarai, located in south Delhi. This area is urbanized, and a home to a large number of people. The area is populated, poorly managed and highly vulnerable to disasters. The study area also has two colleges situated near the residential area because of which the area is subjected to a lot of traffic jam. The purpose of choosing this area for this study is its vulnerability to disasters like fire, earthquake and biohazard. The study area has pockets with high rise buildings or ill-designed high-risk areas without specific consideration for earthquake resistance. Moreover, the area lacks proper waste management. It has been identified that the area is a highly vulnerable place when it comes to hazards like fire, earthquake and biohazards. The people living there are in a constant threat for their lives. One of the major problems is that the community lacks dedication and determination, which has been tested through a schedule and observation method, to change their circumstances and bring about a change in the area that would benefit them and their families.

Keywords Vulnerable · Earthquake · Fire hazard · Biohazard and waste management

V. Sanoria (✉)
School of Education, Christ University, Bangalore, India
e-mail: vinekasanoria@gmail.com

Chandrakanta
Department of Geography, Shaheed Bhagat Singh College, University of Delhi, New Delhi, India

V. R. Sharma and Chandrakanta (eds.), *Making India Disaster Resilient*,
https://doi.org/10.1007/978-3-031-50113-5_14

Introduction

Disasters have been visiting every part of the globe at one time or the other. The world is becoming increasingly vulnerable to disasters. From earthquake to flood and famines, mankind is even more threatened by the forces of nature now more than ever. Disasters can strike at any time, at any place. Nearly three million people worldwide may have been killed in the past 20 years due to disasters such as landslides, earthquakes, floods, avalanches, cyclones, etc.

Throughout history, hydro-meteorological disasters have caused large-scale destruction and still continue to do so. According to the United Nations International Strategy for Disaster Reduction, a hydro-meteorological hazard is a process or phenomenon of atmospheric, hydrological or oceanographic nature that may cause loss of life, injury or other health impacts. The number of hydro-meteorological disasters has only gone up.

Over time there has been a radical change in the types of disasters that affect human beings. Initially famines, wars or epidemics affected the population of a particular country drastically; nowadays however, hydro-meteorological disasters cause massive destruction. On 2011, Japan was hit by a 9.0 magnitude Earthquake, that triggered a deadly 23-foot tsunami in the country's north. According to the official toll, the disaster left 15,839 dead, 5950 injured and 3642 missing. More recently, Hurricane Harvey that hit Texas on August 25, 2017, caused 125 billion in damage according to the National Hurricane Centre. That is more than any other disaster in US history except Hurricane Katrina, which caused 161 billion in damages.

India because of its geographic location is vulnerable to a number of natural hazards, particularly flooding, cyclones, drought, extreme heat wave, landslide, wildfires and earthquakes (Khullar, 2017). According to the National Disaster Management Authority, 1600 people every year are killed due to flooding. In 2017 alone, more than 1200 people died in flood-related incidents, as per the official estimates reported by the state government. In 2013, India experienced the double shock of severe flash floods in the state of Uttarakhand and Cyclone Phailin in the state of Odisha. The flooding and associated landslide in Uttarakhand caused over 4000 deaths and 661 million damage and loss, resulting in the country's worst disaster since 2004 tsunami. In India, vulnerability to natural hazard is exacerbated by high population density and growth in urban areas and coastal areas.

Delhi being the capital of the country lies in Seismic Zone IV, but the destruction here will be more than the areas belonging to zone V region like Himalayas, as the built-up area is widely spread due to which the population density is high (Iyengar & Ghosh, 2004). Another problem the entire city of Delhi faces is the problem of garbage disposal, since the density of population is high, tons of waste is generated on daily basis by the people of the city. The city generates more garbage than it can ever dispose.

The area selected for the present study is **Chirag Delhi** and **Sheikh Sarai** which is in South Delhi. It is located in a highly urbanized and populated residential area. The study area (Map 14.1) also has two colleges—Shaheed Bhagat Singh College

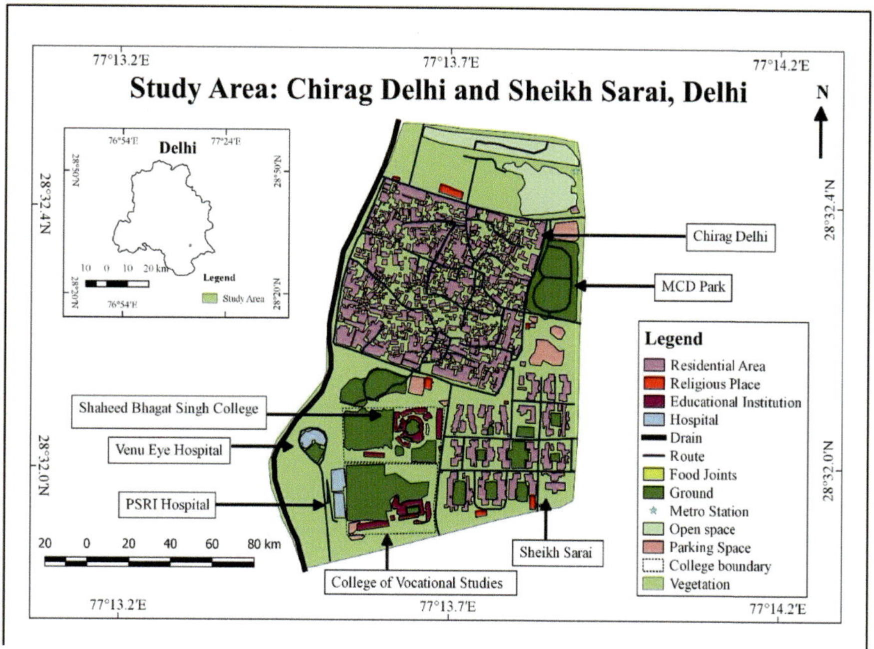

Map 14.1 Study area. *Source* Created by author using QGIS, March 2019

and College of Vocational Studies. Also, it includes two hospitals, PSRI Hospital and Venu Eye Institute & Research Centre. The area also includes a metro station—Chirag Delhi metro station which is located on the Magenta Line of the Delhi Metro. The station was opened for public on 29 May 2018.

The area is clustered and is often subjected to traffic jams. The area is congested and densely populated which increases the disaster vulnerability in the study area. It has pockets with high rise buildings or ill-designed high-risk areas exist without specific consideration of earthquake resistance. The main purpose of choosing this area for this study is its vulnerability to disasters like fire, earthquake and biohazard and to study the effectiveness of disaster management plans by the concerned authorities. The clustered houses make this area extremely prone to disasters.

Objectives

1. To explore the causes of potential hazards, i.e., fire hazard, biohazard and earthquake in the study area.
2. To assess the risk of these hazards in the study area.
3. To assess the socio-economic and physical vulnerabilities to these hazards in the study area.

4. To explore the community awareness and participation as well as the government and non-government initiatives for disaster management present in the study area.
5. To propose a preparedness plan on the basis of the HRVC analysis for the study area.

Database and Methodology

Database

The present study is based on both secondary and primary data.

Primary data has been gathered through field survey including personal interviews which were carried out with the help of a schedule method. Observation method was also used for collecting the data. For the field data, photographs were also taken.

Secondary data has been gathered through Census report (SRS Statistical Report 2018) and State Government official website (https://delhi.gov.in/). The information regarding the disasters faced by India and specifically Delhi was extracted from the official websites of National Institute of Disaster Management (NIDM), National Disaster Management Authority (NDMA) and Indian Meteorological Department (IMD).

Methodology

The respondents for the study were chosen using Random Stratified Sampling Method[15]. A total of 120 respondents were chosen for the study. Composition of Residents is represented using Table 14.1. The data was gathered with the help of a schedule which had both open-ended and close-ended questions. MS Excel was extensively used to carry out the analysis of data. It was used for data compilation and tabulation. This data was then used to reveal the pattern that is highlighted through graphical representation such as bar graphs, pie charts, etc. The maps were digitized with the help of Q.GIS 2.18.3 software.

Literature Review

A comprehensive review of the earlier research studies on earthquake, fire hazard and biohazard conducted in Delhi is highly essential and imperative to identify the problem area related to the present research study.

It has been found in studies that the study area has been inspired by Rajasthani culture in terms of streets and urban space which is why the area is congested with narrow streets (Raj Rewal, 1970). A special analysis has been done to understand the

Table 14.1 Composition of respondents

Area name	Male	Female	Total
Chirag Delhi East	26	5	31
Chirag Delhi West	12	4	16
Chirag Delhi Metro Station	20	3	23
Sheikh Sarai	18	4	22
PSRI hospital	7	4	11
college of vocational studies	1	1	2
Shaheed Bhagat Singh college	4	5	9
Venu eye hospital	4	2	6
Total	92	28	120

Source Field survey, March 2019

seismotectonics of the area (Chauhan, 1975). This helps in getting a better understanding of this area in case an earthquake occurs. The narrow streets add to the vulnerability of the residents. Another research on seismic risk study (Srivastava & Roy, 1982) seeks to give idea about causes of earthquake. The scheme estimates seismic vulnerability of existing building stock quantitatively and qualitatively. Delhi lies in Seismic Zone IV and is liable to earthquakes of 6 to 7.5 Richter scale which falls under a category of high-risk zone as stated by NDMA. A framework guideline has been presented for the construction of houses, forming an integral part of micro-zonation studies that are being taken up for Indian vulnerable cities (Iyengar, 2000).

The area is also highly prone to a fire hazard. Unplanned urbanization and uncontrolled population has intensified the problem further. It was recognized that the greater proportion of poor quality housing, inadequate planning, monitoring and control in metropolitan cities, which become overpopulated, lead to a greater number of fires and other urban disasters (Mukta Girdhar, 2013).

Biomedical waste management is a rising concern as the waste generated is severely harmful to the health of living beings. The waste generated is much more than the waste that can be disposed. About 10–25% of biomedical waste is hazardous. The hazardous part of the waste presents physical, chemical and/or microbiological risk to the general population and healthcare workers associated with handling, treatment, and disposal of waste (Datta et al., 2018). It was found out that the healthcare institutions contribute a large share of biomedical waste. The awareness and practice levels of healthcare workers of a tertiary hospital were studied regarding biomedical waste management and it was highlighted that some major areas of deficit found were about knowledge regarding number of biomedical waste categories, mercury waste disposal and definition of biomedical waste (Bhagawati et al., 2015).

Results and Discussion

In the present study, it was found out that the study area is a congested area which is prone to many hazards such as fire, earthquake and biohazards. This study includes the assessment of socio-economic status of the respondents which helps in highlighting the vulnerability of an individual to disasters. To assess the vulnerability of disaster in Chirag Delhi and Sheikh Sarai, data from 120 respondents have been collected through schedule method. The number of respondents chosen is sufficient to carry out a good micro-level study. Out of the total respondents, 78 per cent were males, whereas the rest 22 per cent were females. The ratio of females is much lower as compared to the males because the females were reluctant to communicate and also because of the fact that the state has a low sex ratio.

It is observed that majority of the people belong to low-income group. These people are more vulnerable than the people from middle or high income groups as the low income group cannot investment on themselves and hence, get more exposed to disasters. It was found out that most of the respondents' houses are more than twenty years old. The ruins of Mughal Period can be seen below in Plate 14.1. Through this we can get an idea about the year of construction of some buildings. This implies that in case an earthquake of medium or high intensity occurs, then these houses will be the first ones to collapse.

Domesticated animals which include livestock and pet are an important factor of assessing the social vulnerability. Animals are dependent on humans so they are vulnerable in case of a disaster such as fire or earthquake. The people whose income is dependent on their livestock will be affected badly in case their livestock is affected due to a disaster.

Transport facilities have always been a capacity in case of a disaster. It helps in reaching the emergency facilities like hospital, police station faster. The respondents who do not own any vehicle are more vulnerable as they are dependent on their neighbors or on the public transport.

Warning before a possible hazard can help in reducing a person's vulnerability and casualties cause by the disaster. It was found out that most of the respondents did not receive any warning from any agency about a disaster. On the other hand, very few did receive a warning about a potential disaster from agencies. This shows how less concerned are the authorities. Temporary shelter during a disaster is extremely important as it will help in rehabilitating people that have been directly affected by the disaster. Most of the people feel that there is no provision for temporary shelter. Through this we can tell how unaware the residents of the study area are. In case of any disaster-related emergency, they would not know where to go in order to save themselves.

Fire vulnerability

The study area is highly vulnerable to fire (Map 14.2). This is due to the fact that the houses are closely built which makes the area extremely congested. The roads are narrow which makes it more vulnerable. Fire in one building will result in harming the surrounding buildings as well.

Plate 14.1 A view of Chirag Delhi. *Source* Clicked by Author, March 2019

Fires are mostly man-made disasters caused by negligence, poor maintenance. Increased usage of electricity, LPG and hazardous chemicals has resulted in an increase of fire hazard potential. The number of fire incidents increase drastically during summers due to the increase in the usage of air conditioners which leads to overheating of cables and other electrical components leading to short circuit. Also, during summers all the combustible material is dry and catches fire with relative ease. In the area, most of the respondents believe that short circuits are the main cause of fire incidents because there have been many cases in the past regarding the same (Fig. 14.1).

Most of the respondents agreed that the wires of the electric poles (Plate 14.2) should be covered with proper wiring, and if the wires are hanging loose, then they should be given immediate attention.

A large number of people are not aware about a fire station nearby which is a troubling sign because it clearly increases the vulnerability of the people. According to them it took more than 40 minutes for a fire brigade to reach the spot. This is because of the fact that the area is very congested and the roads are narrow which makes it difficult for the fire brigade to reach the spot as shown in Plate 14.3.

Another major problem in the area is that people are not aware about any active safety systems present in their locality. Most of the respondents have not heard

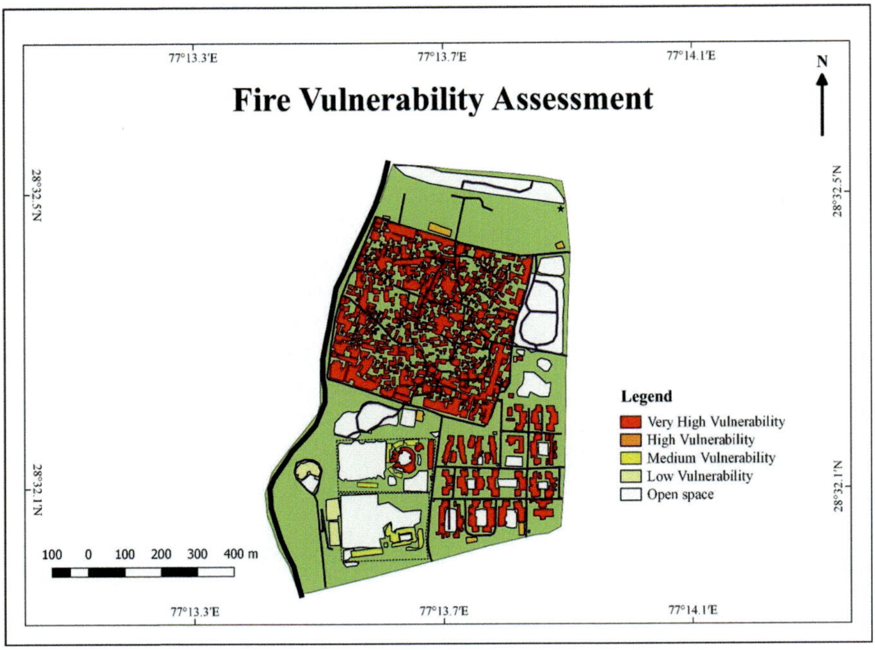

Map 14.2 Fire vulnerability assessment. *Source* Created by Author using QGIS, March 2019

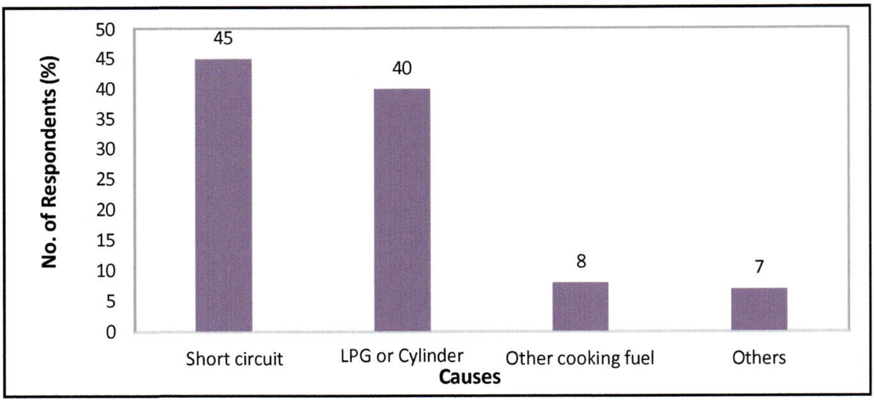

Fig. 14.1 Causes of fire hazard. *Source* Field survey, March 2019

about any awareness programs for pre-disaster management. This shows their awareness levels regarding disaster management. Few of the respondents have attended programs related to disaster management which were conducted in their schools, and some of them were conducted by the government.

The area is extremely prone to a fire hazard so it is of paramount importance that people understand their vulnerability and work toward reducing it. The government

Plate 14.2 Electric pole. *Source* Clicked by author, March 2019

Plate 14.3 Narrow streets of Chirag Delhi. *Source* Clicked by Author, March 2019

Plate 14.4 A poster at Chirag Delhi metro station. *Source* Clicked by Author, March 2019

authorities are trying to spread awareness regarding fire hazards. Plate 14.4 shows a poster put up by the Delhi Metro Rail Corporation during the Fire Safety Week quoting "Fire Prevention is Better than Fire Fighting" at the Chirag Delhi Metro Station. Although this does not reveal how to prevent a fire but it somehow does the job of making people aware that they should at all costs try to prevent a fire than fighting it later. The government should also put extra efforts in reducing the vulnerability of the area. Everyone should be made aware of the do's and don'ts in case a fire breaks out. Fire extinguishers should be installed in the locality, and mock drills should be conducted once in a while.

Earthquake Vulnerability

Delhi lies in a very sensitive seismic zone IV (Map 14.3); it experiences earthquake almost every year. From the study, it was found out that how differently people react during an earthquake and how many people are aware and prepared in case an earthquake occurs. Among the various reactions, the most common were running out of the building, not knowing what to do, hiding under a bed/table and standing next to a wall or under the door frame (Fig. 14.2).

Seismic hazards are the intrinsic natural occurrences of earthquakes and the resulting ground motion and other effects (Bansal et al, 2009). The type of hazard depends on the strength of seismic activity, along with such factors as local topographic and built features, subsurface geology and ground water. Many non-structural components in buildings such as furnishing, equipment, electrical and mechanical fixtures, architectural features, storage cabinets, shelves, and glass may pose hazards when they slide, tip over, fall or collapse during an earthquake. In the study area, a large proportion of houses are not built according to the National Building Code (Fig. 14.4). Also, the vast majority of people do not have a household emergency plan (Fig. 14.3) which is a major problem during or at the time of an earthquake (Map 14.3).

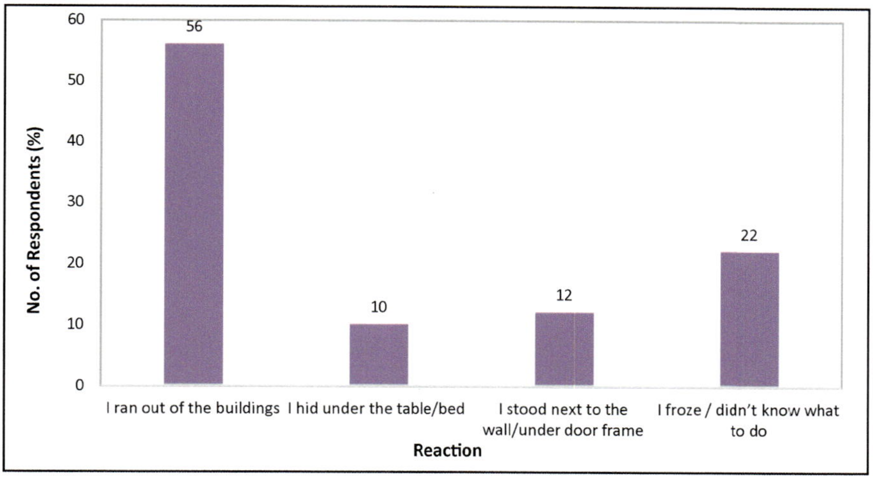

Fig. 14.2 Instant reaction. *Source* Field survey, March 2019

Fig. 14.3 Household emergency plan. *Source* Field survey, March 2019

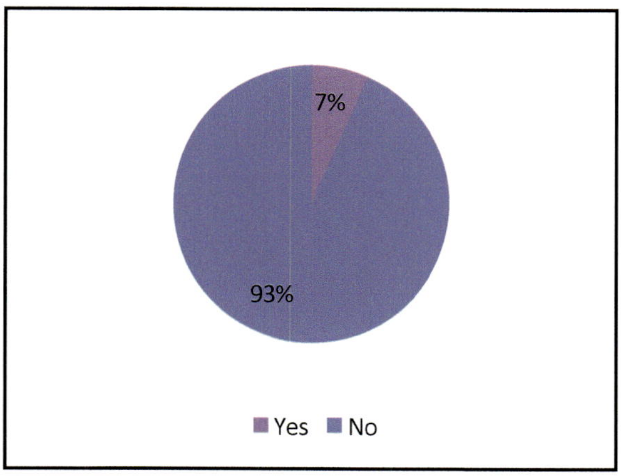

Another interesting observation is that not a single person in the area has disaster insurance, probably because they are not aware about such insurance policy. Whatever the reason may be, it is important that people are aware about such policies so that they can be prepared for a disaster in the future.

Biohazard Vulnerability

Cleanliness in the study area is a major problem. There are various open garbage dumps, and the garbage is scattered here and there. The area is not at all clean and because of this people suffer from a number of diseases, mainly malaria, diarrhea, food poisoning and chikungunya (Fig. 14.5). This could be attributed to improper garbage disposal in the area.

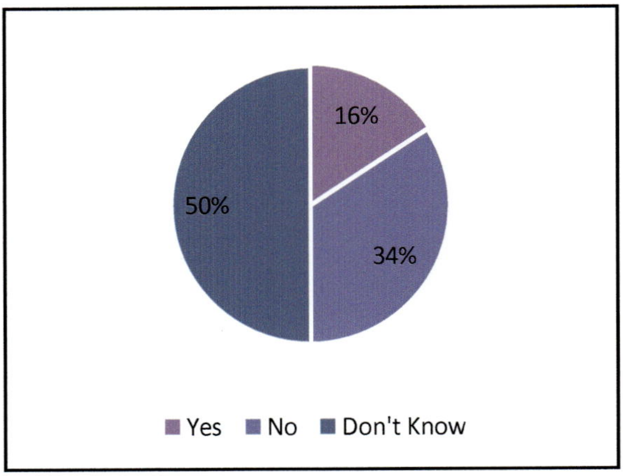

Fig. 14.4 Construction as per the national building code. *Source* Field survey, March 2019

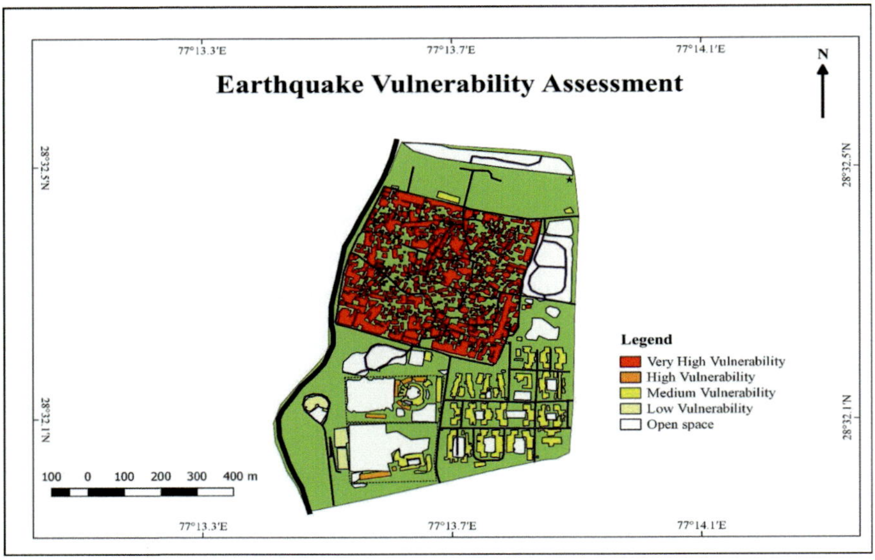

Map 14.3 Earthquake vulnerability assessment. *Source* Created by author using QGIS, March 2019

There are multiple causes of health problems in the study area. The major causes which were identified were improper garbage disposal, lack of sanitation and water logging and mosquitoes in the area. Few respondents agreed that polluted water supply and uncleansed drainage are the reasons because of which people fell sick.

It was found out that the expenditure on health for majority of the people is less than five thousand rupees and for some it is more than that. There are also people who do not spend anything on their health because there are various government schemes

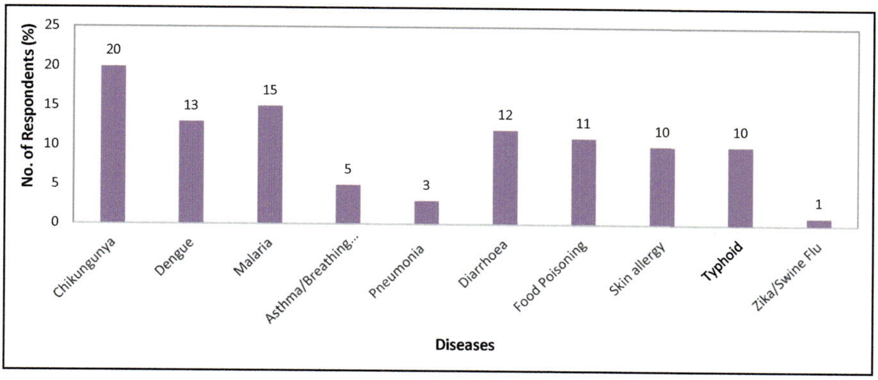

Fig. 14.5 Illness in the family. *Source* Field survey, March 2019

(one of them being CGHS) that provide free medical treatment to the people (Map 14.4).

A walk around the area and you will be exposed to a number of garbage dumping sites (Plate 14.5). The area between Shaheed Bhagat Singh College and College of Vocational Studies is a garbage collecting site, and it is just besides the main road, a large number of people use that road. It is impossible to cross it without covering

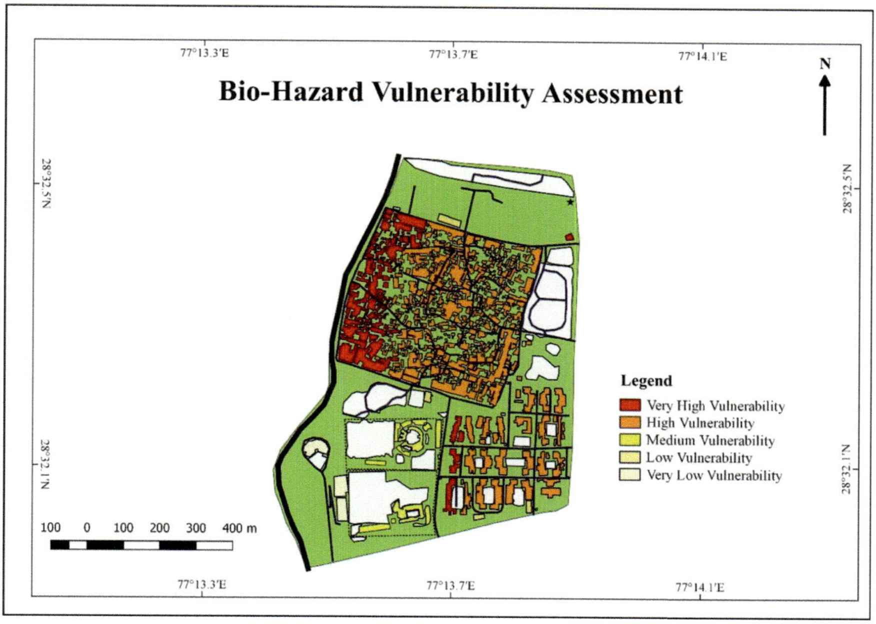

Map 14.4 Biohazard vulnerability assessment. *Source* Created by author using QGIS, March 2019

Plate 14.5 Garbage dumping site of Sheikh Sarai. *Source* Clicked by author, March 2019

your nose. It is sites like these that are a breeding ground for diseases; hence, it is imperative that these sites are exterminated from the area.

There are many medical facilities, government and private, which are accessible to the study area. The hospitals which are accessible to the study area are Pt. Madan Mohan Malviya Hospital, Aam Aadmi Mohalla Clinic, Max Hospital, City Hospital, PSRI Hospital, Venu Eye Hospital and many more.

The part of Chirag Delhi near the drain is also highly vulnerable to biohazards. This area always has a foul smelling air and the drain is a home to many viruses. Therefore, special attention is needed to eradicate this problem or else the people living there will continue to suffer. Biohazard Vulnerability Assessment of the study area is represented using Map (14.4).

Evacuation Plan in Case of Fire

- If fire or smoke is detected, the door should be touched by the back of hand to check whether it is hot. If not, then the door should be opened cautiously and then one should proceed calmly.
- The building should be evacuated.
- Dial 101 or the nearest fire station.
- If there is no way out of the room or building, stay in and seal the door with wet towels or clothes until assistance arrives.
- If there is no trouble in evacuating the building then the red arrows should be followed which will lead to the nearest emergency exit.
- Narrow lanes should be avoided.
- If one emergency exit is blocked then the next nearest one should be used.

- If there is no option left other than escaping through smoke then crawl low to the exit. The head should be kept 1 to 2 feet above the floor. The air will be the cleanest there.
- After the area is evacuated, the assembly area should be used for gathering the people (indicated by white in Map No. 5).
- Parks and parking areas should be avoided for gathering.
- Checking of injuries should be done if any. Stop, drop and roll technique should be used if clothes catch fire. The nearest hospital should be called in case of injuries.
- Going back inside should be avoided until it is safe to do so. The fireman should be informed in case any problem occurs.

In Case of Earthquake

- If indoors, sturdy furniture or an inside wall should be used to take cover. Face should be covered. Kitchen area should be avoided at all costs.
- If outdoors, taking shelter near buildings, utility wires and streetlights should be avoided. Step in a doorway. Face and shoulders should be covered in order to protect from falling debris or glasses.
- After the earthquake has passed, the building should be evacuated calmly and safely.
- The red arrows leading toward open spaces and grounds should be followed.
- Letting the aftershocks pass by is necessary before going back indoors.
- The nearest hospital should be called in case of any emergency.
- Going back should be avoided until it is safe to do so.
- Evacuation (Map 14.5) must be installed at places with public gathering.

Conclusion

The study area is a very congested area with narrow streets which is home to numerous people, out of which most of them are migrants who have come here either for job opportunities or educational facilities. Few of the houses are hundreds of years old which clearly states that they are not built according to the National Building Code. The area is subject to increasing population due to which there is an illegal vertical construction.

This area also lacks proper waste management because in this area the waste is produced in large quantities and the waste keeps piling up moreover, the garbage pickers do not come on a regular basis to pick up the garbage. Since the dumping sites are open here, it creates very obnoxious smell which is hazardous for the people living around, there is an urgent need to cover these dumping sites.

The drainage system here needs improvement because during rainy season water logging occurs which leads to the development of breeding grounds for mosquitoes and causes diseases like dengue, malaria and it even contaminates the drinking water.

It has been identified that the area is a highly vulnerable place when it comes to hazards like fire, earthquake and biohazards. The people living there are in a

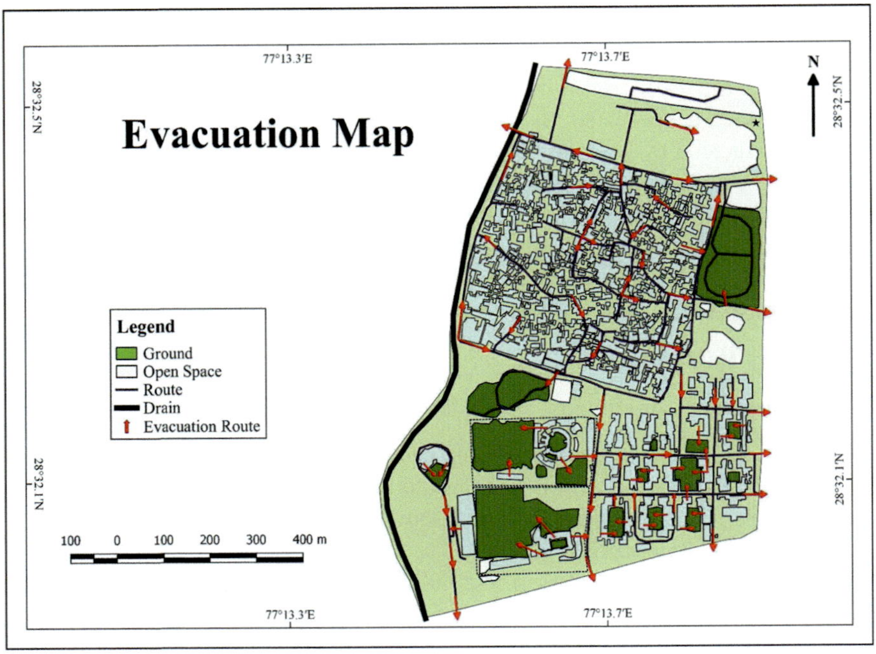

Map 14.5 Evacuation map. *Source* Created by author using QGIS, March 2019

constant threat for their lives. One of the major problems is that the community is not even aware about it or the community is aware but does not want to do anything. It can be seen that the government authorities also do not play any role in reducing the people's exposure to these disasters. The people cannot even complain to the government authorities as they have illegally constructed their houses. The people lack awareness when it comes to these potential hazards, let alone saving themselves during such circumstances.

We can conclude by saying that the study area being a congested region is in dire need for an efficient evacuation plan. The plan must exhibit an evacuation route which can be used during the times of disaster. This plan will make it easier for the residents to vacate the place in no time. This will also help in reducing casualties. The evacuation plan must be placed somewhere where it is accessible to the people.

Acknowledgements I would like to thank the Department of Geography of Shaheed Bhagat Singh College that provided me with this opportunity to conduct this survey.

References

Bansal, B. K., Singh, S. K., Dharmaraju, R., et al. (2009). Source study of two small earthquakes of Delhi, India, and estimation of ground motion from future moderate, local events. *Journal of Seismology, 13*, 89–105. https://doi.org/10.1007/s10950-008-9118-y

Bhagawati, G., Nandwani, S., & Singhal, S. (2015). *Awareness and practices regarding bio-medical waste management among health care workers in a tertiary care hospital in Delhi.*

Datta, P., Mohi, G. K., & Chander, J. (2018). *Biomedical waste management in India: Critical appraisal.*

http://delhi.gov.in/wps/wcm/connect/5366420040f86897b402be7a69ad66a3/10_Disaster+Management+in+Delhi.

http://e.duac.org/images/pdf/2%20Chirag%20Delhi.pdf

http://kantclinicvikaspuri.com/biological-hazard/

https://www.downtoearth.org.in/coverage/delhi-is-earthquake-prone-12655

https://www.researchgate.net/publication/258700119_1960_Delhi_earthquake_Epicentre_depth_and_magnitude

https://www.academia.edu/9603408/The_Heterotopic_Space_of_Chirag_Delhi

https://ndma.gov.in/en/

https://www.hindustantimes.com/delhi-news/chirag-dilli-s-legacy-lost-in-maze-of-lanes/story-jqXzpm3HDRlVM8OE47OFqM.html

https://www.pipldelhi.com/east-delhi-crisis-of-garbage-landfill-site/

Iyengar, R. N. (2000). *Seismic status of Delhi megacity.* Current Science Association.

Iyengar, R. N., Ghosh, S. (2004). *Micro-zonation of earthquake hazard in greater Delhi area.*

Khullar, D. R. (2017). *India: a comprehensive geography.* Kalyani Publishers.

Chapter 15
Surface Deformation Modelling Using C-Band SAR Data—A Case Study on Shimla Town, Himachal Pradesh, India

C. Prakasam, R. Aravinth, Kanwar S. Varinder, and B. Nagarajan

Abstract SAR datasets play an active role in monitoring the active deformation not only due to its high radiometric resolution, but also increased temporal resolution (12 days) and greater swath coverage. The European Space Agency (ESA) provides SAR dataset through its twin constellation of satellites Sentinel 1A and 1B which has a huge impact in data analysis not only for deformation studies, but also in other thrust areas and earth related studies. The study area chosen for the deformation modelling is Shimla Municipal Corporation. According to Seismic Vulnerable zone classification by India, Shimla is present in Zone IV with high vulnerability. The region is also highly prone to disasters such as earthquake, landslides, flash floods, improper building construction due to natural and anthropogenic causes. The deformation rate has been calculated between 2014 and 2015. The datasets were acquired from Sentinel Data Hub organized by ESA. Four C—Band Sentinel—1A SLC TOPSAR data at ascending orbit with dual polarization has been used for deformation analysis. Precise orbit files were applied to the data and then each successive data was coregistered as Master and Slave Image. Three interferograms were formed and used for the analysis. These images were then subject to further pre-processing such as topographic phase removal and phase filtering to smoothen the interferograms. The unwrapped image was acquired by analysing the image through SNAPHU in Linux environment. The output received was converted into displacement data through band

R. Aravinth (✉)
Institute of Environment Education and Research, Bharati Vidyapeeth University, Pune 411043, India
e-mail: aravinthraja662891@gmail.com

B. Nagarajan
National Centre for Geodesy, Indian Institute of Technology, Kanpur 208016, India

C. Prakasam
Department of Geography, School of Earth sciences, Assam University, Diphu Campus (A Central University), Diphu, India
e-mail: cprakasam@gmail.com

K. S. Varinder
Department of Civil Engineering, Chitkara University, Baddi 174103, India

V. R. Sharma and Chandrakanta (eds.), *Making India Disaster Resilient*,
https://doi.org/10.1007/978-3-031-50113-5_15

math into mm units. The results indicate that the surface has subsided nearly−33 to−100 mm/year between 2014 to 2015 and 2016 to 2017. The surface has an upliftment of 39 to 45 mm/year between 2015 and 2016. The result obtained in this study from measuring active monitoring of ground deformation and rate will help stake holders and government authorities to take necessary steps in constructing built-up lands and other environmental related projects.

Keywords Sentinel—1A · Interferograms · Phase unwrapping · Phase filtering · Displacement rate · Range Doppler terrain correction

Introduction

SAR Interferometry (INSAR) is a technique used to detect and measure active ground surface deformation using Radar images of the earth surface acquired from multiple satellite platforms at high precision, wide swath coverage and high temporal frequency (Lanari et al., 2004; Massonnet & Feigl, 1998; Simons et al., 2007). Over the last two decades, the ability of Synthetic Aperture Radar (SAR) to map active ground deformations at mm accuracy caused by various natural and anthropogenic factors have become rather important. Relying on SAR imageries for surface deformation mapping compared to other imageries can be attributed to many factors, namely (i) advances in the sensor and performance of satellite system to produce data at an increasing spatial and temporal resolution along with larger swath coverage to produce maximum data in a single flight. (ii) Recent findings in developing new methods to analyse surface deformation by greatly reducing the noise and atmospheric error from the Radar data. (iii) Increased computational capabilities through parallel processing, big data analytics and cloud computing. European Space Agency's Copernicus initiative on Sentinel missions opened the door for newer possibilities for data analysis not only relying on Radar dataset, but also combining radar data with high spatial resolution and DEM model to produce real-time high accurate results. ESA's initiative Sentinel—1 mission consists of constellation of twin satellites Sentinel -1A and Sentinel—1B that were launched during April 2014 and April 2016, respectively. Both of the satellites share a high temporal coverage as much as 6 days for each satellite (12 days for both 1A and 1B). This high temporal coverage makes highly suitable candidate for interferometric and ground deformation studies. Sentinel 1 data has a regional scale applicability with systematic and rapid product delivery with less than three hours from data acquisition. The Sentinel 1 data are freely accessible to science community and also public and private domain (Lanari et al., 2015). Zhang et al., (2007) studied the 3D-based landslide deformation and application of deformation modelling for effective geological mapping.

The method encompassed in this research is Differential Interferometric Synthetic Aperture Radar (DINSAR) based ground deformation analysis. DINSAR technique is one of the promising method in existence for active mapping of deformation through its high coverage area and increased accuracy (Liu et al., 2016). Classical DINSAR is used to obtain the relative motion between two images. It uses phase difference and interferogram properties of the RADAR to calculate the relative ground motion along the Line of Sight (LOS) of the sensor (Grandin 2015; Crosetto et al., 2011). Compared to DINSAR, other techniques were also developed to predict the ground deformation such as Multi-temporal INSAR (MTINSAR), Polarimetric INSAR (PSINSAR), CAESAR and GEOS-ATSA, etc. (Ferretti et al., 2011; Fornaro et al., 2015). Authors like Gama et al. (2017), Poursanidis and Nektarios (2017), also studied application of SAR dataset for subsidence in mining and landslide monitoring.

In this paper, we have attempted to exploit the advantages of multi-temporal SAR data to detect ground deformations. We have coregistered multiple SAR data and the required interferograms for the selected years was acquired. The interferogram was then filtered and unwrapped to produce a displacement result using phase to displacement tool. The results presented in this paper show the effectiveness of the SAR data to provide near real-time deformations over a wide area.

Objectives

- To create stacked interferograms of the study area.
- Derive phase filtering and unwrapping based on the successive interferograms.
- To estimate surface deformation along the study area through phase to displacement modelling using MCF factor.
- To estimate temporal change of deformation along the study period from 2014 to 2017.

Study Area

The study area chosen to model the surface deformation is Shimla Municipal Corporation located in Shimla district of Himachal Pradesh. The geographical extent of the Shimla town is 77° 0′ and 78° 19′ East longitude and 30° 45′ and 31° 44′ North latitude (Fig. 15.1). The Shimla Tehsil is divided into Shimla Urban and Shimla Rural. Most of the settlements and industrial areas falls within the Shimla Corporation. The average elevation of the Shimla district ranges from 300 to 6000 m. Maximum extent of the Shimla Corporation falls built-up lands and the latter fall under forest and slopes. The terrain is highly undulating and rugged with steep slopes and escarpments (1). Shimla's population has been steadily increasing since the start of nineteenth century. The population of Shimla Tehsil was 13,960 in 1901 and 144,975 in 2001. According to 2011 Census, the total population of Shimla town itself is 169,578 (18). Shimla is one of the tourist hotspots in Himachal region. Nearly five million domestic and foreign Tourists visited Shimla during 2005. Even though these tourists bring a boon

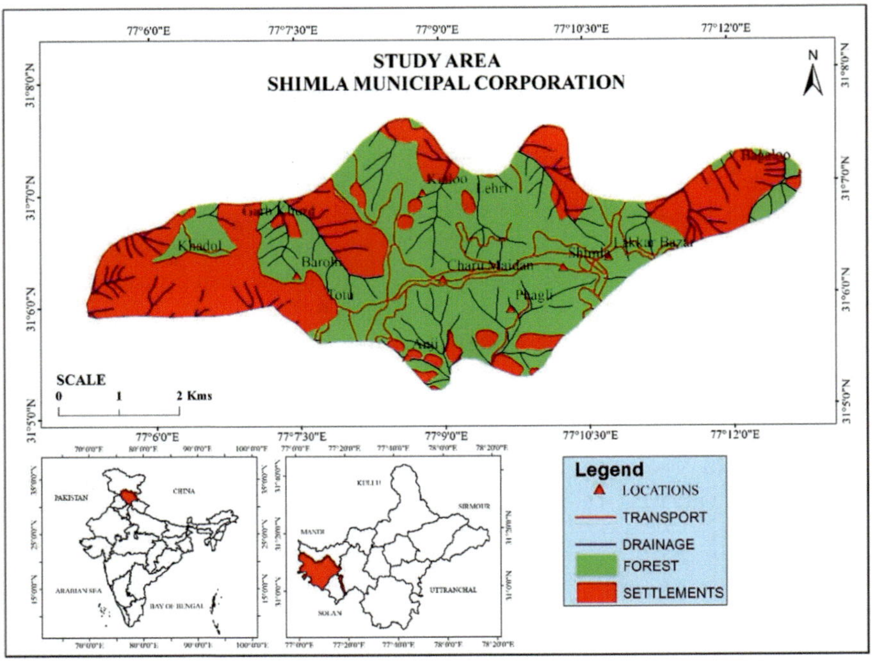

Fig. 15.1 Study area

to economic stability of Shimla tourism, it also proves to be a negative impact for ecological degradation. Given the rapid growth of population and steady increase in tourist visit, it has proven to be a major source for many ecological and environment related problems. Some of the major problems are as land degradation, landslides, land subsidence, unplanned building construction, soil erosion, soil creep and forest fire, etc. Some of the main sites of landslides and land subsidence are Lakkar bazar, Bharari road, Mall road, NH 2, Bypass road and Sanjauli–Jakhu road (19).

Data Used and Methodology Data Used

The base map of the study area was prepared from Survey of India Toposheets (SOI) 53E/04 and 53E/08 respectively. Table 15.1 depicts the details of the datasets used for base map preparation and subsidence mapping. Deformation analysis was carried out using Sentinel—1A C-band SAR data. The Sentinel 1 satellite funded by European Space Agency (ESA) is a constellation of two satellites Sentinel—1A and Sentinel—1B which were launched on February 2014 and 2016 respectively. The primary products of Sentinel—1A sensor are offered in SM, IW and EW (Geudtner et al., 2014). These products are offered in both single polarization

Table 15.1 Details of the datasets used for base map preparation and subsidence mapping

S. No.	Data	Source		Date		Resolution	
1	Toposheets	SOI				1:50,000	
S. No.	Data	Source	Polarization	Orbit	Swath	Date	Resolution (cm)
1	C—Band Sentinel 1—A	ESA	VV	Ascending	IW—2	15/10/2014	5.6
2	C—Band Sentinel 1—A	ESA	VV	Ascending	IW—2	28/09/2015	5.6
3	C—Band Sentinel 1—A	ESA	VV	Ascending	IW—2	10/06/2016	5.6
4	C—Band Sentinel 1—A	ESA	VV	Ascending	IW—2	08/05/2017	5.6

and dual polarization. The interferometric mode is widely used in many land applications throughout the world (Torres et al., 2012). The orbital swaths if the IW mode is 250 km with a spatial resolution of 5 m * 20 m. IW catches data in three subswaths (IW1, IW2, IW3) encompassing entire swath using Terrain Observation with Progressive ScanSAR (TOPSAR). TOPSAR acquires data in a series of burst by cyclically switching the antenna beam different subswaths. This mode of data acquisition provides larger swath widths and enhanced radiometric performance when compared to traditional strip mode (SM) SAR data by reducing the signal-to-noise ratio and avoiding scalloping and azimuth—carrying ambiguity effects (Xixi et al., 2018).

In this study, Sentinel—1A C-Band SAR data acquired in Interferometric mode (IW) along the descending orbit between 15/10/2014 and 08/05/2017 were used to detect subsidence long the Shimla town.

Methodology

The acquired data were processed in ESA's Sentinel Application Platform, an open-source tool box for the processing of SAR datasets (Liu et al., 2018; Lv et al., 2014). Details of the datasets used for base map preparation and subsidence mapping are represented in Table 15.1. The Sentinel—1A dataset has varying resolution within the subswaths of the image. The datasets are pre-processed before the coregistration process (Raspini et al., 2018; Song and Jiang 2016). The second step involves in coregistering two different temporal datasets, one being the Master SLC image and the other being Slave SLC image. The coregistered image is then processed to get

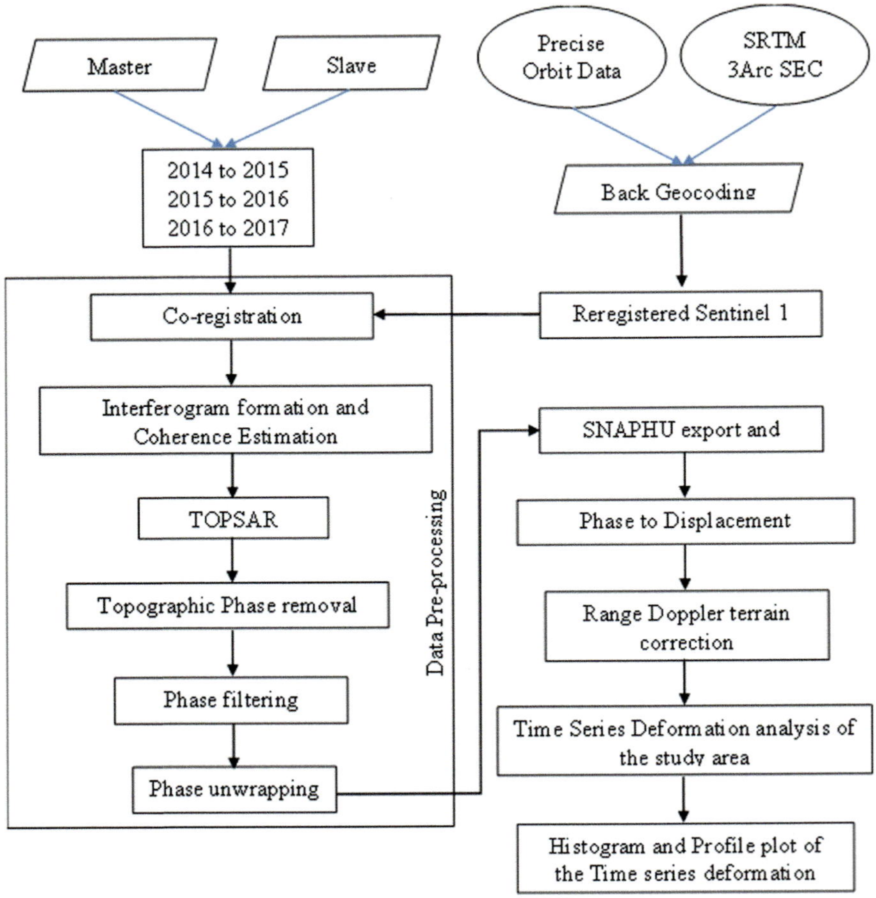

Fig. 15.2 Methodology for deformation analysis. *Source* (authors)

interferograms of the temporal datasets. Subsequent steps are followed such as Topographic deburst, Topographic phase removal, Phase filtering and Unwrapping of the interferograms. Finally, the interferogram is processed using Band math to achieve the displacement value of the study area. More detailed steps of the analysis are given in Analysis part of the study area. Figure 15.2 indicates the overall methodology for the deformation analysis.

Results and Discussion

The acquired datasets are imported into SNAP platform for pre-processing. The sentinel—1A SLC data acquired in IW mode is a multi-sized product where bands have varying resolution and sizes. In order to get the interferograms, the products are first resampled to the specified subswath within the IW mode to get a uniform resolution over the image. The products are resampled using the desired reference band as a source product. In our study area, the entire Shimla town falls under subswath of IW 3. Hence, the particular band acquired within the subswath used has a reference product. Nearest neighbourhood classification is used as a resampling method for classification. In order to perform the interferometric processing, the input products should have two or more SLC images coregistered of the same location for different temporal periods. The coregistration is executed with the help of Sentinel 1 TOPS coregistration tool. The registration accuracy of the Sentinel 1 SAR imagery needs to be 1 cm. To achieve that, one image is kept has a Master Image and the other as Slave Image. For example, to estimate displacement between the year 2014 and 2015, 2014 is considered as a Master Image and 2015 is considered a Slave Image (Fig. 15.3). The phase difference of slave image will be matched to the phase difference of Master Image. The orbital vectors are adjusted for each image through Sentinel precise orbit vectors. The image is coregistered with the help of SRTM DEM to calculate for topographic variations. The RGB combination of Coregistered image will reflect green and yellow colour where the area is successfully Coregistered, and red wavelength will indicate the absence of one or more data in that particular area. Next step of the process is to calculate the interferogram between the coregistered images. The interferogram is formed by cross multiplying the Master and slave Images. The resulting amplitude represents multiplication of both the images, while phase represents the variation in phase difference between the images (Fig. 15.4).

Fig. 15.3 Image coregistration

Fig. 15.4 Interferogram formation

The difference in travel path of the wavelengths between two successive images represents the interferometric phase of the images. The phase difference in interferogram can be attributed to five different factors (ESA'S Sentinel website).

$$\Delta_{\varphi} = \Delta_{\varphi}\text{flat} + \Delta_{\varphi}\text{elevation} + \Delta_{\varphi}\text{displacement} + \Delta_{\varphi}\text{atmosphere} + \Delta_{\varphi}\text{noise}$$

- Δ_{φ} flat = flat earth phase
- Δ_{φ} elevation = topographic variation
- Δ_{φ} displacement = interferometric phase
- Δ_{φ} atmosphere = atmospheric contribution to interferometric phase
- Δ_{φ} noise = phase noise due to temporal change of the scatters, different look angle and volume scattering.

The value of coherence is also estimated along with interferogram. High coherence data reflects bright image and area with low coherence reflects dark grey to black image. In the above image (Fig. 15.5), area present in the Shimla town has high coherence and areas surrounding Shimla have less coherence. Once interferogram is created, the next step is to do a TopoDeburst. The processed interferogram from TOPSAR IW and EW SLC mode will have 3 swaths and 5 swaths respectively. The information is gathered with a series of burst for each image where nearly 50–100 samples overlap for each image. The deburst tool brings the complex image into one single image for further processing. The next step is to do topographic phase removal which accounts for variation in elevation present in the earth surface. The topographic phase is removed with help of SRTM 3ARCSEC GDEM. Phase filtering is a necessary pre-processing technique that will reduce the residual error during the

Fig. 15.5 Coherence estimation

phase unwrapping increasing its accuracy (Fig. 15.6). The method used for phase filtering was put forth by R. M. Goldstein and C. L. Werner known as Goldstein phase filtering (Goldstein & Charles, 1998). The retrieved filtered image is then unwrapped to relate it to the topographic height. The unwrapped phase variation between two points provides the measurement of altitude variation along the Line of Sight (LOS). The filtered image is unwrapped through the phase unwrapping tool in RADAR operations (Fig. 15.7). The unwrapped image is then exported through SNAPHU export for deformation analysis. The exported image is then processed in Linux command terminal to get the final processed unwrapped image. The unwrapped image will not have spatial coordinates within it. In order to get spatial coordinates wrapped, interferogram is used through SNAPHU import. The final step of the analysis is to estimate the displacement along the study area using phase to displacement tool. The distance can be distorted due to terrain variation and tilting of the sensor. This error is removed through terrain correction methods. The acquired displaced image is then terrain corrected through Range Doppler terrain correction method. This will compensate the distortions so that the geometric variations within the image will be as close as to the real world (Fig. 15.8).

Ground Subsidence in the Study Area

The data used for modelling subsidence and Upliftment in Shimla Municipal Corporation is acquired from Sentinel—1A Interferometric mode (IW) in Dual Polarization along the ascending orbit. The subsidence is calculated along the Line of Sight (LOS)

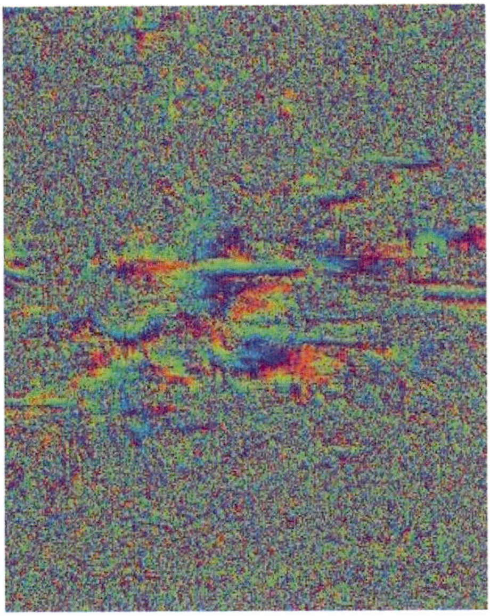

Fig. 15.6 Phase filtering

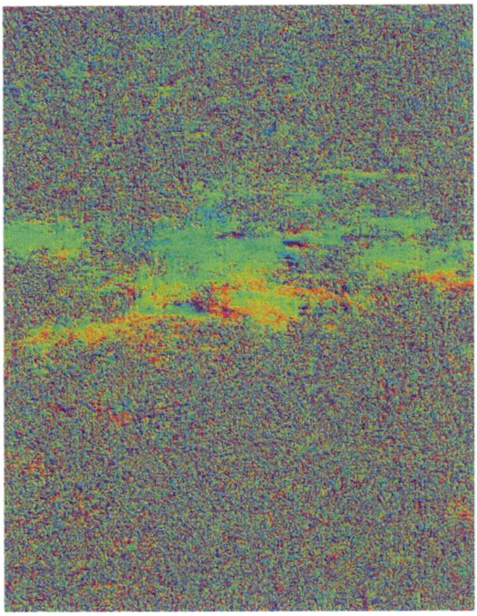

Fig. 15.7 Unwrapped interferogram

of the sensor. Based on the observations both in the year 2014–2015 and 2016 to 2017, the land has subsided from − 31 to − 100 mm/year. Between the year 2014 to 2015, the ground has subsided between − 47 to − 100 mm/year (Fig. 15.9) and in the year 2016 to 2017, the ground has subsided − 31 to − 96 mm/year (Fig. 15.12). For the

Fig. 15.8 Displacement along the study area

year 2014–015, the subsidence rate between − 62 and − 76 mm/year covers a larger extent along the middle part of the Shimla Municipal Corporation (Fig. 15.10). The least amount of deformation ranging between − 47 and − 54 mm/year have been found to be along the eastern part of Shimla. The highest amount of deformation between − 77 and − 100 mm/year has been located along the western part of the Shimla (Fig. 15.11a–c). For the year 2016 to 2017 (Fig. 15.12a–c), the least amount of displacement occurred between − 31 and − 47 mm along the eastern part of the Shimla Municipal corporation. Area with moderate level of subsidence − 48 to − 62 mm/year has been found along the middle part of observed Municipal Corporation to a greater extent (Fig. 15.9). The highest level of subsidence has been observed along the western part of the Shimla Municipal Corporation.

Regions surrounding the Khodal area has suffered the maximum deformation between the year 2014–2015 (Fig. 15.10c, d). The high deformation rate is high due to the change in surface features along the particular area within a short span of time. Fig. 11.16 represents the overall deformation occurred in Shimla Municipal Corporation and regions surrounding Shimla. From the results, a profile plot has been developed to define the variation in displacements between the individual pixel (Fig. 15.13). The profile signifies the minimum and maximum deformation occurred between the pixel rate of 1500 to 2000 and 3500 to 4000. Fig. 15.14a, b shows the region of extreme deformation in the study area. While Fig. 15.14c represents the area with lowest amount of deformation near the built-up lands.

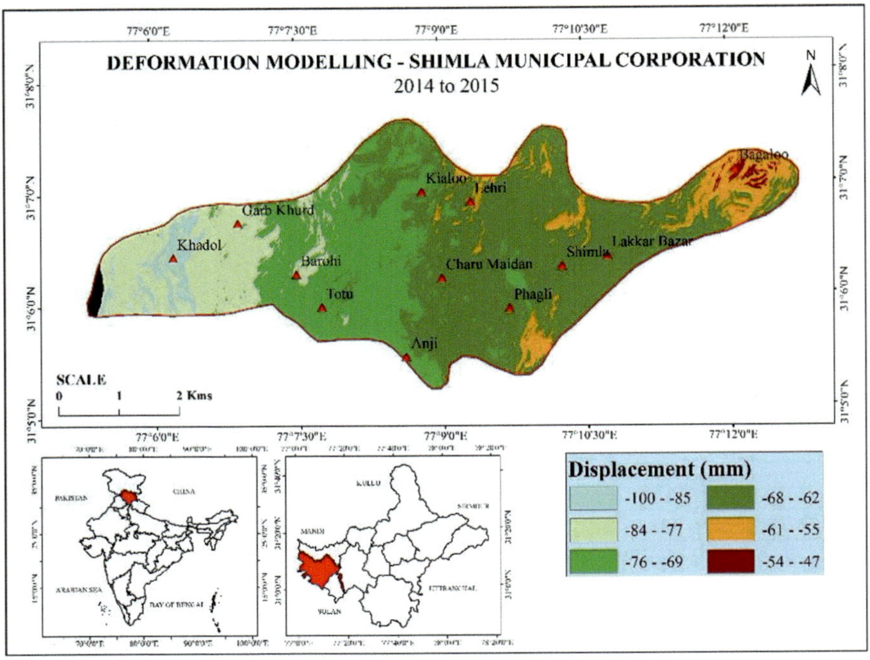

Fig. 15.9 Deformation modelling along Shimla Municipal Corporation for the year (2014–2015)

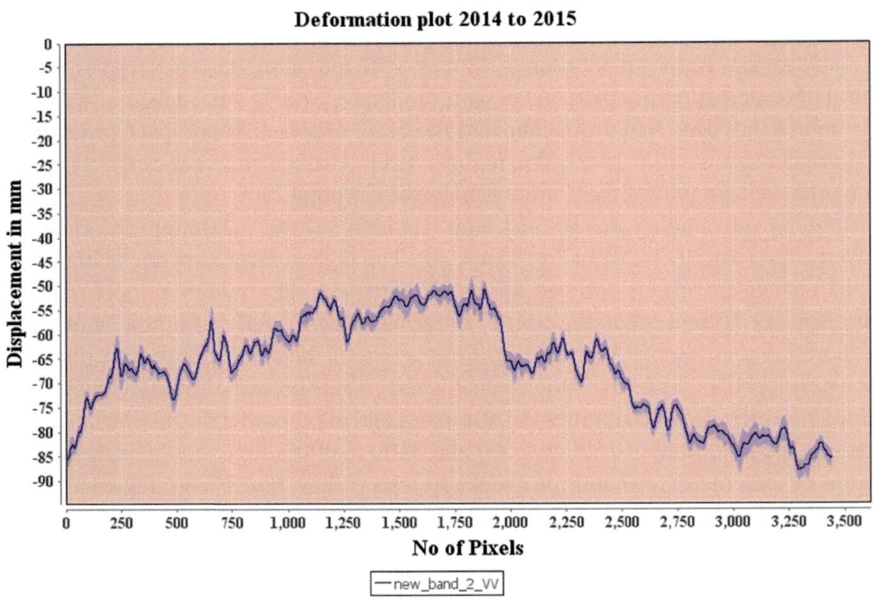

Fig. 15.10 Overall deformation rate along the study area (2014–2015)

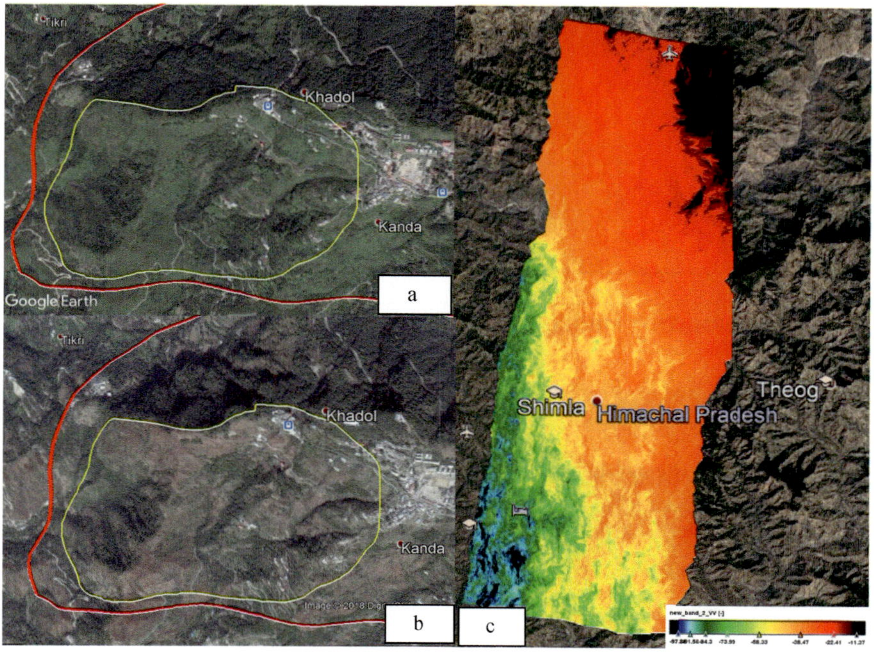

Fig. 15.11 **a** and **b** Represent the area in and around Khadol were the highest amount of deformation has occurred, **c** represents the overall deformation rate in Shimla town and other surrounding areas overlaid in Google Earth imagery

Ground Uplift in the Study Area

Between the year 2015 and 2016, there was upliftment in the study area along the LOS of the sensor. Most of the Municipal Corporation falls under 45–52 mm/year. Between 53 and 58 mm/year, the highest amount of upliftment was found scattered all around the study area. The least amount of change ranging from 39 to 44 mm/year was observed in small patches surrounding the region of Bagloo (Fig. 15.12a). A profile plot (Fig. 15.12b) has been generated for the study area, which reveals that the lowest amount of subsidence has occurred between the 1000 and 1500-pixel frequency and highest amount of deformation occurred between 4500 and 5000.

Some of the highly deformed areas along the Barohi and surrounding region have been given in figure (Fig. 15.17a–c). The temporal variation between these images show that there is a reduced forest cover along the study area within the analysed time frame, while Fig. 15.15 represents the deformation modelling and Fig. 15.16 represents the deformation rate along the study area from 2015 to 2016 along the entire IW—2 mode of the sensor. The area indicated by the arrow mark indicates a smooth terrain, which is due to the loss of image information in that particular region.

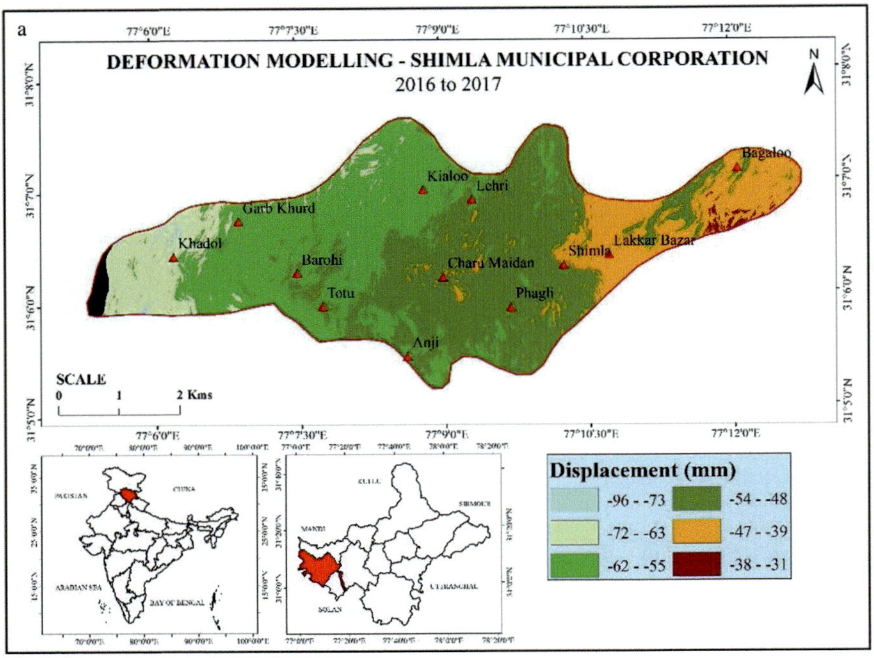

Fig. 15.12 **a** Deformation modelling along Shimla Municipal Corporation for the year (2016–2017), **b** and **c** represent the highly deformed region (Khadol) in the study area. This process is attributed to the ever-changing terrain in that area due to natural process such as climatic factors, **d** and **e** represent the less deformed area, the region along the built-up lands in the study area

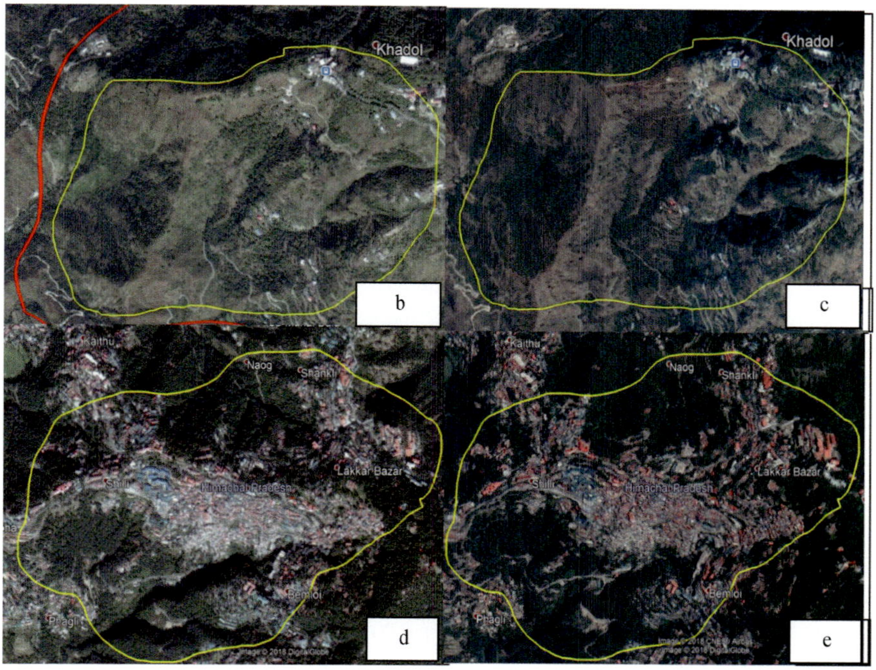

Fig. 15.12 (continued)

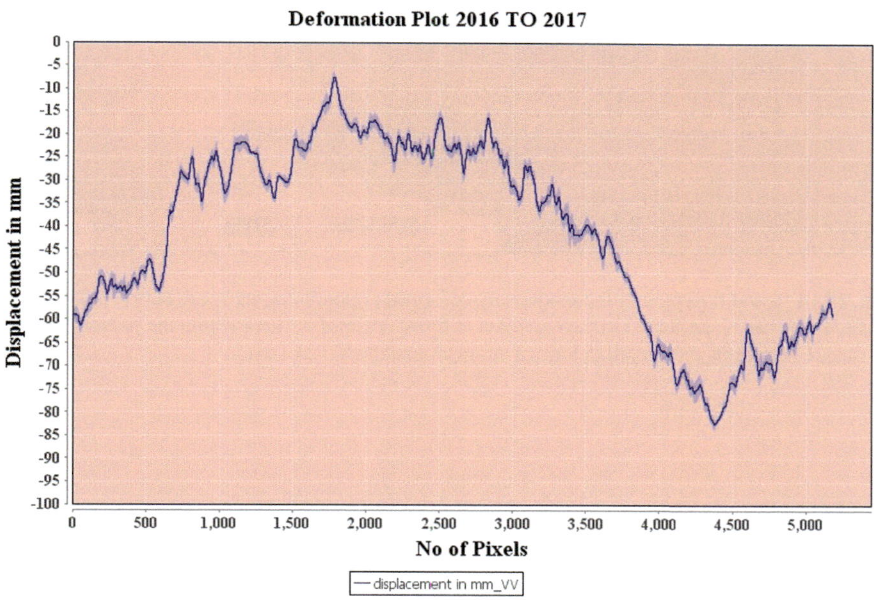

Fig. 15.13 Represent the overall deformation rate along the study area

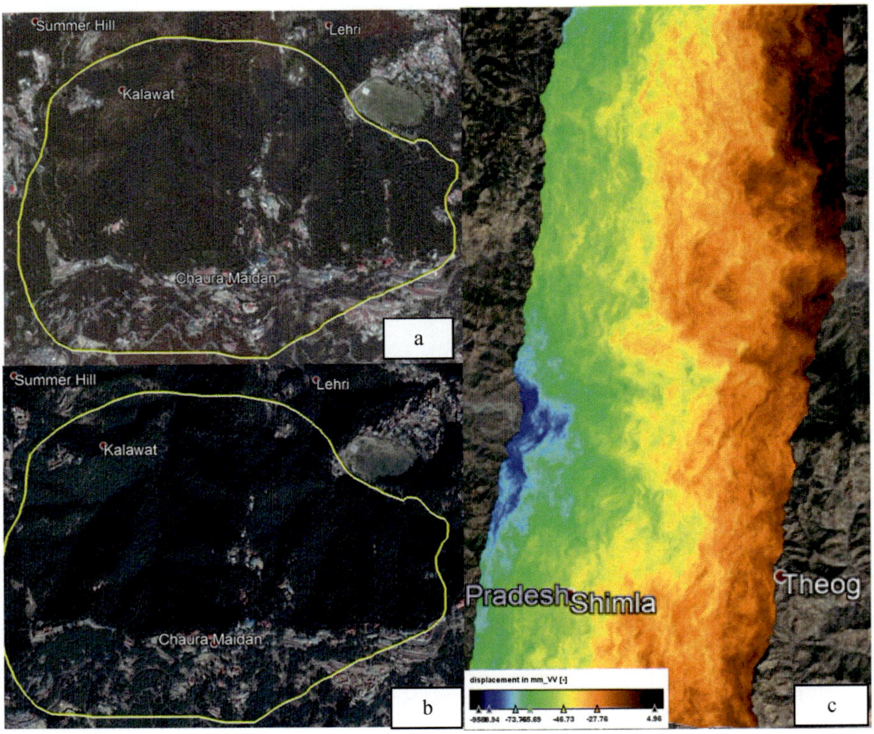

Fig. 15.14 **a** and **b** Represent the moderately deformed region in the study are between 2016 and 2017. These can be attributed to the reduction in forest cover of the area within the particular span of time, **c** represents the overall deformed are in around the Shimla region

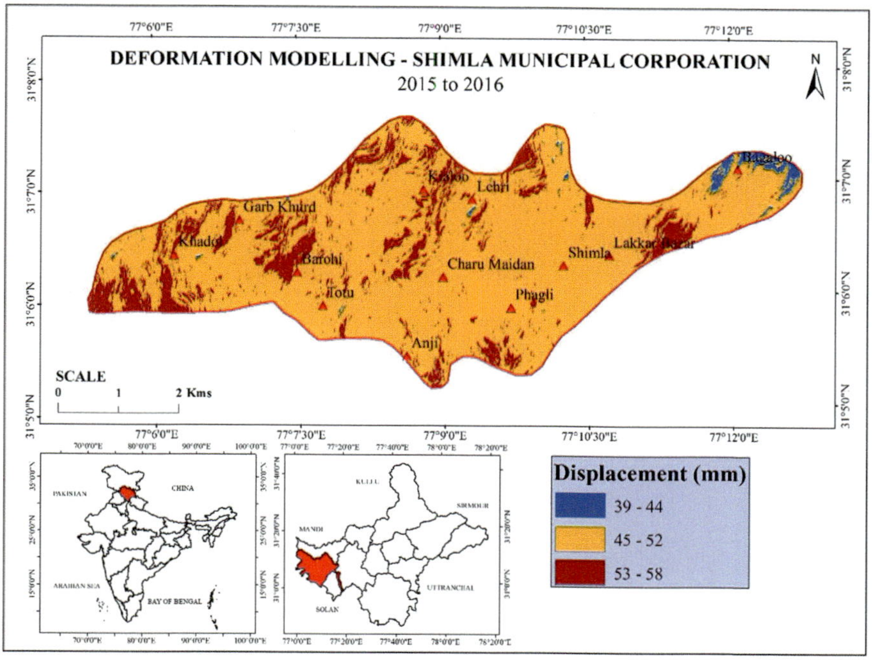

Fig. 15.15 Deformation modelling along Shimla Municipal Corporation for the year (2015–2016)

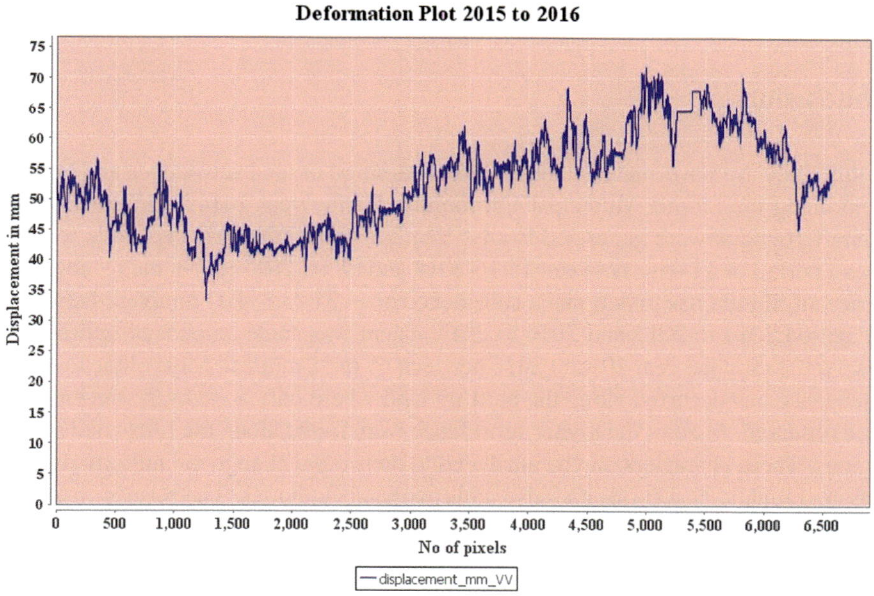

Fig. 15.16 Overall deformation rate along the study area (2015 to 2016)

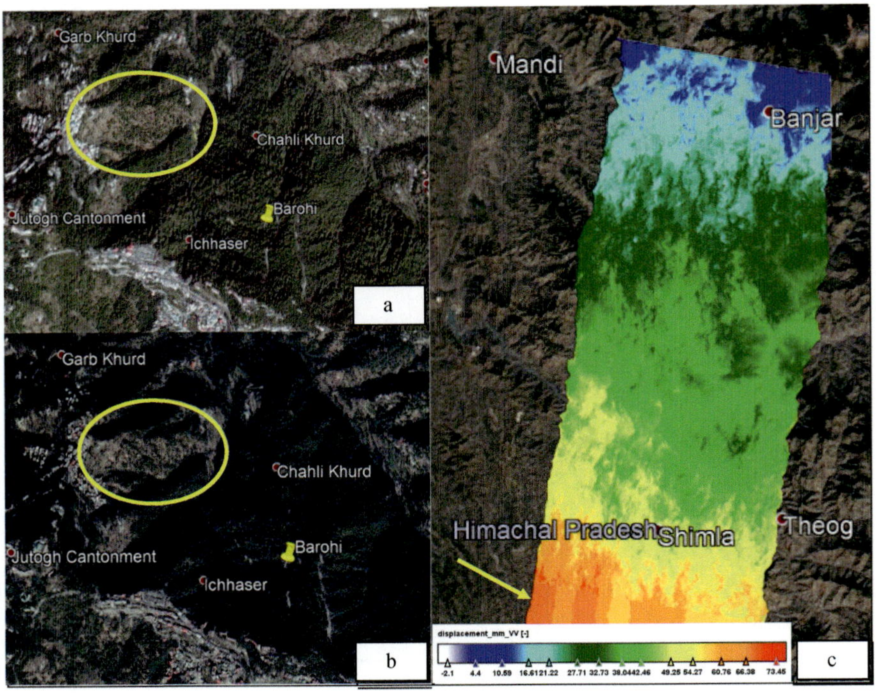

Fig. 15.17 **a** and **b** Represent the areas with higher amount of deformation between the year 2015 and 2016, **b** shows the extent of deformation along the Shimla region and surrounding areas

Conclusion

In this study, we have analysed the temporal variation of land subsidence and upliftment along the Shimla Municipal Corporation for the year 2014–2017. A total of 3 interferograms were generated from 4 Sentinel—1A TOPSAR IW mode with a revisit period of 12 days between 15/10/2014 and 08/05/2015. From the results, it's clear that Shimla has undergone a subsidence of − 31 to − 100 mm/year between the period 2014 to 2015 and 2016 to 2017. There was some amount of upliftment observed during the year 2015 to 2016 between 45 to 52 mm/ear. Most of the least of subsidence has occurred along the built-up land where only a slight movement has been observed. Most of the higher subsidence were found along the forest and slope regions. These characters maybe attributed to their rapid change in landform due to varying climatic conditions and also anthropogenic activities. The results obtained in the study will help in modelling the deformation pattern and rate which will prove to be a valuable input during the constructional and reformational activities. This data will also play an important role in identifying areas that are more prone to disasters such as landslides and floods and take necessary precautionary measures and practices.

Acknowledgements The research work done is a part of NRDMS-DST-funded research project. We would like to express our sincerest gratitude to NRDMS-DST, GOI, New Delhi, India, for funding this research project.

We would like to thank CSIR, New Delhi, GOI, for sponsoring the SRF—Direct, scholarship for pursing research work.

References

Crosetto, M., Oriol, M., María, C., & Bruno, C. (2011). Spaceborne differential SAR interferometry: Data analysis tools for deformation measurement. *Remote Sensing, 3*(2), 305–318. https://doi.org/10.3390/rs3020305

Feigl, K. L., & Massonnet, D. (1998). Radar interferometry and its application to changes in the earth's surface. *Reviews of Geophysics, 36*(4), 441–500. https://doi.org/10.1029/97RG03139

Ferretti, A., Fumagalli, A., Novali, F., Prati, C., Rocca, F., & Rucci, A. (2011). A new algorithm for processing interferometric data-stacks: SqueeSAR. *IEEE Transactions on Geoscience and Remote Sensing, 49*(9), 3460–3470. https://doi.org/10.1109/TGRS.2011.2124465

Fornaro, G., Verde, S., Reale, D., & Pauciullo, A. (2015). CAESAR: An approach based on covariance matrix decomposition to improve multibaseline-multitemporal interferometric SAR processing. *IEEE Transactions on Geoscience and Remote Sensing, 53*(4), 2050–65. https://doi.org/10.1109/TGRS.2014.2352853

Gama, F. F., Cantone, A., Mura, J. C., Pasquali, P., Paradella, W.R., dos Santos, A.R., & Silva, G.G. (2017). Monitoring subsidence of open pit iron mines at Carajás province based on SBAS interferometric technique using TerraSAR- X Data. *Remote Sensing Applications: Society and Environment,* 8(September), 199–211. https://doi.org/10.1016/j.rsase.2017.09.001

Geudtner, D., Torres, R., Snoeij, P., Davidson, M., & Rommen, B. (2014). Sentinel-1 system capabilities and applications. *International Geoscience and Remote Sensing Symposium (IGARSS),* 1457–1460. https://doi.org/10.1109/IGARSS.2014.6946711

Goldstein, R. M., & Werner, C. L. (1998). Radar interferogram filtering for geophysical applications. *Geophysical Research Letters, 25*(21), 4035–4038.

Grandin, R. (2015). Interferometric processing of SLC Sentinel-1 tops data. *Proceedings of [FP1] FRINGE'15: Advances in the Science and Applications of SAR Interferometry and Sentinel-1 InSAR Workshop,* 2220–34. https://doi.org/10.5270/Fringe2015.116

Lanari, R., Mora, O., Manunta, M., Mallorquí, J.J., Berardino, P., & Sansosti, E. (2004). A small-baseline approach for investigating deformations on full-resolution differential SAR interferograms. *IEEE Transactions on Geoscience and Remote Sensing, 42*(7), 1377–1386. https://doi.org/10.1109/TGRS.2004.828196

Lanari, R., Berardino, P., Bonano, M., Casu, F., De Luca, C., Elefante, S., Fusco, A., Manunta, M., Manzo1, M., Ojha1, C., Pepe, A., Sansosti, E., & Zinno, I. (2015). Sentinel-1 results: SBAS-DInSAR processing chain developments and land subsidence analysis. *2015 IEEE International Geoscience and Remote Sensing Symposium (IGARSS),* 2836– 2839. https://doi.org/10.1109/IGARSS.2015.7326405

Liu, X., Wang, Y., & Yan, S. (2018). Interferometric SAR time series analysis for ground subsidence of the abandoned mining area in North Peixian using Sentinel-1A TOPS data. *Journal of the Indian Society of Remote Sensing, 46*(3), 451–461. https://doi.org/10.1007/s12524-017-0708-4

Liu, Y., Mingsheng, L., Xuguo, S., Lu, Z., & Cory, C. (2016). Potential loess landslide deformation monitoring using L-Band SAR interferometry. *Geo-Spatial Information Science, 19*(4), 273–277. https://doi.org/10.1080/10095020.2016.1258202

Lv, X., Yazici, B., Zeghal, M., Bennett, V., & Abdoun, T. (2014). Joint-scatterer processing for time-series InSAR. *IEEE Transactions on Geoscience and Remote Sensing*, 52(11), 7205–21. https://doi.org/10.1109/TGRS.2014.2309346

Poursanidis, D., & Nektarios, C. (2017). Remote sensing, natural hazards and the contribution of ESA sentinels missions. *Remote Sensing Applications: Society and Environment, 6*, 25–38. https://doi.org/10.1016/j.rsase.2017.02.001

Raspini, F., Bianchini, S., Ciampalini, A., Del Soldato, M., Solari, L., FabrizioNovali, S. D., Conte, AlessioRucci, Ferretti, A., & Casagli, N. (2018). Continuous, semi-automatic monitoring of ground deformation using Sentinel-1 Satellites. *Scientific Reports, 8*(1), 1–11. https://doi.org/10.1038/s41598-018-25369-w

Simons, M., & Rosen, P. A. (2007). Interferometric synthetic aperture radar geodesy. *Treatise on Geophysics, 3*, 391–446. https://doi.org/10.1016/B978-044452748-6.00059-6

Song, Z., & Jiang, H. (2016). A high accuracy coregistration method for TOPSAR interferometry. *2015 IEEE International on Signal and Image Processing*, 133–36. https://doi.org/10.1109/SIPROCESS.2016.7888238

Torres, R., Snoeij, P., Geudtner, D., Bibby, D., Davidson, M., Attema, E., & Potin, P., et al. (2012). GMES Sentinel-1 mission. *Remote Sensing of Environment, 120*, 9–24. https://doi.org/10.1016/j.rse.2011.05.028

Zhang, K., & Zhang, S. (2007). 3D Visualization of landslide deformation. *Geospatial Information Science, 10*(1), 67–70. https://doi.org/10.1007/s11806-006-0133-2

Part V
Disaster and Gender

Chapter 16
Gendered Spaces, Climate Change and Resilience in a Squatter Slum of Global South

Mamta Sharma, Jag Mohan, and Anjana Mathur Jagmohan

Abstract Climate change is affecting the cities of almost all the countries of Global South. Slums, in the already over-spilling cities, are at the receiving end of the worst impacts of disasters resulting from the climate change. Women and children are the most affected in times of disasters and also the most neglected. The paper studies how women in the slum Bela Gaon adapt to disasters—in this case floods and rising waters of River Yamuna. The study delves into the relationship between education, financial status of women and the condition of houses, strategies of facing disasters and division of spaces within the community. Findings point out to a deep relationship between women empowered by financial independence and education to spaces divided by gender, house structure and losses incurred due to climate change and disasters.

Keywords Gendered spaces · Women empowerment · Climate change

Introduction

With the global climate continuously warming, constantly increasing temperatures are causing climate and unprecedented disasters around the globe. Most affected by these climate-induced disasters are the poor in the cities of Global South. Flash floods, droughts and high temperatures first affect the poor living in urban squatter slums. Cheap building material, uncemented floors poor weak roofs and walls make every spell of heavy rains, river level a life-threatening situation for the dwellers. Often it is the women and children of the household who are the worst affected,

M. Sharma
Aditi Mahavidyalaya, University of Delhi, New Delhi, India

J. Mohan
Department of Geography, Aditi Mahavidyalaya, University of Delhi, New Delhi, India

A. M. Jagmohan (✉)
Department of Geography, Dyal Singh College, University of Delhi, New Delhi, India
e-mail: anjanamathur@dsc.du.ac.in

V. R. Sharma and Chandrakanta (eds.), *Making India Disaster Resilient*,
https://doi.org/10.1007/978-3-031-50113-5_16

due to malnutrition and weak immunity. Women, who are constrained both by the space and vulnerable economic situation and taking care of the household goods and the kids, find themselves in the worst situation at the times of the hazards and also because of the roles they play in growing food and providing for the energy and water needs of their families and because they comprise a large number of the poor communities that depend on natural resources for their livelihood.

The Terminology

Gendered spaces are areas in which a particular gender of people and particular types of gender expression are considered welcome or appropriate, and other types are unwelcome or inappropriate. As a critical concept in the geography of gender, gendering of spaces is an important means by which social systems maintain the organization of gender. They reinforce particular ways of being a man or being a woman and can maintain the relationships between men and women.

In cities of the Global South, the common division of spaces is **private-parochial-public** (Jabeen, 2019). The traditional view in most **Indian cities** is that private space belongs to women, while public space belongs to men, while parochial space is utilized by women for common activities.

The gendering of spaces helps to reinforce a culture's gender norms (Tyler & Cohen, 2010). People who are in the "wrong" spaces are subject to punishments of various sorts, from teasing to violence. Which spaces receive which gender sends messages about the proper character and duties of a person of a given gender (e.g., the wilderness is a masculine space, so men should be strong and tough). And gendered spaces can enable the dominant gender to retain control of valuable resources, as when important informal mentoring or making of business deals occurs at an all-male country club.

Climate change refers to the rapidly changing climate mainly due to human activities. Increasing temperatures, sea level rises, changing patterns of precipitation and more frequent and severe extreme events are expected to have largely adverse effects on key determinants of human health, including clean air and water and adequate shelter.

The effects of climate on human society and human's ability to mitigate and adapt to them are mediated by social factors, including gender.

Resilience refers to the capacity both to cope and to adapt. While coping focuses more on the disaster at the moment, adapting means a change in thinking and actions post-disaster.

Squatter settlement is any collection of buildings where the people have no legal rights to the land they are built upon. The people are living there illegally and do not own the land. They provide housing for many of the world's poorest people and offer basic shelter.

Global South is an emerging term which refers to countries seen as low and middle income in Asia, Africa, Latin America and the Caribbean by the World Bank. These nations are often described as newly industrialized or in the process of industrializing. Global South does not necessarily refer to geographical south (Wikipedia).

Conceptual Framework

The Global South cities are at a junction where industrialization and tertiary sectors are experiencing a boom and agricultural sector is facing a negative growth rate. The other developmental issues include the gigantic problem of population explosion, rural–urban migration and the resultant urban population explosion as well as almost regular occurrence of natural disasters in which thousands of lives are lost. Disasters resulting from climate change are mostly affecting the poor and the deprived sections of the cities. Even within the squatter slums where the poor live, there is a differential impact of disasters on monetary lines and on gender lines. Men, women, poor and very poor cope differently, and have different strategies of adaptation and resilience from climatic extreme and disasters. It is this aspect of climate change and disaster management which forms the base of this study.

Aim and Objectives of Study

- To understand how the division of spaces occurs between genders in a squatter slum of a Global South city;
- To understand the relationship between gendered spaces, resilience to climate change and adaptation.

Study Area

The chosen squatter slum is located in the Yamuna bed of Delhi and is quite spread out in more than three clusters near Rajghat Thermal Power Station. The area is very much prone to floods every monsoon, and every time there are torrential rains and any other time when water level of Yamuna rises due to release of water from Hathni Kund Barrage in Haryana, upstream. The slum has houses in three clusters with kuchha roads coming down to the river bed to all the three clusters. None of the houses have any access to municipal water supply and sewer facility.

The area has more 100 families, and many of them have been living in the slum for many decades. The economic condition of all the slum dwellers varies and so does their educational status. The oldest residents have been here for more than 60 years. Some of them are the fourth generation of migrants from Hapur and Meerut (Uttar Pradesh) and have slowly prospered with time. The elders in the slum narrate that there were quite many families living here till about 25 years back when many of them were given 25 m^2 residential plots in Bawana (outskirts of Delhi) for house construction. The rest of the remaining families have stayed back in anticipation of plot allotment. Leaving the slum shall take away the chance of free allotment of residential plots, feel the present residents.

Bela Gaon, situated in the Yamuna bed off the Ring Road bypass in Central Delhi, being very close to the central Delhi, offers immense advantage to the slum dwellers in terms of accessibility for jobs and schools for children. This is reflected in a diverse profile in terms of jobs done by men and women. While the elderly ladies and men stay at home, many young ladies work in the samadhis of leaders nearby as attendants, many are house attendants in Daryaganj, and many work in the shops of Old Delhi, Sadar Bazar and Darya Ganj. There were many drivers of commercial cargo vehicles who also owned the vehicles. There were many dairy farm owners who had a couple of caretakers for their milk providing cattle. At the same time many men who were totally illiterate worked as daily wagers in Sadar Bazar, pulling rickshaws and carrying freight on their backs.

Methodology

In order to attain the aims and objectives of the study the variables which are to be studied included home ownership, household headship, location of the house, construction material used, economic activities engaged by the people of the slum and also individual and household income.

The main technique used for the study is observation and participatory method.

By participatory methodology we mean a *research style*, "an orientation to inquiry" (Reason & Bradbury, 2001, p. 1). Participatory research or PR presents people as researchers in pursuit of answers to questions encountered in daily life.

The study also used many other techniques. These include semi-structured questionnaire, oral testimonies and story collection as a foundation for collective analysis and photograph stories.

Separate interviews were conducted to obtain useful information about the rights and responsibilities specific to women, spaces dominated by women and men and also by both.

The slum was divided on the basis of income of the families and further analyzed for house type, construction material used and also the socio-cultural background of dwellers. Typical house types were sketched and photographed.

Gendering of Spaces

The spaces that the higher income group are used to are the private and parochial spaces. Bedroom is a private space, corridors and verandahs are the parochial spaces, and roads are public spaces. While private spaces are essential for sleeping, parochial spaces are meeting places for known people, and public spaces are for the unknown people with whom there is little or no interaction.

In squatters where there is already a space crunch, there is an unseen battle for spaces. It was observed that higher the income of the household, more was the space occupied and more was the input of the woman of the household in house building material and house plan, etc.

Also women dominate the private and sometimes the parochial space, while men dominate the public spaces.

Women use the parochial spaces for meeting with the other women and also sometimes for community work or for self-employment.

Gender and Climate Risk

Gender is one of the primary drivers of differential climatic vulnerability. The ability to take anticipatory and reactive actions in response to a climate change disaster depends immensely on the status of woman in the society. A woman in a squatter slum who is solely dependent on the male member for the earnings and has a least say in the house plan and house building material. An earning woman is also able to gather support from other women of the slum cluster in preparing for any eventuality arising out of disaster. It was found that an economically empowered woman, however meager her earnings might be, gives her an empowered vision to judiciously use and multiply the available resources and use them efficiently during the times of crisis.

Climate-Induced Risks in the Study Area

The Yamuna bed area is located in such topography that whenever there is a slight change in the water level, the place gets flooded. Even half an hour of rain submerges many shanties in the village. Therefore, torrential rains, occurring very often due to climate change, have been creating havoc in Bela Gaon. Along with it, the area has started to face immense heat waves as well as very high humidity levels which occur due to its close proximity to the river. The temperatures inside the shanties rise to very high levels making it very difficult to bear.

Presence of a lot of standing water around the village from the river results in mosquitoes along with many other insects which result in the prevalence of malaria and dengue fevers which many times prove fatal.

Absence of potable water makes the people drink water pumped up through tube wells, forcing the people to drink the local hand pump water, which is affected by the seeping chemical effluents from industries upstream and results in numerous waterborne diseases like cholera and diarrhea.

According to the residents, the extreme temperatures, flooding and drinking water scarcity are the main problems faced.

Literature Review

Gender has been one of the favorite topics of researchers in the field of Sociology. However, it is fairly recently that geographers and disaster management specialists have discovered the importance of gender in geographical spaces, resource management and even during disasters. While the spaces of the two genders have started to mix and overlap in the cosmopolitan culture of the higher income groups of the cities of Global South, it is the spaces in the unplanned settlements in these very cities which demands urgent study. The whole life of women in slums in these cities is spent negotiating the small 10 × 10 ft. shanties and three feet gullies surrounding them. A lot of sociologists and recently gender experts have started to intricately study these space divisions. We shall be needing all these definitions and criteria for identification of the space divisions in squatter slums of the Global South cities. Some of the studies dealing exclusively with gendered spaces are given below.

Nakahal (2015) in this paper examines gendered urban spaces that are shaped under/by the capitalist patriarchal system. According to her, architectural standards recreate gender, racial and class hierarchies, just as local cultural productions reinforce specific notions of women-as-space. As a result, we are left with an unchallenged reproduction of gender binaries and a reinforcement of what women are "supposed" to be and do. In the end, this paper attempts to disrupt these binaries and hierarchies through relocating our bodies and rewriting gender in space.

Beebeejaun (2017) in her article "Gender, Urban Space and the Right to Everyday Life", suggests that a fuller recognition of the contested publics that coexist within the contemporary city and the gendered mediation of everyday experiences could enable planners and policymakers to undertake more inclusive forms of intervention in urban space.

Juran (2012), in his article "The Gendered Nature of Disasters: Women survivors in Post- Tsunami Tamil Nadu", argues that women confront human rights gaps during "normal" times and that such pre-existing inequalities are simply reified and magnified in times of disaster. These contentions are upheld by providing a theoretical review of gender and disaster, a survey of actual accounts of gender and disaster across space and by buttressing the literature with examples from post-tsunami Tamil Nadu. The aim of this article is to analyze salient gender-based issues in a specific post-disaster context and to add to the discourse on gender and disaster writ large.

Enarson (2002) in her article, "Through Women's Eyes: A Gendered Research Agenda for Disaster Social Science", suggests how analysis of the gendered terrain of disaster both develops disaster theory and fosters more equitable and effective disaster practice.

Denton (2010) in her article "Climate change vulnerability, impacts, and adaptation: Why does gender matter?" argues that if climate change policy is about ensuring a sustainable future by combining development and environment issues, it must take into account the interests of all stakeholders. The Global Environment Facility and the Clean Development Mechanism of the Kyoto Protocol can play a role in ensuring sustainable development, provided they are implemented in a way that does not disadvantage women and the poor.

Articles dealing with climate change disasters in urban areas include an article by Chatterjee, Monalisa (2010) "Slum dwellers response to floods in megacities of India". Ms. Chatterjee emphasizes that people living in informal settlements instead have to employ a combination of structural means and complex networks of assistance to recover from floods. Based on the results deduced from data collected with the help of household surveys in the slums of Mumbai, the study demonstrates the types of coping strategies used by slum dwellers and the changing characteristics of these mechanisms under the influence of global change processes in megacities.

Wilby and Keenan (2012) in their paper "Adapting To Flood Risk Under Climate Change" review steps being taken by actors at international, national, regional and community levels to adapt to flood risk from tidal, fluvial, surface and groundwater sources. They refer to existing inventories, national and sectoral adaptation plans, flood inquiries, building and planning codes, city plans, research literature and international policy reviews. They also distinguish between the *enabling* environment for adaptation and specific *implementing* measures to manage flood risk.

Brown et al. (2012) in their paper "From practice to theory: emerging lessons from Asia for building urban climate change resilience" aim to capture and analyze emerging experiences, lessons and tensions evident from several years of work underway through the Asian Cities Climate Change Resilience Network, a network of secondary cities in South and Southeast Asia that have engaged in a process to analyze vulnerabilities and plan and implement measures to address them. With the support of the Rockefeller Foundation and numerous partners, these cities have identified more than 59 specific resilience-building measures, of which 23 are being implemented. Through this work we see 10 critical urban climate change resilience action areas that cities must consider in order to strengthen their ability to anticipate, prepare for and respond to the types of sudden and slow onset impacts.

Reed et al. (2013) in their paper "Shared learning" for building urban climate resilience—experiences from Asian cities" consider how resilience thinking and, in particular, its emphasis on learning have been applied in 10 cities in Vietnam, India, Thailand and Indonesia. This article suggests that deliberate, strategic intervention by facilitators may contribute to more transformative change on behalf of equitable, socially just outcomes—and thus cautions against seeing urban climate vulnerability as a technical challenge, or shared learning as a "toolkit" for building resilience.

Findings

Gendering of Spaces in Bela Gaon

There is a clear difference between the prosperous and poor slum dwellers when building material, space occupied the dwelling, and amenities owned are concerned.

The prosperous families in the slum had more than one dwelling space according to their needs. The elderly widowed family members had one shed each to themselves which contained apart from a bed, storage space for valuable household items. The elderly women have more say in the household decisions along with the sons. Daughters-in-law had comparatively lesser say in the house.

Space Usage in a Prosperous Family Dwelling Set

See (Fig. 16.1).

The building material used for the dwelling is much better and stronger than other dwellings. The big structure visible in Fig. 16.2 is that of the elderly widowed mother. Its insides are in Fig. 16.1. The dwelling has a good bed with a ceiling fan on top. It doubles up as storage for other valuables owned by the family like the washing machine, a scooty, family jewelry, etc. (Fig. 16.3), which is stored in the metal trunks. The ceiling of the dwelling is strong and has many single logs. The floor is also better. The walls are made of wooden doors and planks procured from outside. There are also concrete pillars on the corners of the dwelling (Fig. 16.4).

The dwelling in Fig. 16.5 is that of the ėlderly widowed grandfather. It has a stable bed, a cooler and also serves as storage space for drinking water. The material of the dwelling is strong with metal sheets as walls and thick canvas layers as roof. Figures 16.6 and 16.7 show the third dwelling unit of the family. This does not have any storage but has a refrigerator, a television and a cooler for the son and his family. The room is comparatively less crowded and has a good height.

Cooking is done in rains in the son's bedroom (Fig. 16.8). Otherwise in sunny days, cooking is done on a cooking gas stove outside son's bedroom in open (Fig. 16.9). Son's dwelling also has the bathing space opposite to it. Although just a ramshackle shed with walls and no roof, they consider it a privilege to have a bathing space (Fig. 16.10). The dwelling has no toilet, and everyone defecates in the open in the agricultural fields around them.

Fig. 16.1 Inside view of elderly widow women room

Fig. 16.2 Outside view of elderly widow women room

Fig. 16.3 Storage area

Women and Domestic Space in a High-Income Household

Domestic space includes the space within the household meant for sleeping and eating. Generally the space is utilized only during the night times, and outside men do not come inside. The prosperous household surveyed had three dwelling units, but all three of them were used only at night for eating and sleeping. Even at these times, they were used more by women and lesser by men of the house.

The women of the household have more space in their dwelling for themselves, for their kids and also for storing, dressing, grocery and vegetables.

The elderly men avoid going into the dwelling units of young sons, grandsons and daughters-in law.

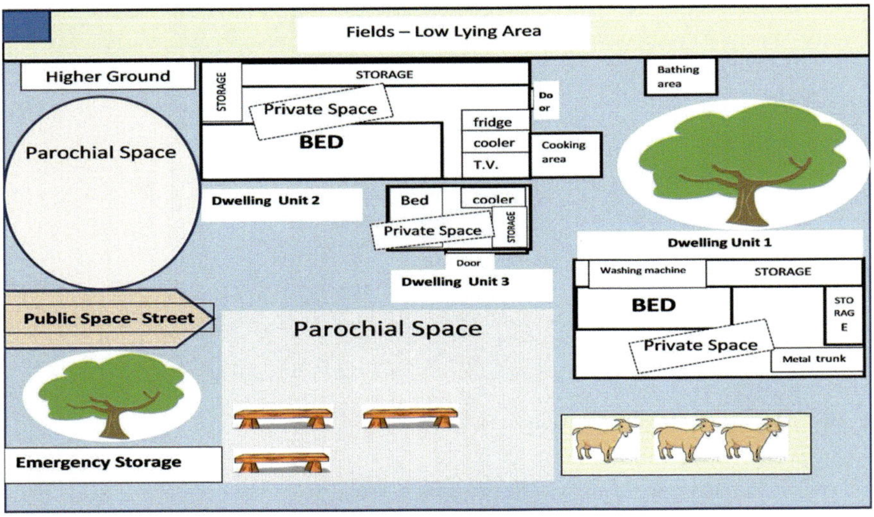

Fig. 16.4 Sketch of low-income household dwelling unit in Bela Gaon

Fig. 16.5 Dwelling of widowed grandfather

Fig. 16.6 Inside view:the third dwelling unit- Son's bedroom

Fig. 16.7 Inside view: third dwelling unit

Fig. 16.8 Son's bedroom: inside view

Fig. 16.9 Son's bedroom: outside view

Fig. 16.10 Bathing area

Women and Parochial Space in a Low-Income Household

The space outside the house is generally used by the women of the household to meet other women and also to work with a common activities like cow dung cakes drying, storing important items which cannot be taken when the area is flooded. In Bela Gaon, it is this parochial space that is the space used by the elderly as well as the young women of the household for all the activities—cutting vegetables, tying domestic animals (Fig. 16.12), smoking hookah, shredding animal green fodder and just sitting idle.

The parochial spaces have many charpoys (wood framed bed with plastic woven net) used by all the women for interacting with each other (Fig. 16.11). This is the most frequented place by both men and women (Fig. 16.12).

Fig. 16.11 Interaction point

Fig. 16.12 Domestic animal

Public Space

Streets and passages comprise of public spaces. These spaces are generally avoided by non-earning women (Fig. 16.13).

These spaces are the least frequented spaces by both men and women in the slum. These are used only for going to another house or for going out of the area.

Space Usage in a Low-Income Family Dwelling

Figure 16.14 shows a set of low-income family dwellings. Even in a slum, there are low-income families who earn less than the others. The main advantage that these shanties give is very little or no rent to be paid for them. These shanties are owned by the residents themselves who are not able to afford to buy better material for walls and roofs. The women in these dwellings do not work and have little say in

Fig. 16.13 Village passage

the household matters. Figure 16.15 shows a low-income shanty. There is a flimsy curtain which is tied up during the day and is pulled down during the night and serves as one of the walls. The other walls are also very weak and made of thatch tied together (Fig. 16.16).

The household has an open clay stove (Fig. 16.17) which is just at the boundary of the dwelling. Rains make cooking difficult and often the family does not eat during the rains. The kitchen material is stored inside (Fig. 16.18). The common space outside the dwellings is used by the family to tie the cows owned by the male member's employer, who himself lives in the city (Fig. 16.19).

The dwelling unit of the low-income household consists of just one unit (Fig. 16.20) as it is difficult to build another unit and more material required to build it. The space has a bed, storage space and cooking space all fitted together.

Fig. 16.14 Low-income family dwellings

Fig. 16.15 Low-income shanty

Fig. 16.16 Outer walls

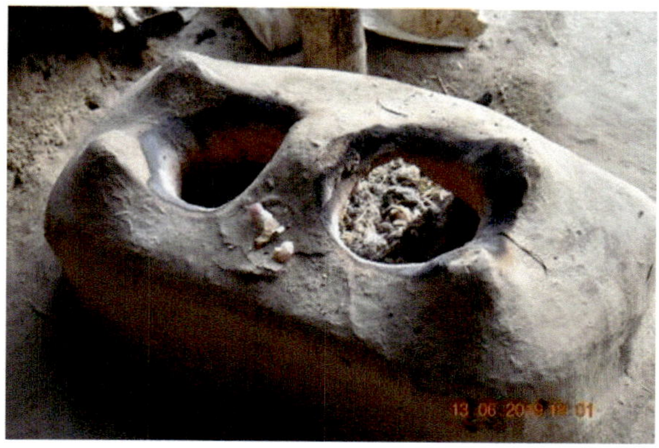

Fig. 16.17 Open clay stove

Fig. 16.18 Kitchen

Fig. 16.19 Common space

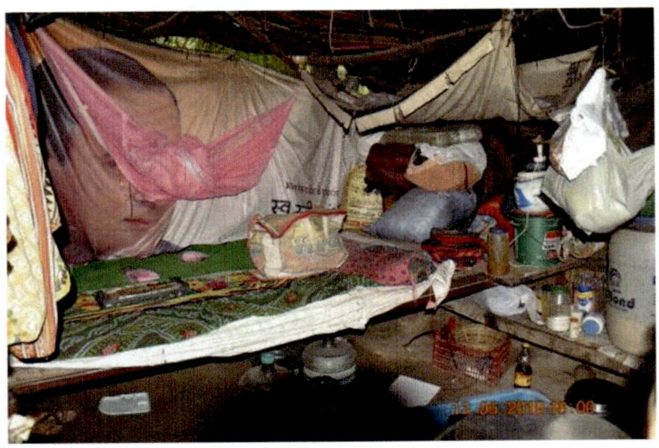

Fig. 16.20 Single unit of a low-income household

Domestic Space Use in Low-Income Households

Domestic space in a low-income household is the single dwelling unit owned by the household. Although the space inside the dwelling is almost the same it is the amenities owned which differ. A low-income household, built with flimsy material, is devoid of any refrigerator, cooler, cooking gas, scooty, washing machine, etc. All the belongings are stored in bags or hung from the ceiling. Women dominate the dwelling insides, while men go out to work. There were no elderly men and women found in the low-income households. Elderly had been moved to their original villages in Uttar Pradesh because of space crunch.

The women have very little space for themselves in the house. There is a cramping of space for storing grocery, vegetables and clothes. Cooking space is either absent within the house or is covered all the times except when cooking. Within the household the women only have their bed where they sit or sleep.

Parochial Space in Low-Income Households

The space outside the dwelling units was the space used by women of all the households to interact and talk to each other. They also used the space for any economic activity which could be done from home. In this case, one of the households was owned by a milk man who was also taking care of his employer's cows. He used the space for tying cows and their calves.

The public space was the narrow kuchha road behind the dwelling units which mostly remained deserted and was used only for commuting purposes (Fig. 16.4).

Analysis

Gendered Spaces and Climate Resilience

1. There is a marked difference in the dwelling units of a high-income household and a low-income household. Financial security allowed the women of the household to purchase amenities like cooking gas, scooty and truck using which were able to

    a. Procure better material for walls and roofs of the dwelling;
    b. Switch cooking place from outside to inside;
    c. Shift easily to higher places in times of flooding;
    d. Lock some material which does not spoil in water in a shed;
    e. Women were well prepared to face rising waters and also the diseases associated with it;
    f. Children of the household were comparatively better fed and had no signs of malnutrition or diseases. Thus they were less likely to be affected by waterborne and mosquito bite diseases;
    g. About 20 L jars of bottled water for domestic use were found in all the higher income households. The women found it better to buy bottled water by paying a cost rather than risk the lives of their families by drinking local hand pump water. Hand pump water was used only for bathing and cleaning and also for cattle.

2. The lower income household women, in contrast, have a lesser say in the household matters.
3. Lesser income household has

    - Very flimsy and weak building materialwhich has made the women and children vulnerable to extreme heat, malarial outbreak and cholera;
    - Lesser amenities exposed them to climate changes like floods;
    - Rains force the women not to cook;
    - Difficulty in shifting to higher ground in times of flooding;

- Loss of material goods during floods;
- The children of these households suffered from malnutrition and were most likely to suffer cholera, typhoid, diarrhea and other diseases associated with rising flood waters;
- Lower income families could not buy bottled water for drinking and cooking. As a result they are forced to consume the locally available water from tube wells and hand pumps.

Women and Adaptation to Climate Change

There exists a gender divide and an economic divide in adaptation to climate change and disasters in the Bela Gaon squatter slum.

The high-income women are more equipped to deal with floods, diseases and disasters. It is the availability of money, and a different perspective to look at things and situations due to exposure to the outer world gives the women more confidence to deal with situations.

The higher income group women also had more knowledge about the vaccinations available for children, facilities provided by the government for the under privileged, oral rehydration for children in case of cholera and continuous nausea. The richer women also bought bottled water from the market when the children fell ill.

In order to cope with the immense heat generated during the summers and also during heat waves, the women found it essential to have a room cooler or a desert cooler and also a ceiling fan to improve the temperatures inside.

The lower income group women have a comparatively lesser say in the household matters. Lesser money in hands make them dependent on the men of the household much more for pocket money, household expenses as well as decisions about purchases, building material and preparedness for disasters, disease and flooding. The women also had very little knowledge about vaccinations, cholera and malaria treatments and oral rehydration. Table fans or a single ceiling fan was found in all the lower income households to tackle the temperatures. Some houses lacked the fourth wall making its residents extremely vulnerable to the elements of weather and also to attack by mosquitoes and animals. Thus, they are more vulnerable to disasters and climate change.

Conclusions

Spaces—public and private, for centuries, have been symbols of gender divide. Be it in developed countries, where wilderness (public spaces) is considered a domain of men and homes (private) are where women have to prove themselves; or in the countries of Global South, where homes and especially bedrooms and kitchen are considered private spaces for women, while drawing rooms and corridors are parochial spaces, and roads and space between houses are considered public (Human Geography, e-textbook), (Gender, Climate Change and Health, WHO).

The same gender divide translates into women, power and decision-making in households. An earning women has more say in the decisions of the household and occupies comparatively more space in the house. The better-income households in the slum had the women not only occupying more but also better space—one having ceiling fan just above the sleeping areas.

Disaster when it comes does not move away from the houses of the richer persons but its impact is different for well of and poor. Heat wave, floods and waterborne diseases affect the houses and the people not equipped to deal with them. The comparatively well-off women and their families reach out more to government agencies for help during the times of crisis. Their awareness about various entitlements also raises their living conditions.

Financial independence makes the women of the household more aware about the developments around them, medicines and other ways of coping with rising waters and mosquito menace.

Gender is an important factor while planning for climate-induced disasters and developing resilience. Economic independence as well as education plays an important role in readiness for climate change extremes, floods and diseases.

In order to prepare a society ready to face any disaster, the first step has to be empowering women—through education and self-sustenance. It is the women who bring the society to life—literally and metaphorically. Empowering them through education and skills shall definitely result in an empowered society, which shall be ready to face any climate change. Equal space in the household and equal say in household matters are the beginnings any citizen of the world—whether from rural or from urban areas—must adopt for his own better future as well as that of the society.

References

Beebeejaun, Y. (2017). Gender, urban space and the right to everyday life. *Journal of Urban Affairs, 39*(3).

Brown, A., Dayal, A., & Del Rio, C. R. (2012). From practice to theory: Emerging lessons from Asia for building urban climate change resilience. *Environment and Urbanization, 24*(2), 531–556.

Chatterjee, M. (2010, April). Slum dwellers response to flooding events in the megacities of India. *Mitigation and Adaptation Strategies for Global Change, 15*(4), 337–353.

Denton, F. (2010). Climate change vulnerability, impacts and adaptation: Why does gender matter. *Gender and Development, 10*(2)

Enarson, E. (1998). Through women's eyes: A gendered research agenda for -disasters social science. *Disasters, 22*(2), 157–173.

Jabeen, H. (2019). Gendered space and climate resilience in informal settlements in Khulna City, Bangladesh. *Environment and Urbanization*, *31*(1), 115–138.

Juran, L. (2012). The gendered nature of disasters: women survivors in post Tsunami Tamil Nadu. *Indian Journal of Gender Studies, 19*(1), 1–29. Sage

Nelson, V., & Meadows, K. et.al. (2002). Uncertain predictions, invisible impacts and the need to mainstream gender in climate change adaptations. *Gender and Development, 10*(2).

Neumeier, B. (2013). Gender and urban space. *Gender Forum, An Internet Journal For Gender Studies*

Rains, E., & Krishna, A. et al. (2018). Urbanization and India's Slum Continuum: evidence on the range of policy needs and scope of mobility. *International Growth Centre, London School of Economics and Political Science, London.*

Reason, P., & Bradbury, H. (2001). *Handbook of action research-participative enquiry and practice.* Sage.

Reed, S. A. et al. (2013). Shared learning" for building urban climate resilience—experiences from Asian cities. *Environment and Urbanization, 25*(2), 393–412.

Sultana, F. (2014). Gendering climate change: Geographical insights. *The Professional Geographer*, *66*(3).

Tyler, M., & Cohen, L. (2010). Spaces that matter: Gender perormativity and organizational space. *Organization Studies, 31*(2), 175–198.

Wilby, R. L. (2012). Adapting to flood risk under climate change. *Progress in Physical Geography: Earth and Environment, 36*(3), 348–378.

Part VI
Human Aspects: Impact, Vulnerability and Governance

Chapter 17
Community Participation in Disaster Risk Reduction: A Case Study of Chamoli District, Uttarakhand

Suman Das and Ashis Kumar Saha

Abstract The Himalayan terrain is highly fragile and presently faced with the dilemma of maintaining a balance between development and ecology and environment conservation with changing climate resulted into increasing number of disasters. In the recent past, cloud burst and flash floods events are on rise in case of 2013 Kedarnath Tragedy, which resulted into huge loss of life and property. Hence, there is a need for a systematic linkage between disaster risk reduction and climate change adaptation to advance sustainable development. Therefore, if the affected communities are involved in the process of disaster risk assessment and reduction, the impact of disaster can be minimized to a great extent, and therefore developing a suitable community-based disaster management technique is required. This paper presents a preliminary finding on community engagement in disaster risk assessment, management and strengthen the capacity of communities to cope up with the disaster involving the concept of community-based disaster management (CBDM) by using local knowledge at Kandey and surrounding villages in Chamoli district of Uttarakhand. Emphasis was placed on collection of disaster experience and skills development of locals in the identifying and characterizing of various hazards in villages with special emphasis on landslide and preparedness for coping with disaster. The methodology has basically proposed in two stage: assessing of community vulnerabilities, capacity of multiple hazards in selected settlement by community engagement and participation characterized by training and skills development for multi-hazards risk assessment and disaster risk mitigation. Participatory rural appraisal (PRA) and participatory mapping have been effectively used to identify risk elements and coping capacity of the community in case of any future event. Finally, a disaster management plan has been prepared through community participation in the form of simplified maps showing safe shelters, escape routes, medical facilities, etc. The

S. Das (✉)
Department of Geography, Kirori Mal College, University of Delhi, New Delhi, India
e-mail: sumandas@kmc.du.ac.in

A. K. Saha
Department of Geography, Delhi School of Economics, University of Delhi, New Delhi, India

V. R. Sharma and Chandrakanta (eds.), *Making India Disaster Resilient*,
https://doi.org/10.1007/978-3-031-50113-5_17

study provides the example of the utility of community-based disaster management approach for solution toward the effective mitigation of landslide and associated hazards especially in developing countries.

Keywords Community-based disaster management (CBDM) · Community participation · Participatory rural appraisal (PRA) tolls · Disaster risk analysis · And Participatory mapping

Introduction

Himalayan terrain is highly susceptible and vulnerable to natural disasters due to its unique geo-climatic conditions. Earthquakes, cloud burst, flash floods and landslides have been a recurrent phenomenon. In the past, cloud burst and flash flood events are on rise resulted into immense loss of life and property, which may be attributed to extensive anthropogenic activities in the mountain region and climate change in some extent. Despite the developing scientific and technological advances, it has been tough to reduce the impact of natural hazards as experienced in case of 2013 Kedarnath Tragedy. As a result, there is a need for a systematic linkage between disaster risk reduction and climate change adaptation to advance sustainable development. Therefore, if the affected communities are involved in the process of disaster risk assessment and reduction, the impact of disaster can be minimized to a greater extent. So, developing a suitable community-based disaster management (CBDM) technique is required (Chen et al., 2006). Participatory mapping is a "bottom-up approach" which allows the people to create maps for all, in contrast to the traditional "top-down approach", that is relying on those with the resources and power to create maps which will benefit the people either directly or indirectly. Mapping of disaster is useful for policymakers, planners and residents for the collection of knowledge, administration of municipality services, establishment of boundaries and empowerment of those involved with the land. Local knowledge of a spaces can be used to complement existing formal data sets, to inform planners and policymakers and to empower residents to envision improvements of their spaces. It is also important to compile and understand local knowledge in order to understand complex communities (Warner, 2015).

A CBDM approach aims to lessen their socially constructed vulnerability by involving local communities as active participants in a disaster program (Anderson, 2003). There is also a broadening consensus of CBDM, e.g., it is a cost-effective method to train and educate communities about risks they may face, provide them access to resources and knowledge and to develop community-based preparedness and mitigation programs (Prakash, 2013b). This approach has arisen as a complement to structural mitigation and even certain types of non-structural mitigation programs like building codes, land use, development regulations, etc. (Anderson & Malcolm., 2013). It is perceived that an aware, informed and prepared community is better ability to cope up with disaster and help to minimize the impact of disaster. Community-based risk assessments are valuable measures to support the risk evaluation and mitigation process and to find appropriate risk reduction options such as land use planning, early warning systems, preparedness, awareness-building activities and suitable adaptation strategies (Glade & Crozier, 2005; Kunzler et al. 2012).

Study Area

The study area includes Kandey, Bamyala, Dogri, Nelgadora, Anandpur and Tangsha villages of Dasholi Tehsil in Chamoli district (Fig. 17.1). The study area locates in the northeastern part of the district respectively. The region covers of about 459 hectares in Garhwal Himalayas. The study area is extent from 25° 05.6″ N to 30° 25′ 16.7″ N and from 79° 17′ 32.8″ E to 79° 17′ 31.6″ E longitude. The average height of the villages in the study area ranges between 1500 mt and 1700 mt above mean sea level.

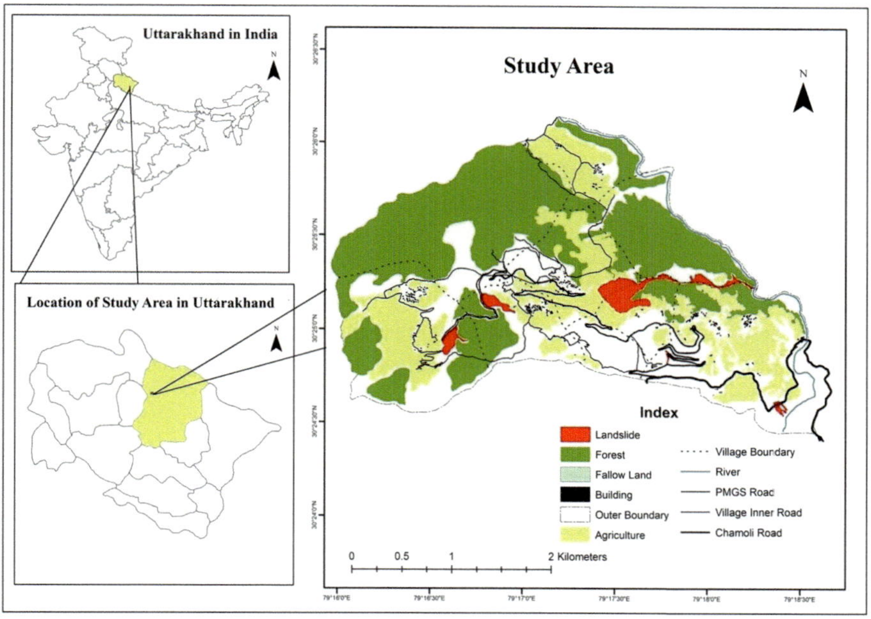

Fig. 17.1 Study area

Data and Methodology

Disaster risk reduction (DRR) is a set of exercise carried out to minimize the vulnerabilities and disaster risks in a community and avoid (prevention) or to limit (preparedness and mitigation) the hostile impact of any hazards. Although numerous scientific and technological methodologies exist for hazard identification, risk assessment, monitoring and control and mitigate; yet the community is hardly involved and benefited by their application (Andreson et al., 2011; Das, 2015). Therefore, an endeavor has been made to involve and use the local community's experience and knowledge in handle with the issue of disaster risk assessment and management while applying scientific fundamentals of disaster management in a broader sense. The database used for the study is solely based on primary survey conducted among the village community in the study area using participatory rural appraisal tools, viz. community-based participatory mapping, focus group discussion, etc. The field survey was conducted during February 24, 2014, to March 12, 2014.

This study is based on a combination of anthropological research methods, including participatory observation and structured, as well as semi-structured interviews and geographical methods of visualization and participatory mapping. All data generation was conducted in the context of long-term fieldwork. The procedure for community-based disaster risk management has following steps. At the initiation of this process, some significant actions have been taken to initiate the work in a systematic way. It includes formation of task forces for supporting the program in disaster assessment and mitigation at local level. After the formation of the group a brief orientation program was organized. The main objectives of the program have been explained to the community along with procedure implement and brief idea of the expected results. It is expected that if the residents of local communities clearly understood the importance of the study and what may they could gain from the involvement, it would be much easier to recruit potential participants. After the orientation program the process of community-based disaster risk management has been conducted in two stages as discussed below.

Phase I: Hazard, Vulnerability and Capacity Assessment

It emphasized on gather, compilation and analysis of information and mapping process for the purpose of assessing the hazards, vulnerabilities and capacities within a locality (Das, 2015; Patterson et al., 2010). Information about the past disaster events in a locality could be collected from disaster experiences gathered by the local people or by their ancestors. It is designed to reveal disaster history of the local community and acknowledge the descriptions of an emergency situations by reviewing of past disaster events and personal experiences (Das et al., 2017; Holcombe et al., 2012). Through exchanging information, participants could have an idea about the natural hazards in their community like frequency of occurrence of various hazards, areas likely to be most affected, magnitude and intensity of the hazard, spatial extent, duration, seasonal pattern, existing warning system, etc. These past disasters experiences could be further used to help participants in identifying the hazards at local level which threaten their community from time to time (Das et al., 2017).

Phase II: Preparedness and Management Plan

As most hazards cannot be controlled, steps should be taken in advance to lower the impact of a hazard or anticipated the stress at reducing the vulnerability of the exposed communities (Andreson & Holcombe, 2013). Prevention and preparedness target on increasing the resilience of a community and protect the assets and livelihoods from a prevailing hazard. Appropriate actions need to be integrated into the plan. These measures are among the most necessary elements of disaster risk reduction (DRR) as prevention to cut down the losses experienced and reduce the costs of recovery (Das et al., 2017; Prakash, 2013a). Disaster preparedness plan is made based on involving local community and its local resources so that community can cope up during the time of disaster hit or unless the external resources came into existence and minimize from the disaster loss (Mori et al., 2005). The aims of disaster preparedness plan are to

- Ensure that appropriate systems and logistic are in place to provide quick and effective response during and post-disaster.
- Prepare the local community to cope up the disaster situation in the first 24 h or till then outside help has not reached.
- Set up an emergency resources and operations center with the facilities and functions such as available emergency resources as well as local resources, list and contact information of skilled and trained human resources, information bulletin and data on past, existing and potential disasters; a set of the CBDM plan, its schedules and changes with time; instruments, module for training of disaster management, etc. Emergency center should also carry out activities like celebrating disaster risk reduction day or conducting mock drills or exercises.

For successful risk mitigation, involvement of local communities in the decision-making is equally important (Pearce, 2003). In this study some structural mitigation has been suggested to the village panchayat for reducing the top-ranked hazards village-wise, and the mitigation for low-rank hazards was developed with the help of community resources and knowledge. Therefore, the local community is involved, planned and implemented the mitigation of various hazards. So, Phase-II involves application of the information obtained through the exercises in Phase-I for disaster risk reduction at local level.

Finding and Discussion

Hazards Assessment

During orientation program the villagers identified many types of hazards and risks that are prevalent in the villages along with main threat from the landslide. Therefore, these hazards and risks were also considered in the community response study. Information collected from the participant villagers was examined, and all the prevailing hazards were listed by the participated community. The hazards identified by the community were landslide, loss of livelihood due to wild animal intrusion, attacking on human being by wild animal, flash flood and water logging, accidental death or injury during fodder collection, hail storm, cold wave and forest fire that were the major hazards of the villages, among them most of the hazards occur in rainy season. Water logging occurs in rainy season area due to poor drainage system, increasing the probability of flash flood. Hail storm and cold wave occur during the winter season from November to March. Hazards and land use/cover are strongly related to each other. Forest fire occurs in the region because both of natural and manmade causes. During the pick summer season most of the forest fire occurs due to the natural cause but during the winter the forest fire resulted from the practice of slash and burning process. Thus, we can see that the village in Dasholi is in the grips of one form of hazard or other all year round.

Therefore, the data collected in local communities and supplemented by the participants on past disaster and their severity, frequency and impact and stresses are discussed and tabulated. Both the frequency and severity are classified as *low, medium and high,* and their corresponding score is also assigned (Table 17.1). A hazard with a high frequency and severity is denoted as 3, medium frequency and severity is denoted as 2 and low frequency and severity is denoted as 1. Finally, a cumulative of all the different types of hazards is obtained to show the cumulative risk from all identified hazards in all the villages.

From hazards frequency and severity analysis it has been concluded that forest fire, accidental deaths and attack on humans by wild animals are all low to medium frequency and severity hazards in all the villages. Landslide is a high frequency and high severity hazard in villages Kandey, Bamyala, Nelgadora and Anandpur. On the other hand, flash flood is an important hazard only in Dogri village. Hail storm is a high frequency and high-intensity hazard in all villages except Nelgadora and Anandpur, for these village tough frequency is high but severity is low because of less agricultural lands are available to them. Thus, it can be seen that all the villages of the study area are under the influence of one kind of hazard or other.

It is also clear from the cumulative score of frequency and severity of various hazards that Bamyala and Dogri villages rank highest in both the frequencies which has a cumulative point of 19 as well as severity which has a cumulative point of 17 of hazards. Kandey village with a frequency of 17 and severity of 14 comes next followed by Nelgadora and Anandpur (frequency of 16 and severity of 13). Tangsha village has the least hazard risk, with a hazard frequency of 15 and hazard severity of

Table 17.1 Severity and frequency of different hazards

S. No.	Villages	Hazards frequency and severity	Hazards								Total score
			Landslide	Loss of livelihood due to wild animal Intrusion	Attacking on human being by wild animal[a]	Flash flood and water logging	[a]Accidental death or injury during fodder collection	Hail and storm	[a]Cold wave	Forest fire	
1	Kandey	Frequency	High	High	Low	Medium	Low	High	High	Low	17
		Severity	High	Medium	Low	Low	Low	High	Low	Low	13
2	Dogri	Frequency	Medium	Medium	Medium	High	Medium	High	High	Low	19
		Severity	Medium	Medium	Low	High	Low	High	Low	Low	14
3	Nelgadora	Frequency	High	Low	Medium	Medium	Low	High	High	Low	16
		Severity	High	Low	Low	Medium	Low	Medium	Medium	Low	13
4	Bamyala	Frequency	High	High	Medium	Low	Medium	High	High	Low	18
		Severity	Medium	Medium	Medium	Low	Low	High	Medium	Low	14
5	Anandpur	Frequency	High	Medium	Low	Medium	Low	High	High	Low	16
		Severity	High	Medium	Low	Low	Low	Medium	Medium	Low	13

(continued)

Table 17.1 (continued)

S. No.	Villages	Hazards frequency and severity				Hazards					Total score
			Landslide	Loss of livelihood due to wild animal Intrusion	Attacking on human being by wild animal[a]	Flash flood and water logging	[a]Accidental death or injury during fodder collection	Hail and storm	[a]Cold wave	Forest fire	
6	Tangsha	Frequency	Medium	Medium	Low	Medium	Low	High	High	Low	15
		Severity	Medium	Medium	Low	Medium	Low	Medium	Low	Low	12

Indicators:
High frequency hazard: One that has occurred at least twice in 3 years. The occurrence of the hazard appears to have increased in recent decades.
*Medium frequency hazar*d: One that has occurred 1 to 2 times in the last 5 years.
Low frequency hazard: Hazards that have occurred only once in the last 5 years. Highly severe hazard: One that has caused severe damage in past events.
Medium severe hazard: One that has caused moderate damage in past events.
Low severe hazard: One that has not created a situation where the affected family or community has been able to cope and manage using their own capacities and resources.
Risk score: High—3, medium—2, low—1,
[a]The foot note clarifies the weightage methodology for different hazards

12. Thus, from the above analysis table it can be concluded that Bamyala and Dogri villages have the highest hazard risk. Kandey, Nelgadora and Anandpur have high to moderate risk, and Tangsha has the least hazard risk of all the villages studied.

Assessing Vulnerabilities

Assessing vulnerability involved identifying village boundary and physiographic features (i.e., drainages, slopes, ground cracks, etc.), exploring the problems, risk areas and placing this information on a map. The participants had to highlight all this information on a detail map called as a "community sketch". It should also include the areas of past disaster locations, their effects, locating the vulnerable buildings and potential hazardous areas along with, identification of areas where attention should be given (Das et al., 2017). Secondly, both the qualitative and quantitative information are crucial in assessing present and future vulnerabilities. Much of this information has been gathered at community level.

Vulnerability analysis has been done in six villages of Dasholi using human population, land, crop, livestock loss and infrastructure damages as five indicators (Fig. 17.2). From qualitative vulnerability analysis, it is clear that human population and human infrastructure are at a greatest risk from landslide in all the villages except Dogri and Tangsha where the agricultural land suffers major threat from the landslide hazards. Hazards such as intrusion by wild animals affect livestock and grain, but their danger to human life is minimal; but the hazards is another prime concern for all the villagers as the crops are damaged by the wild animal which is a major loss of livelihoods for the villagers. Flash flood affects all the six indicators in village Dogri because the villages locate at down slope, but in the rest it has only a limited impact. Accidental death or injury during fodder collection affects human population, but its effects range from medium to low and that too is most common in Dogri as the slope is steep in this village. Hail storm is very injurious to crops, and this is clearly seen in the table. Finally cold wave and forest fire are not considered very great threat and vulnerability because it varies generally low or medium.

The qualitative vulnerability summarizes and represents the vulnerability of different elements identified by the community in the table form (Table. 17.2). In the above vulnerability analysis, an element with a high vulnerability is denoted as 3, medium vulnerability is denoted as 2, and low vulnerability is denoted as 1. Finally a cumulative of all the different types of hazards is obtained to show the vulnerability in all the villages (Fig. 17.3), and it is categorized into three classes, namely—(A) < 5, (B) 5–9 and (C) > 9.

Therefore, a cumulative vulnerability of all hazards in each village has been calculated, and it was found that Kandey, Nelgadora and Dogri villages rank the highest in population vulnerability, all the village except Nelgadora which rank low come under medium vulnerability in loss of agricultural land and crop grain losses because of less agricultural land availability among the people, and people involve in others economic activity. In case of livelihood loss, the village Tangsha has low

Fig. 17.2 Vulnerabilities of various hazards: 1. Active landslide and settlements, 2. crack development house due to landslide, 3. loss of agricultural land due to landslide, 4. damaged road posing threat to commuters, 5. damage done by water seepage during heavy rain, 6. women carrying over-burden grass and fuel wood/fodder from the forest

vulnerability, and all other villages come under medium vulnerability. Lastly the vulnerability of infrastructure damage is low for Bamyala and Tangsha village next followed by medium vulnerability for rest of villages. The total vulnerability score for all the villages which clearly indicates that Kandey, Nelgadora and Dogri have the highest vulnerability, Bamyala and Nelgadora villages have medium vulnerability, and Tangsha has the least vulnerability of all the villages studied.

In case of quantitative vulnerability analysis, it presents these facts on the basis of data. It is found that all villages have a great threaten from landslide as landslide resulted in loss of 335 nails (local measurement unit of agricultural land) of agricultural land and 26 house. It is also found that the threat from the landslide on human population and infrastructure is high on villages like Kandey, Anandpur and Nelgadora. Whereas the major threatens for Dogri and Tangsha from landslide are on the agricultural land, agriculture land became vulnerable for Bamyala village due to recently activated landslide along the prime road of the village. The probability of damaging the road from landslide is important for all the villages. Loss of livelihood by wild life intrusion is counted as Rs. 20,500 annually. The hazards attacking on human being by wild animals and accidental death or injury during fodder collection record nine and fifteen cases in past three years, respectively. The vulnerability is assessed based on the past experience of the villagers. The accidental death or injury is mainly caused because the road in village which is used for going into forest is in poor condition; in some places, it is prone to slide, and also the slope of the forest is sliding prone and women of villagers walk with a huge load on their back in such steep slope. The probability loss from the hazards water logging and flash flood is high for Dogri village.

A total amount of nearly Rs. 23,000 as annually loss occurs by the water logging or flash flood problems. A total of 29 house 104 human lives are identified as vulnerable

Table 17.2 Vulnerability analysis

Qualitative categorization (based on livelihood asset)

Villages	Vulnerable elements				Hazards					Total score
		Landslide	[b]Livelihood loss due to Wild animal intrusion	[a]Attacking on human being by wild animal	Water logging and flash flood	bAccidental death or injury during fodder collection	Hail and storm	Cold wave	Forest fire	
Kandey	Human population	Medium	Low	Low	Low	Low	0	Low	Low	8
	Agricultural land	High	0	0	Medium	0	0	0	0	5
	Crop grain	High	High	0	Medium	0	High	Medium	Low	12
	Livestock loss	Low	Low	0	Low	Low	0	Low	Low	6
	Infrastructure damage	Medium	0	0	Low	0	Low	0	Low	5
Dogri	Human population	Low	Low	Low	High	Medium	0	Low	Low	10
	Agricultural land	High	0	0	Low	0	0	0	0	4
	Crop grain	High	High	0	0	0	High	Medium	Low	12
	Livestock loss	Low	Low	0	High	Low	0	Low	Low	8
	Infrastructure damage	Low	0	0	High	0	Low	0	Low	6
Nelgadora	Human population	High	Low	Low	0	Mcdium	0	Low	Low	9

(continued)

Table 17.2 (continued)

Qualitative categorization (based on livelihood asset)										
Villages	Vulnerable elements				Hazards					Total score
		Landslide	[b]Livelihood loss due to Wild animal intrusion	[a]Attacking on human being by wild animal	Water logging and flash flood	bAccidental death or injury during fodder collection	Hail and storm	Cold wave	Forest fire	
	Agricultural land	High	0	0	0	0	0	0	0	3
	Crop grain	Low	High	0	0	0	High	Low	Low	9
	Livestock loss	Medium	Low	0	0	Medium	0	0	Low	6
	Infrastructure damage	High	0	0	Low	0	Low	0	Low	6
Bamyala	Human population	*Low*	*Low*	*Low*	*0*	*Low*	*0*	*Low*	*Low*	*6*
	Agricultural land	*High*	*0*	*0*	*0*	*0*	*0*	*0*	*0*	*3*
	Crop grain	*High*	*High*	*0*	*Low*	*0*	*High*	*Low*	*Low*	*12*
	Livestock loss	*Low*	*Low*	*Low*	*0*	*Low*	*0*	*0*	*Low*	*15*
	Infrastructure damage	*Low*	*0*	*0*	*0*	*0*	*Low*	*0*	*Low*	*4*

(continued)

Table 17.2 (continued)

Qualitative categorization (based on livelihood asset)										
Villages	Vulnerable elements				Hazards					Total score
		Landslide	[b]Livelihood loss due to Wild animal intrusion	[a]Attacking on human being by wild animal	Water logging and flash flood	bAccidental death or injury during fodder collection	Hail and storm	Cold wave	Forest fire	
Anandpur	Human Population	High	Low	Low	Low	Low	0	Low	Low	9
	Land	High	0	0	Medium	0	0	0	0	5
	Crop Grain	High	High	0	Medium	0	High	Low	Low	13
	Livestock loss	Low	Low	0	Low	Low	0	0	Low	5
	Infrastructure Damage	High	0	0	Low	0	Low	0	Low	6
Tangsha	Human population	0	Low	Low	0	0	0	Low	0	3
	Agricultural land	High	0	0	Low	0	0	0	0	4
	Crop grain	High	High	0	Low	0	High	Low	Low	12
	Livestock loss	0	Low	0	0	0	0	0	0	1
	Infrastructure damage	Low	0	0	0	0	Low	0	Low	3

Indicators:
High: The hazard occurs frequently and is extremely damaging. Communities suffer large losses, and recovery takes a long time.
Medium: The damage caused by the hazard is less, but still significant. Recovery requires external support.
Low: The damage caused by the hazards is negligible, and community can cope up with that hazards with its internal support.
Zero: The hazard either did not occur or if it did, it caused negligible damage.
Vulnerability score: *High*—3, *medium*—2, *low*—1.
[a]Peoples may be injured or died based on the past experience.
[b]Based on past experience on crop loss by wild animal intrusion

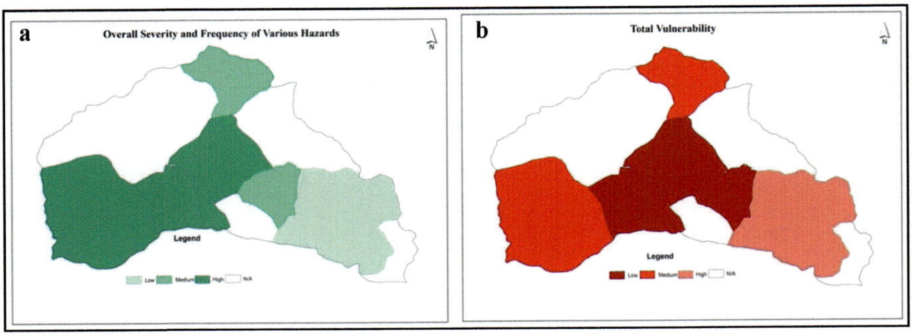

Fig. 17.3 **a** Severity and frequency of various hazards, **b** level of vulnerability in villages

to these hazards, among them 12 h are from the Dogri village. The loss of agricultural land is also high in this village, nearly about thirty nails per year. The vulnerability from the hazards of hail storm is mainly on the crops. The estimate loss from such hazards is calculated as Rs. 91,000, if it happens during the sowing time of crops, the expected loss is much higher, and therefore, alternative livelihoods should be created among the villagers, especially during the season of such hazards. There are two kinds of vulnerability from cold wave: one is on the crops during the time of November and March; as in the month January and February no crops have been cultivated. Secondly these specific hazards have specially threatened the old age people of the village. The forest fires have zero vulnerability as the villagers practice to control the forest fire away from the settlement, and also no such loss has identified from the past time.

Capacity Assessment

The third step is assessing capacities of the local communities in disaster management. Some of the necessary capacities are institutional, such as access to humanitarian support, trained personnel, trained health workers and issues of governance. Other capacities are linked with material inputs such as radio communication, emergency food supplies and rescue equipment that are assessed for each village (Islam et. al., 2014). Preparedness, awareness and the existence of an evacuation plan at community level help to limit losses and save lives.

The capacity assessment at different villages identified by the community is summarized and represented in Table 17.3. It was observed and interpreted that on all counts of disaster preparedness, the villages in Dasholi lack capacity. Thus, for example, early warning and communication of disasters is low in all the villages, while only Tangsha village has direct access to basic services and motorable road for vulnerable population. No village has any Emergency Community Relief Fund or any Community Food Storage Facility that can be used at the times of disaster. Relief material and distribution is identically low in all the villages, while there is no rescue equipment or trained rescue worker in any village. Thus, it can be concluded from this table that no proper disaster management facility is available in the villages.

Risk Assessment

Each hazard experienced in both the past and the present may not affect every village to the same extent. Some village may have been extremely affected, while others remained almost untouched toward a hazard. In order to reflect the reality of this situation, community-wise disaster risk assessment (Fig. 17.4) of identified hazards and stresses is therefore necessary (Prakash, 2013a). Disaster risk analysis of the selected villages is done by overlay analysis using the following formula:

$$\text{Disaster Risk} = (\text{Hazards} * \text{Vulnerability})/\text{Capacity}$$

From the overlay analysis it was observed that villages Bamyala, Kandey and Nelgadora have high-risk zone, villages Dogri and Anandpur have medium-risk zone, and Tangsha comes under the low-risk zone.

In all the six villages, people are aware of landslide risk because landslides occur very frequently in the area and cause substantial damage to life and property. In the study area, perception of risk is formed mainly from experience of the actual damaging landslide events in recent past, from the loss of agricultural land, forest, houses and livelihoods. From the hazards risk prioritization, it was found that villages Kandey, Bamyala, Nelgadora, Anandpur and Tangsha have given first priority to landslide hazard risk. These villages have lost their agricultural land and houses because of landslide damage and the rest one village namely Dogri ranks second because of these villagers have only loss their agricultural land by landslide. Loss of livelihoods due to wild animal intrusion mainly damaging the crops by monkey, wild boar and others animals is another important hazard for the villagers. The villages Kandey, Bamyala, Anandpur, Tangsha and Nagar have been given second priority, and villages Dogri and Nelgadora given third priority to this hazard, respectively. The hazard attacking on human being by wild animals is given fourth priority by Kandey, Bamyala and Anandpur villages and fifth rank by Dogri, Nelgadora and Tangsha villages, respectively. The attack on villagers occurs when women go in long distance within the forest to collect the grass and fuel wood where there is more risk of attacking by forest animals like tiger, leopard, bear, wild boar on women or sometimes attacking on children who are staying alone in house. Water logging and flash flood are the major problems in Dogri village, and it is given first priority. According to villagers there is a drain which overflows during the rainy season with huge velocity and high water discharge from the upper slope. During this time crack has been developed on wall of the house and also the water flowed beneath our houses. Two houses were destroyed in the year 2013 during rainy season as the water blow up with them. Village Nelgadora was ranked second, and villages Anandpur

Table 17.3 Capacity assessment of the village

S. No.	Capacity indicator	Villages					
		Kandey	Dogri	Nelgadora	Bamyala	Anandpur	Tangsha
1	Early warning and communication	L	L	0	L	L	L
2	Rescue equipment and tools (ambulance, fire extinguisher, drilling equipment, rope, stairs, etc.)	0	0	0	0	0	0
3	Trained rescue and first aid workers, etc.	0	0	0	0	0	0
4	Access to natural resources water, land, etc.	M	H	L	M	M	M
5	Temporary shelter	L	L	L	L	L	M
6	Community food storage facilities	0	0	0	0	0	0
7	Relief materials and distribution mechanisms	L	L	L	L	L	L
8	Community, political leaders' awareness about disaster	M	L	L	L	L	M
9	Access to service providers (agriculture office, health office, post office, market, etc.)	M	M	L	L	M	H
10	Access road to vulnerable community	M	M	L	M	M	H
11	Emergency community fund	0	0	0	0	0	0
12	Diversified options for livelihoods of vulnerable	L	L	L	L	L	L
	Total score	12	12	7	10	11	15

Indicators:
Capacity score:
High (H) = 3, Medium (M) = 2, Low (L) = 1, 0 = zero capacity among the villagers. High = community can cope up with the disaster with its internal resource base. Medium = community is prepared but external support is required.
Low = community is aware but not prepared.
Zero = community have neither aware nor prepared.

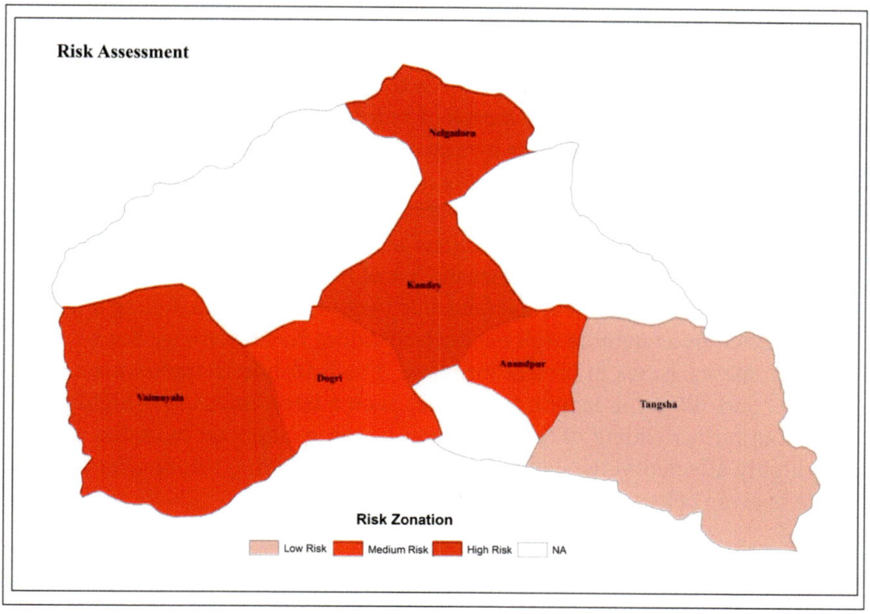

Fig. 17.4 Disaster risk assessment for selected villages

and Tangsha third and fifth by Kandey village, respectively, but the village Bamyala ranked eight to this disaster as it is located on top of the hill. The next hazard, accidental death or injury during fodder collection, is given sixth by Kandey, Dogri and Tangsha villages and fifth by Anandpur and Bamyala villages, and Nelgadora ranked seventh for this hazard. Hail and storm is another important hazard which makes a huge loss of cultivated crops every year. Villages Bamyala, Kandey and Anandpur give third rank to this hazard, and villages Dogri, Nelgadora and Tangsha given forth rank to these hazards. The hazard cold wave was given sixth rank by the Nelgadora and Bamyala, the rest of villages given seventh rank to this hazard. The hazard forest fire which was caused due to both natural and manmade processes resulted into loss of forest cover for the villagers. This hazard ranked eighth by all the villagers except Bamyala which ranked seventh as the villagers practice forest fire for boosting agricultural activity.

Disaster Mitigation Plan

In this study some structural mitigation has been suggested to the village panchayat for reducing the top-ranked hazards village-wise, and the mitigation for low-rank hazards was developed with the help of community resources and knowledge.

From all the above mentioned hazards it is identified that landslide, flash floods, hail and crop loss due to animal intrusion are the most important hazards which continuously threaten the life and livelihoods for the villagers. Land use planning is the most economical and effective means of reducing hazards losses; i.e., frequent damage from landslides can be avert by evacuating areas that continue to have slope failures. Frequent landsliding areas can be also prioritized for landslide monitoring and mitigation. Further development should be avoided over the area showing paths of debris flow slides, and immediate actions must be taken to protect those slopes from further sliding in order to minimize the risk of life and properties located nearby (Jaiswal & Westen, 2013). The small landslide over the study area, falling along the road is protected by a retaining wall but the landslide at Kandey village has not taken any appropriate measures to protect the slope. The landslide is approximately 350 m in height and 300 mt in length, to stabilize such a huge landslide is highly cost-effective, but further sliding of the slope must be mitigating; otherwise the villages may be highly affected in the coming year. Some engineering solution for stabling the slope is given below:

1. Grass turf: The process involves modification of the slope below the friction angle, and then use of turf to cover the slope. The advantage of this method is that it does not require periodic maintenance over a long period, and since it involves flattening of slope, it is more reliable in containing shallow mass movements. As a disadvantage, this method may not be feasible if the slope is steep and where flattening of the slope is difficult. The use of grass turf is not very common in the study area but for such a huge landslide it may proof as cost-effective technique.
2. Concrete retaining wall: This is the most common mitigation measure used in the study area. This measure involves constructing a concrete retaining wall at the toe of the cut slope to stabilize slope failure. Figure 5.15 shows the concrete retaining structure constructed to stabilize a cut slope failure. The method is suitable along the road where availability of space permits construction of a retaining wall. The average design life of a retaining wall is 15 years. The advantage of this method is that it also does not require continuous maintenance. But, as a disadvantage since the maximum height of the structure is 3 m, it may not be suitable for containing a large landslide.

However, before selecting any mitigation strategy, the actual cost–benefit analysis has to be performed by trained professionals, using the real estimated values of costs of measures, investment period, maintenance of costs and benefit rates, etc. One type of measures may not be suitable for the entire area; rather, several different combinations of mitigation measures can be considered depending on the type of elements at risk such as some slopes can be stabilized using grass turf while some using retaining wall or even cost–benefit analysis of option such as alteration of bridges, bioengineering solutions, measures like soil nailing, shot creating, etc. can be carried out. Nevertheless, in all analysis, information from risk models will form an important input.

For water logging and flash flood mitigation proper drainage network should be developed over the area and diversion of water through making newly drain is required to check the hazards. Community should monitor and clean the drain time to time so that water can flow smoothly without any obstruction. The other hazards can be mitigating with the help of community awareness and livelihood security.

Disaster Preparedness

It assumes that an informed, aware and prepared community will be better able to cope with disaster than otherwise. For various hazards risk reduction community is prepared on the basis of various age groups. The major four groups have been created namely—school children, young people, women and old age people.

The following activities are performed for the disaster preparedness:

1. Formation of relief and rescue team: Despite all the efforts disaster may continue to inflict upon the society, so the community can face the remnant disaster in a planned way; a relief and rescue team has been prepared in each village. The main work of the rescue team is saving the life of people unless the external help will reach. The team also participate and monitor the work done by the community for hazards mitigation. For emergency medical facilities, a list of local cab driver of each village has been prepared, so that, during the time of disaster it can be used for sending the victim to hospital.
2. Mapping alternative routes and safe shelters: The second stage consists of identification of alternative evacuation routes and safe shelter places drawn by the people based on their local knowledge of the terrain. Community sketch map prepared by participate community shows evacuation routes in the villages. The community from each village have identified a temporary shelter where they can stay during the disaster hit or unless the re-habitation program is achieved. The community has identified their school building, community center and temple as temporary shelter during the disaster time. The identified distance between the emergency evacuation route and the temporary shelter is between the 10 and 15 min ranges. In Kandey village the school building has been identified as shelter during disaster hit. For Dogri village people have been identified their temple as safe shelter during disaster hit. In the village Bamyala though people do not face any major risk from landslide but people have identified the community hall as the safe shelter during the time of disaster hit. The village Nelgadora has also identified their community hall as a shelter during disaster hit but the major problems have no direct access to motor able road. The villages have 1.5 km far away from the motor able road, though the community have identified the temporary shelter nearby but to provide emergency medical facilities to the villagers became difficult. Lastly the village Nagartangsha has identified their panchayat bhavan as a shelter during the disaster hit. The village has direct access to motor able

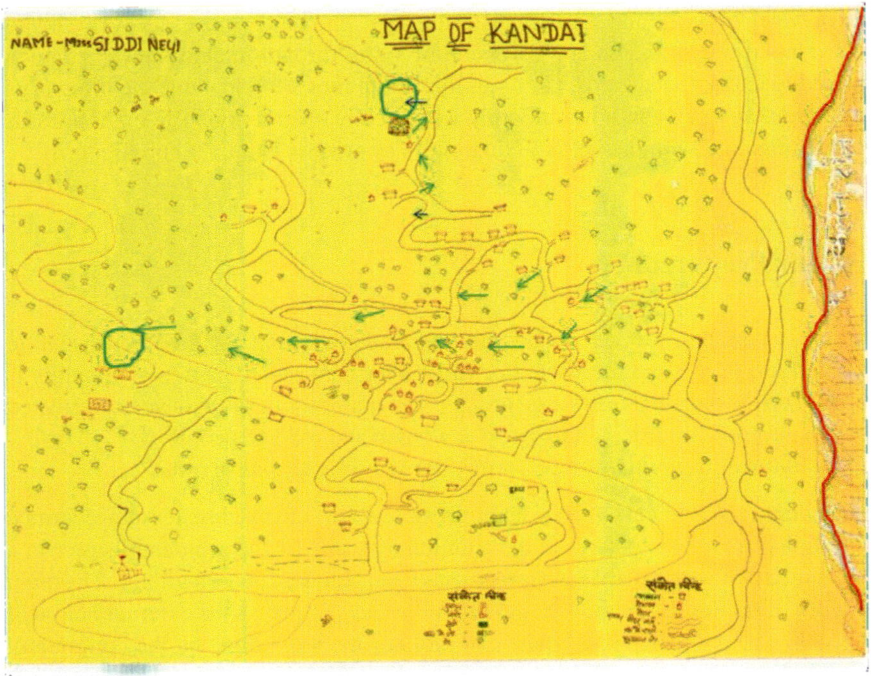

Fig. 17.5 Emergency route plan for Kandey village

road and primary health center in their village. Figure 17.5 depicts the emergency route plan for Kandey village.

3. Awareness and Skill Development: In third stage awareness about the major hazards like landslide, flash flood and water logging is given, so that community can understand how to identify the characteristics and predict the landslide occurrence (Fig. 17.6). The participants in awareness program are divided in four groups, namely school children as they are the future generation to the village, it helps them to understand the various hazards problems in their village. Second young men as they are the people who will serve as rescue and relief operation during a disaster hit in the villages. Third and fourth groups are women and old aged people, respectively, who are the most vulnerable group during a disaster hit. A list of what to do and not is prepared for various hazards and discussed with each group.

   The awareness about how to predict the occurrence of hazards has been created among the participated community so that they can issue a warning system during the pre-disaster based on the forecasting of rainfall, received from the TV, radio or newspaper. It will help the community to prepare and keep the valuable item in a safe place before the disaster hit. Women who are more vulnerable to accidental death or injury during the time of fodder collection from the forest have created a group to go into the forest so that they can protect each other. For farmer's

Fig. 17.6 Disaster preparedness and management. 1. Community participation for preparing vulnerability, hazards and emergency route sketch map, 2. Formation of disaster relief and rescue team for villages, 3. Disaster risk reduction and awareness program among school children, 4. Among women, 5. Among old age people, 6. Villagers preparing retaining wall as crack develops on houses due to landslide

alternative livelihood like animal husbandry, handicraft and handlooms should be promoted to minimize the loss from hail and storm or wild life intrusion can be minimized.

4. Identification of representative: The final work in this process is to identify a representative within each community who can liaison with the local officers and emergency management personnel during pre- and post-disaster phases. The political representative from each village has been selected as the community representative.

Recommendation

The recommendations are based on the experience gained throughout the study and other works carried out in this field. Some of the recommendations for mitigating the hazards over the study area are given below.

- The most important measures to be taken are the preparation of slope stabilization by grass turfing over the sliding area immediately so that further sliding of land can be stopped.
- The practice of intensive agriculture (paddy cultivation) above the crest of main landslide.
- Villagers are being encouraged to leave paddy cultivation and adopt plantation agriculture to check the further landslide.
- A proper drainage network should be developed within the slide area so that soil cannot be saturated and minimize the occurrence of further landslide.

- To check the water logging and flash flood the first and foremost mitigation measure is drainage correction. This involves maintenance of natural drainage channels, both micro- and macro- in vulnerable slopes.
- In Dogri village the diversion of water by making newly drainage channel is highly required so that houses of villagers can be saved.
- For emergency purposes, during the times of disaster, a community fund needs to be established with the help of NGOs and government agencies.
- Regular monitoring and management by specialized agencies and local volunteers should be done to ensure full preparedness during disasters.
- A primary health center with basic facilities needs to be established in every village of the study area.

Conclusion

This research aims at establishing systematic mapping of sustainable environmental knowledge for disaster risk reduction in local level. It was shown how visualized forms of local knowledge may contribute to an improved adaptation for a disaster resilience community. Participatory mapping may allow for more efficient, sustainable and pluralistic forms of risk and hazard management and a greater acceptance by the local population. However, an important finding of our research is that these bodies of local environmental knowledge are rarely taken into account by professionals in disaster control. During the whole work it was found that

- Most people of the villages are aware of landslide disasters, and they accept the fact that their area is prone to landslides.
- They are aware of the cause of landslides (i.e., very high rainfall), and they know that the period between July to September is the most problematic and the time of occurrence of other hazards.
- Most of the people have lived in the study area all their life and their ancestors for many generations, and in spite of the fact that they have experienced or witnessed multiple disasters, they do not want to leave their native place.
    - People tolerate the risk because of certain benefits such as working in nearby district headquarter and well fertile land for cultivation.
    - They have no emergency preparedness plans and insurance coverage of properties for a disaster.
    - They depend on the local government and local organizations such as rescue operation unit, a non-governmental organization, for post-disaster help and mitigation.
    - The low economic status of most communities does not permit them to make another house in other safer place or carry out additional investment for reinforcement of their houses in order to protect them from landslide damages.

So, it can be told that community-based disaster management is a novel method of combating disasters. As seen in the Kedarnath disaster many people lost their lives due to lack of awareness and information. Community-based disaster management (CBDM) gives communities the training required to protect themselves from disasters and the adverse effects of climate change. It provides the skills and equipment for a small team to carry out an emergency first response when a disaster has taken place. This study will help in capacity building at local level to cope up with the disaster specifically to the fragile Himalayan environment and reducing or minimizing the effect of disaster loss.

References

Anderson-Berry, L. J. (2003). Community vulnerability to tropical cyclones: Cairns, 1996–2000. *Natural Hazards, 30*(2), 209–232.

Anderson, M. G., Holcombe, E., Blake, J. R., Ghesquire, F., Holm-Nielsen, N., & Fisseha, T. (2011). Reducing landslide risk in communities: Evidence from the Eastern Caribbean. *Applied Geography, 31*(2), 590–599.

Anderson, M. G., & Holcombe, E. (2013). *Community-based landslide risk reduction: managing disasters in small steps*. World Bank Publications.

Chen, L. C., Liu, Y.C., Chan, K. C. (2006). Integrated community-based disaster management program in Taiwan: A case study of Shang-An village Natural. *Hazards, 37*(1–2), 209.

Das, S. (2015). Community based disaster risk management: Managing disaster in small steps. *International Journal of Multidisciplinary Research and Development, 2*(1), 8–11.

Das, S., & Saha, A. K. (2017). Community based disaster management for climate change adaptation: A conceptual model. In V. S. Negi, (Ed.), *Climate Change: Perspective in 21st Century* (pp. 155–163). Research India Press.

Glade, T., & Crozier, M. J. (2005). *Landslide hazard and risk: Concluding comment and perspectives Landslide hazard and risk*, (pp. 767–774). Wiley.

Holcombe, E., Smith, S., Wright, E., & Anderson, M. G. (2012). An integrated approach for evaluating the effectiveness of landslide risk reduction in unplanned communities in the Caribbean. *Natural Hazards, 61*(2), 351–385.

Islam, M. A., Chattoraj, S. L., Ray, C. P. (2014). Ukhimath landslide 2012 at Uttarakhand, India: Causes and consequences. *International Journal of Geomatics and Geosciences,*(4), 3–544.

Jaiswal, P., & van Westen, C. J. (2013). Use of quantitative landslide hazard and risk information for local disaster risk reduction along a transportation corridor: A case study from Nilgiri district India. *Natural Hazards, 65*(1), 887–913.

Künzler, M., Huggel, C., & Ramírez, J. M. (2012). A risk analysis for floods and lahars: Case study in the Cordillera Central of Colombia. *Natural Hazards, 64*(1), 767–796

Mori, M., Hosoda, T., Ishikawa, Y., Tuda, M., Fujimoto, R., & Iwama, T. (2005). *Landslide management by community-based approach in the Republic of Armenia*. Japan International Cooperation Agency, Government of Armenia.

Prakash, S. (2013a). Capacity development for landslides risk reduction in India. In *Landslides: Global risk preparedness* (pp. 369–383). Berlin, Heidelberg: Springer.

Prakash, S. (2013b). Education, training and capacity development for mainstreaming landslides risk management. In *Landslide science and practice* (pp. 257–264). Berlin, Heidelberg: Springer.

Pearce, L. (2003). Disaster management and community planning, and public participation: How to achieve sustainable hazard mitigation. *Natural Hazards, 28*(2), 211–228.

Patterson, O., Weil, F., & Patel, K. (2010). The role of community in disaster response: Conceptual models. *Population Research and Policy Review*, *29*(2), 127–141.
Warner, C. (2015). Participatory mapping: a literature review of community-based research and participatory planning. In*: Social Hub for community housing, faculty of architecture and town planning technion.* Cambridge, Massachusetts: Massachusetts Institute of Technology.

Chapter 18
Regulatory Framework for Regional Cooperation on Disaster Risk Reduction (DRR) in India and Globe

Upma Gautam and Deeksha Bajpai Tewari

Abstract All member states of South Asian Association for Regional Cooperation (SAARC) are prone to recurring natural disasters as well as disasters induced by humans. South Asian countries have experienced a wide-ranging geological and hydro-meteorological disasters may it be severe droughts, floods, cyclones, landslides, earthquakes or tsunamis causing not only irreparable loss to life but also shattering the economy of these member states. Huge population, poverty and corruption make the region more vulnerable and augment the magnitude of any disaster manifold. The impact of these disasters is not restricted within the physical and political boundaries of these states but is discernible across the borders. Though in almost all the SAARC countries there has been a successful mapping of the legislative and institutional framework in the region for the purpose of disaster risk reduction, but the major lacunae lie in the inadequate cooperation amongst the countries at the regional and international level. Sustainable synchronisation, alliance and mainstreaming of a comprehensive legal and institutional framework are required to be adopted and implemented in this region which ensures a rapid response mechanism dealing with disasters providing not only adequate and timely relief but also assistance which is more humanitarian.

Keywords SAARC · Disaster risk reduction · Natural disaster · Poverty · Corruption · Disaster management

---

U. Gautam
University School of Law and Legal Studies, Guru Gobind Singh Indraprastha University, New Delhi, India
e-mail: upma@ipu.ac.in

D. B. Tewari (✉)
Department of Geography, Dyal Singh College, University of Delhi, New Delhi, India
e-mail: deekshabtewari@gmail.com

V. R. Sharma and Chandrakanta (eds.), *Making India Disaster Resilient*,
https://doi.org/10.1007/978-3-031-50113-5_18

Introduction: Conceptualising the Concept

SAARC as an organisation was founded for various mutually beneficial objectives, amongst which the promotion of the welfare of the people in region thereby improving their standards of living was primary. This objective was sought to be achieved by the mutual collaboration and assistance amongst the member states; thus realising the goal of accelerated development in economic, social and cultural spheres in the region. As the name suggests, this organisation works towards developing and strengthening trust amongst the member states in the region; with the aim of understanding and empathising with each other's problems. Additionally the organisation seeks to develop a framework—legal and institutional—as well as ensuring its successful implementation; thus to minimise the impact of the commonly shared problems. Poverty, corruption, trans-national crimes and disasters, may it be natural or manmade, are some of the common grounds on which the South Asian countries have to come together to build a robust mechanism through mutual regional cooperation and assistance for providing a dignified life to each and every citizen of this region.

Over the past few decades South Asia has seen an immense economic growth; thus ameliorating the health services, access to facilities, infrastructure, etc. The other side of this remarkable story is also a reality of South Asia being a high disaster-prone region, where a high-intensity disaster can set a halt to these progress measures. United Nations International Strategy for Disaster Risk Reduction (UNISDR) defines "disaster" as a serious disruption of the functioning of a community or a society at any scale due to hazardous events interacting with conditions of exposure, vulnerability and capacity, resulting to one or more of the following: human, material, economic and environmental losses and impacts.

In case of South Asia region the term "disaster" has a multidimensional impact in addition to above-mentioned definition. The demographic situation, rapid technological and economic changes, accelerated urbanisation, consequent need for infrastructure development in a high-risk and disaster-ridden environment like South Asia further exacerbate the vulnerability and diminish the resilience of the people and their communities in the region.

It has been more than ten years since the formulation of a comprehensive disaster management in South Asia was undertaken in 2007. Its prime concern was to create and develop a comprehensive disaster management system in South Asia which fulfils the need of a prompt and an effective response in case of any kind of disaster in the region efficiently. This framework was further supported and strengthened by the "SAARC Agreement on Rapid Response to Natural Disasters" in the year 2011, which made it obligatory for the member states for developing and implementing effective measures for disaster risk reduction which included not only mutually assisted emergent arrangements for relief during disasters but also exchange of information pertaining to the same. Time and again it is also been acknowledged by the international community that for successful and effective risk reduction policy—across the globe and in individual country—its systematic integration into the plans, policies and programmes for sustainable development and poverty reduction is a prerequisite, and this further needs to be supported through bilateral, regional and international cooperation.

The present research article is to map out the gaps between the legal and institutional frameworks as are in force at present and their effectiveness as far the implementation in South Asia is concerned. It further articulates the reasons behind the inability of SAARC countries in formulation of an efficacious organisation which is effective in managing disaster in the region. In its conclusion it attempts to suggest feasible alterations which can be made in the legal and institutional frameworks resulting into mechanism which deals with disaster risk reduction and management in SAARC countries in a productive manner with mutual assistance, cooperation and collective self-reliance.

Analysis of Legislative and Institutional Framework for Disaster Risk Reduction: Is South Asia Prepared?

The states which are members of the South Asian Association for Regional Cooperation (SAARC) are highly prone to hydro-meteorological and geological hazards such as floods, landslides, droughts, cyclones, earthquakes, heat waves, avalanches and tsunamis. Amongst the eight member nations India, Pakistan and Bangladesh exhibit the largest losses, which is due to large exposure at risk and the high level of hazards. The various factors that trigger the impact of a hazard include urbanisation, environmental degradation, lack of strong governance, political instability, border disputes, ineffective regional networks and climate change. Moreover, the development and influence of a natural hazard are not restricted to the political boundary of a nation. Floods, earthquakes, forest fires and volcanoes have significant cross-border impacts. Hence, in the present times good governance, regional stability and effective environmental management are required to have minimum impacts of a disaster.

Laws and Policies Regulating Disaster Management in Islamic Republic of Afghanistan

Afghanistan is a party to many international as well as regional conventions which plays an important part in providing international disaster assistance. The nation is also a member of the SAARC Agreement on Rapid Response to Natural Disasters. Following this Agreement, Afghanistan has incorporated a system for making provisions for disaster preparedness, response and for making sundry arrangements for different disaster relief mechanisms.

To develop a system relating to disaster response, preparedness and risk reduction in the time of disaster—natural or manmade—The National Disaster Management

Law ("DM Law") came into force in 2012. Following the enactment of this Act, Afghanistan National Disaster Management Authority (ANDMA) was constituted. It became the responsibility of ANDMA to regulate and coordinate various response and enforcement activities during a disaster. Furthermore, the Act has established commissions at three levels: national, provincial and district. At the national level is the National Disaster Management Commission (NDMC); the sessions of which are conducted in emergency or at the request of ANDMA. The decisions of NDMC are implemented by the ANDMA. The government, social organisations and the people from concerned areas help the ANDMA in implementing the decisions of the commission.

Besides the DM Act, 2012 the National Disaster Management Plan, 2010 (NDMP) has also been developed with a vision to reduce the number of deaths as well as the distress caused by disasters. The National Disaster Risk Reduction Plan (NDRRP) and the National Disaster Response and Recovery Plan (NDRRP) are the two major components of this plan.

Laws and Policies Regulating Disaster Management in People's Republic of Bangladesh

Bangladesh enacted the Disaster Management Act, 2012 to govern the disaster relief activities in a synchronised, robust and effective manner so as to ensure a coordinated mechanism of disaster management amongst the different organisations. The Act provides for several committees including National Disaster Management Council, National Disaster Response Coordination Group and National Level Disaster Management Committee for smooth operation of disaster management activities from national level to field level. Furthermore there exists two separate funding infrastructures: National Disaster Management Fund and District Disaster Management Fund. The responsibility of ensuring coordination amongst the management at different levels lies with the Ministry of Disaster Management and Relief (MoDM&R). The responsibilities of the ministry include: relief and rehabilitation—during and post disaster—of affected people and to execute disaster management programmes at community level.

One of the best frameworks for disaster management is The Cyclone Preparedness Programme (CPP). It incorporates within its ambit early institution of warning, training, education, community participation and preparedness. The most pertinent attribute of CPP is the efforts of the volunteers who constitute a part of the community itself. The training given to these volunteers include: circulating cyclone warnings, conducting rescue operations, giving first aids, awareness about radio communications equipment and evacuation.

Laws and Policies Regulating Disaster Management in Kingdom of Bhutan

The Royal Government of the Kingdom of Bhutan enacted the Disaster Management Act in 2013. The key aim of this Act is to protect human life and property of the people by reducing and managing risk arising out of disaster. Under this Act an authority at national level namely National Disaster Management Authority (NDMA) and at local level namely Dzongkhag Disaster Management Committee (DDMC) were established. Further the Act has established a Department of Disaster Management, as the secretariat to the NDMA. NDMA has various responsibilities such as making strategies and policies for disaster management activities, developing a contingency plan, systems (structural and non-structural), guidelines and procedures, ensuring the creation of an Inter-Ministerial Task Force (IMTF), etc. The IMTF is a group of technical experts from relevant agencies who help in giving technical guidance whilst formulating any disaster management operations. They also give advice for setting up facilities for disaster management and risk reduction activities, etc. DDMC prepares and implements the Dzongkhag Disaster Management and Contingency Plan, monitors and evaluates preventive measures, ensures establishment and functioning of Critical Disaster Management Facility and directs Dungkhag, Thromde and Gewog Disaster Management subcommittees. It can also perform additional functions if directed by the NDMA under the Act.

The Act obligates an association with other countries and international organisations and provides for a mechanism to be built by the NDMA for international cooperation in disaster management.

Laws and Policies Regulating Disaster Management in Republic of Maldives

The Republic of Maldives is committed towards protecting its people from various disasters and the enforcement of global standards such the Hyogo Framework for Action and Sendai Framework for Disaster Risk Reduction. The government enacted The Disaster Management Act in 2015. This Act provides the principles through which the disaster management is governed in Maldives. It constituted the National Disaster Management Authority (NDMA), under the Ministry of Defence and National Security. The main function of NDMA is to provide a framework for disaster risk reduction and management in the country.

Laws and Policies Regulating Disaster Management in Federal Democratic Republic of Nepal

In the year 1982 Nepal enacted its first law towards disaster management, the Natural Disaster Relief Act (NDRA) also known as the Natural Calamity Relief Act (NCRA); however it was replaced by Disaster Risk Reduction and Management Act in the year 2017. Under this Act different authorities are created at the national, state and district level. At the national level a National Council for Disaster Management (NCDM) is constituted, and under it a National Disaster Risk Reduction and Management Authority (NDRRMA) is formulated which acts as a central point for governing disaster management system in the country and formulates strategies for enforcement and supervision of disaster related activities. Similarly there is a provision for State Disaster Management Authority and a District Disaster Management Authority as well.

These authorities are further responsible for informing the government about the areas prone to disasters.

Nepal passed The Local Government Operation Act (LGOA) 2017, to replace Local Self Governance Act (LSGA) 1999. This Act provides for self-governance at the local level. The Act highlights the importance of disaster management at local level by involving the local people and local bodies. Besides formulating, implementing, monitoring and evaluating for disaster risk reduction the Act also provides for coordination between disaster management at federal, state and local level.

Laws and Policies Regulating Disaster Management in Islamic Republic of Pakistan

Pakistan passed The National Disaster Management Act in the 2010. Under the Act an authority was created at the national level: National Disaster Management Authority (NDMA). Furthermore similar authorities called provincial/Regional Disaster Management Authorities (PDMA) were appointed at the provincial level, and District Disaster Management Authorities (DDMA) were made at district levels.

The authority at the national level coordinates and monitors the response in the event of any disaster and lays down the framework for operations in disaster. Moreover, it is the responsibility of the authorities to help in prevention and mitigation of a disaster. Additionally, the Act provides for calling of services of any person, and this person will act as a member and has all the powers which are authorised under the Act.

The Act also authorises the Federal Government to take measures and provide support and assistance if any other nation is affected by a disaster.

Laws and Policies Regulating Disaster Management in Democratic Socialist Republic of Sri Lanka

Sri Lanka passed the Sri Lanka Disaster Management Act, in the year 2005, which established the National Council for Disaster Management. This council is headed by the President and Prime Minister and includes as members the Leader of the Opposition, twenty Ministers, the Chief Ministers of Provinces and five opposition Members of Parliament. Further the Act has constituted The Disaster Management Centre (DMC) which will function under the Council and will be accountable for the execution of the policies proposed by the Act.

Laws and Policies Regulating Disaster Management in Republic of India

The Republic of India enacted the Disaster Management Act in the year 2005. The Act provides for a three-tier model of disaster management. The National Disaster Management Authority (NDMA) functions at the national level. It has the responsibility of drafting policies and approving a disaster management plan. Additionally, it helps in enforcing and implementing the plans and policies for disaster management. Moreover, the Act also provides for an executive body, the National Executive Committee (NEC). The main role of NEC is to help NDMA in executing its miscellaneous functions, implementing the policies and plans and ensuring the compliance of instructions devised by the Central Government for the purpose of disaster management. At the state level, there are State Disaster Management Authority (SDMA) and the State Executive Committee (SEC) which perform similar functions for their respective states. At the district level there exists a District Disaster Management Authority (DDMA).

The provision concerning the National Disaster Response Force (NDRF) deserves a special mention as it has played a pertinent role in providing an efficacious response mechanism in time of an occurrence or in the peril of an incoming disaster.

From the above deliberation, it can be inferred that most of the member states in SAARC have a three-tier model to address the issue of disaster management in the region. The legal provisions of the countries focus upon the coordination and cooperation at federal, state and local level for disaster management. Few of the member states have an institutional framework in place for allocating funds to the organisations working towards the relief operations during a disaster across the political boundaries. Further, the divergent legal framework of the SAARC countries pertaining disaster risk reduction and management lacks provisions for promoting regional cooperation, assistance and cooperative self-reliance, amongst the countries in the region. Since there lies no obligation on the member states to provide assistance during disasters and also in the absence of an overarching regulatory authority entrusted with the jurisdiction over all the member states, the relief work which requires immediate response so as to minimise the impacts of the disaster is adversely affected.

Diverse Regional Models for Disaster Risk Reduction: Mapping a Comparison

Disaster and its impact on developing and developed countries in the back drop of economy loss are incomparable. Limited resources and infrastructure, poverty, corruption and high on population are some of the factors due to which the developing countries are not so well equipped to deal with even minor disasters, leave alone the major ones. On the contrary, the demographic position of South Asian countries makes this region prone to frequent disasters. In past few decades, this region has come to standstill on account of severe natural disasters be it the earthquakes, tsunami, floods, etc. resulting in massive loss of life and damage to property. But what made it worse was disconnected response of the member states of SAARC towards these disasters for minimising the loss or even sense of loss. This disconnect in the region resulted into the situation *firstly,* where either the donor nation of the region assisting the disaster hit nation was bearing the cost or, *secondly,* where the dependence for the financial assistance was wholly on international organisations, national governments and altruistic foundations. Since the frequency of disasters was not only on increase in the SAARC region but was the global phenomenon, the assistance on the international level was inadequate which further added on to the misery in the region.

Since the regional cooperation is a well acknowledged, accepted and implemented model for disaster risk reduction and management, it becomes a prerequisite to compare framework of some important existing regional models in the backdrop of developing an understanding of the provisions outlined therein, for the promotion of cooperation, assistance and self-reliance amongst the member states of the respective region.

African Union (AU)

In order to achieve a better life for the people and to advance coordination and cooperation amongst African states the "Organisation of African Unity" was founded which was later on replaced by the "African Union" having a goal to develop unity and solidarity between the member states and institutionalise the political and socio-economic integration of Africa.

Socio-economic growth of Africa has been debilitated by cataclysmic natural disasters. As a response, to tackle the drought and famine conditions the organisation of African Unity postulated the "Africa's Priority Programme for Economic

Recovery" (APPER). Later on the OAU in association with the United Nations established the "Special Emergency Assistance Fund for Drought and Famine in Africa" (SEAF).

SEAF was entrusted with a task to provide emergency assistance to African Countries in the form of grants and loans. Voluntary contributions are made by member states, organisations and associations, and compulsory contribution to this fund is made by the AU. As the fund is depleting, the maximum grant that can be provided through this fund is reduced from US$ 500,000 to US$ 200,000.

Realising that it was essential for governments to allocate dedicated public funding to enhance the disaster risk reduction at the national level, in the year 2004, "The Africa Regional Strategy for Disaster Risk Reduction" was formulated. This led to the establishment of crucial "Programme of Action for the Implementation of the Africa Regional Strategy for Disaster Risk Reduction" (2006–2015). This emphasised on integrating disaster risk reduction with the emergency response management. Allocation of appropriate funds for emergency response, preparedness and recovery activities were proposed.

"Joint Africa-Arab Fund for Disaster Response" was heralded as an additional financial instrument for addressing natural disasters in the region. Contribution to this fund comes from a multitude of budgetary sources of the African Union, League of Arab States, African and Arab Countries, Civil Society, private sector, international partners and regional and international organisations.

These important developments led the ministers of the African Union to ascertain the feasibility of creating a pan-African disaster risk pool to ameliorate the financing capacity for disaster risk reduction. Thus, the "African Risk Capacity Agency" (ARC) was established. This nodal agency acts as an advanced sovereign risk management instrument that allows African Union Nations to pool resources to provide effective emergency finance in case of a natural disaster thereby making the region self-reliant. The ARC not only provides for a regionally managed funding for emergencies but also facilitates contingency planning of the member nations.

Creation of different funding mechanisms and establishment of specialised insurance risk pool by the ARC have helped AU in taking rapid strides in the field of disaster risk reduction and management. But there is still a lack of financial resources within the AU which limits the implementation of the various strategies adopted. This is seconded by the reality that most of the above-mentioned instruments are formed with the aid of other international entities, for example. The ARC was established by the AU, yet it is supported by the World Food Programme in association with the monetary assistance from the UK, the Rockefeller Foundation and international fund for agricultural development.

Organisation of American States (OAS)

In the year 1948, the Organisation of American States (OAS) was set up to build up cooperation and resolve obscurities amongst the member states. The OAS office for sustainable development and environment (OSDE) through its "Department of Risk Management and Adaptation to Climate Change" was entrusted with the task of disaster risk reduction in this region. Further, the member nations pushed the member states to form the "Inter-American convention" in to develop a better responsive and assistance mechanism towards natural disasters, but, unfortunately, it was a failed attempt. Nonetheless in the year 1995, the "Inter-American Emergency Aid Fund" (FONDEM) was institutionalised for providing socio-technical and economic assistance to any of the member state witnessing an emergent situation due to a natural disaster. This fund was solely based on voluntary and altruistic contributions, and the decisions of the permanent council to allocate the fund were executed by the secretary general.

Further, in the year 2001, a central forum namely, the "Inter-American Committee for Natural Disaster Reduction" (IACNDR) was founded at the OAS, to deal with issues pertaining to natural disasters. The IACNDR is duty bound to implement the Inter-American Strategic Plan for policy on vulnerability reduction, risk management and disaster response (IASP), by seeking funding from the member states as well as international organisations.

In the past years the funding instrument FONDEM has been utilised on various occasions providing assistance to different American countries. In addition to this the "Pan-American Development Fund" (PDF) has also been successful in garnering financial resources through different public and private organisations. It is critical to point out that IACNDR has established effective connections with allied organisations like the "White helmets initiatives", "the Caribbean Disaster Emergency Response Agency" (CDERA) and the "Coordination Centre for the Prevention of Natural Disasters in Central America"(CEPREDENAC) and has been successful in generating considerable funds for disaster risk reduction and management.

European Union (EU)

Regional cooperation was the prime objective with which EU was founded. Further, to promote social and economic development in the region, "Community Action Program" (1997) was initiated which in the year 2001 resulted in the formation of "Community Mechanism to Facilitate Reinforced Cooperation in Civil Protection, Assistance and Interventions". The chief aim of this was to promote cooperation amongst member states during emergencies of trans-boundary nature. Later on this paved way for the establishment of the "European Union Solidarity Fund" (EUSF) in the year 2002, which was chiefly focused for providing not only financial assistance during emergent situations but also to support the efforts of the affected state by contributing a share in their public expenditure for the same.

The prerequisite for the utilisation of the EUSF was an occurrence of "a major natural disaster", which implied by virtue of its definition as a "direct damage" estimated of, over 3 billion Euros at 2011 prices or, more than 0.6% of the Country's Gross National Income.

Understanding the importance of coordination and cooperation in the event of disasters, the community mechanism for civil protection assistance interventions was replaced by the "Community Civil Protection Mechanism". Establishment of this new framework provided a legal sanctity to the previous mechanisms by undertaking the setting up of bodies such as the "Monitoring and Information Centre" (MIC) and "Common Emergency Communication and Information System" (CECIS). The much-needed finance was ensured through the "Civil Protection Financial Instrument" (CPFI), an ancillary of the EU Solidarity Fund. In case of emergencies caused by natural hazards the financial help through this instrument is provided in the form of "grants" or "public procurement contracts".

The capacity of the EU to successfully respond to disasters is ably supported by the "Treaty of Lisbon", by means of incorporating a solidarity clause which guarantees that the Union and its members play a pro-active role when a member state is affected by a natural or a human-induced disaster. In 2013, a significantly integrated and cost-effective approach to EU disaster management was developed. The MIC was renamed as the "Emergency Response Coordination Centre". It is the heart of EU civil protection mechanism and coordinates the delivery of assistance to disaster-stricken countries. ERCC's work in the event of forest fires in Sweden and forest fires in Greece in July 2018 is noteworthy. A budget of around 368.4 million has been allocated for the implementation of the Union Mechanism for the year 2014–20 periods.

The EU with an annual budget of around US$ 1,35 billion assigned for humanitarian operations makes it one of the world's most significant humanitarian aid donors, and because of this the region is applauded for having one of the most evolved disaster management systems with adequate financial resource. EERC is credited to be an effective regional regulatory framework for the EU, because it has at its disposal not just the monetary funds but also a pool of other resources and technology to provide an efficient assistance to the beneficiary nations thereby making the region resilient to disasters.

Association of South East Asian Nations (ASEAN)

The Asia Pacific region faces the greatest risk of disasters of any region worldwide as the data demonstrates that more than 50% of the global disasters mortalities between the years 2004 and 2014 occurred in this region. This region is comprised of some nations which face high level of risk but are less well prepared and others which are less prone to hazards but collectively the region is one of the most disaster prone region of the world. ASEAN has been active in utilising the provisions of International Law to attempt to cooperate in disaster risk reduction and response. Three pillars or "Communities" form the strong backbone of ASEAN, namely political and security, economic and socio-cultural. Disasters fall within the socio-cultural community where "[building] disaster resilient nations and safer communities" is an enlisted objective under social welfare and protection. ASEAN as the regional block has a long history of focused approach towards promoting regional cooperation for the mitigation of the impacts of disasters in the region. ASEAN, in the year 2003 (exactly one year before the tsunami disaster), established the ASEAN Committee for Disaster Management (ACDM), which devised a framework called the Comprehensive Agreement on Disaster Management and Emergency Response (AADMER) and was adopted by ASEAN in Vientiane, Laos, in July 2005 and ratified by all member states. It is a regional treaty that has been identified as the best practice amongst all existing across the globe. What adds to its effectiveness are the facts that, *firstly,* it is a legally binding disaster regulatory framework for the region; and, *secondly,* this document is an ASEAN counterpart which not only incorporates the aims and objectives of Hugo Framework for Action (HFA) but also aims for an effective implementation of these disaster risk reduction strategies in the region for betterment of the same.

Further, in order to facilitate the coordinated response towards disaster and to promote intensified regional framework dealing with disaster risk reduction and management the "ASEAN Centre for Humanitarian Assistance on disaster management" (AHA) was established. In addition to this AADMER provided for the deliberate financial donations by not only member states but also international organisations, regional financial institutions and international donor community, to a fund, namely, "ASEAN Disaster Management Relief Fund" (ADMER Fund). This fund acted as a constructive resource for the functioning finances of the AHA Centre. To add to the credit of member nations, regular contributions by them to this fund acted as an emergency fund and played a key role in disaster relief operations and ancillary actions in consonance with AADMER.

ASEAN's work on disasters is carried out at various levels including ministerial, official and expert/technical levels. In December 2015, the ASEAN became the leading regional block to put forth Vision 2025 on disaster management. This was adopted by ASEAN disaster ministers and Conference of the Parties to AADMER. This vision document placed ASEAN as a pioneer in transforming disaster management landscape in the South-East Asian region and beyond. Though there are limited funds available with the ADMER Fund to fully support and implement all the programmes of ADMER and majorly is dependent on external funding, still this region is amongst very few regional blocks which has inculcated standard operating procedure thereby making provisions for standby arrangements and coordinated operations for minimising the impacts of disaster in the region. AADMER's worldwide acceptance as a leading light for furthering regional cooperation as effective tool in disaster risk reduction and management is evident by the fact of it being the basis for the SAARC in reaching the Agreement on Rapid Response to Natural Disasters (ARRND).

Synergy in SAARC for Disaster Risk Reduction: Issues and Challenges

South Asian countries as a whole are prone towards similar kinds of disasters and also face more or less identical impediments in the implementation of policies and programmes pertaining to disaster risk reduction and management. SAARC as a regional block in its charter of establishment spells out a range of objectives amongst which promotion of regional cooperation for the purpose of economic, social and cultural development is primary and vital. Time and again, the member states have insisted on developing a robust disaster risk reduction and management framework for the region based on the mutual trust and assistance. SAARC Agreement on Rapid Response for Natural Disasters (AARND) is one such framework which was the initiative of SAARC Disaster Management Committee providing a mechanism for prevention, preparation, mitigation and rehabilitation in the region.

As far as South Asia is concerned the aforesaid objectives were aimed to be fulfilled by the establishment of SAARC Disaster Management Centre (SDMC) which is supposed to be centre of excellence in the region responsible for research, information exchange and capacity building in the member states by providing policy advice and requisite services, so as to advance effective disaster risk reduction and management in the region. Further, apart from SDMC, what is required in this region due to increasing rate of natural disasters is the joint immediate response of the member nations followed by an adequate and flexible monetary support from them, for which standby arrangements are to be made for ensuring relief in disaster affected areas.

But the reality is far different and distant from this ideal situation of maintaining a separate fund for disaster management at the regional level, as very few member states have provisions regarding such standby arrangements for relief operations during disasters. It would be pertinent here to state the fact that SAARC region consists of developing countries which do not even year mark separate funds at the national level for immediate response for disasters within the country, and in such a scenario, it is beyond expectation to rely on such countries, for financial assistance in course of a disaster emergency beyond their borders.

Further, though a range of institutions are founded by the SAARC countries for promoting regional cooperation based on regional integration resulting into regional development as a whole, but these institutions are not empowered enough enabling them to be pro-active in the region. For instance, SAARC Disaster Management Centres (SDMC) are capacity building institutions which are reposed with the responsibilities of preparing a road map regarding various aspects of disaster risk reduction and management and sharing good practices of disaster management thereby enhancing the preparedness of the member nations of the region to meet the concerned future challenges. But ironically, these centres have failed to set up a dedicated operation wing or a SAARC Disaster Rapid Action Force (SDMRAF). Further, due to weak internal mechanisms these centres failed miserably during the times of need, may it be Nepal Earthquake, Bangladesh Cyclone or Uttarakhand Flash floods, and individual efforts were made by the SAARC countries for assistance in search, rescue and rehabilitation operations. Therefore, the challenge that lies before these countries is to convert policies, conventions and institutions, promoting regional cooperation in the region, from a mere document to performance. Moreover, in an era where regional cooperation for successful implementation of disaster risk reduction policies and management is an accepted norm in all international frameworks proposed, the investigation of various existing worldwide regional models and their comparison to SAARC reveals the truth that in the SAARC region, regional cooperation amongst the member states, regarding disaster management is a distant reality. This is ironical and also depressive to see that a region that shares such a long and common cultural, historical and geographical lineage is not able to arrive at a common minimum understanding on an issue as cardinal as disaster risk reduction. Though, interestingly, the countries of this region are successfully negotiating and cooperating individually at international forums involved in formulation of various policies regarding disaster risk reduction and thereby helping in making the countries worldwide disaster resilient, but concerns and necessities of SAARC as a region are minimally aired and addressed. It is incredibly vital for this region to transform its approach by conceding the intra-regional differences existing amongst the member states. The major factor which holds back the cooperation in this region for mutual assistance during the times of disaster is the continued political differences amongst the SAARC countries.

In addition to this, to realise this very objective of regional integration in SAARC region two crucial members namely India and Pakistan initially have to integrate amongst each other, as till the time the political differences amongst these two countries are not reduced, the expected regional cooperation in the region for successful implementation of policies pertaining to disaster risk reduction and management is an allegory. Resolution of political tensions amongst these two countries often acts as an insurmountable bottleneck that cannot be meandered through for achieving regional integration in the region. But, taking a cue from the EU, it may be postulated that giving precedence to economic integration over geo-political nuances may pave the way forward for the realisation of a common disaster risk reduction model which is beneficial for all the member states.

Resilient SAARC for Disaster Risk Reduction: Conclusion and Way Forward

"Solidarity", mutual "support" and "assistance" are the key attributes adopting which the SAARC as a region can be resilient to disasters. Though SAARC Agreement on Rapid Response to Natural Disasters is a complimentary instrument to ASEAN Agreement on Disaster Management and Emergency Response and includes same objectives including joint efforts at the regional level regarding disaster risk assessment, reduction, preparedness, response and rehabilitation, but in reality the member nations of SAARC region have not integrated these objectives for the purpose of implementation true to its letter and spirit. Unlike, "ASEAN Disaster Management Relief Fund" (ADMER Fund) and "European Union Solidarity Fund" (EUSF), SAARC region lacks economic cooperation and makes this economically fragmented region incapable not only in dealing with disasters but also in preparedness and rehabilitation. A separate allocated regional fund for SAARC region having regular and proportionate contributions from the member states will enhance the response mechanism thereby resulting into effective disaster risk reduction and management.

Further, legal, social and technological cooperation are effective tools for better preparedness and planning for the region regarding disasters. "Emergency response capacity" and "Emergency response coordination" as per EU model should be the guiding inspirations for the SAARC countries. For the purpose of making the region resilient to disasters, similar institutions should be established which coordinate sharing of human and infrastructural resources, finances, knowledge and technological advancements amongst the member states for the betterment and upliftment of the region ridden by several other serious constraints like poverty, corruption, high population and so on.

Recently, countries of South Asia, namely India and Bangladesh, presented an excellent example before the international community regarding how an effective community preparedness and usage of technology can avert and minimise life loss in case of a disaster, "Fani"—the cyclone which hit one of the poor states of India (Odisha) and Bangladesh. Though life loss was minimised but the infrastructural loss was humongous and rebuilding the lost assets would require intra-country and inter-country cooperation. Despite the fact that India is financially well placed in comparison to other SAARC countries but still this distressful time presents an opportunity to SAARC as a region, to show solidarity and mutual assistance in rehabilitation operations, and if successful in doing so, it would place this region amongst other successful regional models.

Part VII
Summary

Chapter 19
Summary and Concluding Remarks

Vishwa Raj Sharma and Chandrakanta

> National Disaster Management Authority of India and Indian National Policy on Disaster Management vision is... *"To build a safe and disaster resilient India by developing a holistic, proactive, multi-disaster oriented and technology-driven strategy through a culture of prevention, mitigation, preparedness, and response"*.

At this juncture, many parts of the country have been facing various disasters in varying degrees where some regions are heavily vulnerable to hazards. This book is an attempt to trace an in-depth and comprehensive summary of a study on **"Making India disaster resilient: challenges and future perspectives"**. The chapters are organized into various sections in the book, namely case studies related to floods, climate change, and land use, fire and smog hazards, earthquake hazards associated with multi-hazard scenarios, disaster, and gender perspectives, including the prominent last section which is further accompanied by human aspects: impact, vulnerability, and governance to disasters in India.

India is a multi-disaster-prone country where the frequency and intensity of disasters have increased over the past few decades. Due to geo-climatic variations and given socio-cultural and economic settings, India is fronting disasters, and for these reasons, it becomes rather evident as if India is the land of future disasters.

Most of the riverine low-laying plains of India appear to have problems with floods annually. The situation of floods in India is the outcome of both natural and human-induced causes. Thus, there is a prerequisite for integrated flood management and warning system.

In this context, the first point of focus on a disaster preparedness plan needs to be absorbed on local and regional area flood area mapping, management, and monitoring of riverine zone based on satellite imagery and advanced geographical information

V. R. Sharma (✉)
Department of Geography, University of Delhi, New Delhi, India
e-mail: vrsharma2002@gmail.com

Chandrakanta
Department of Geography, Shaheed Bhagat Singh College, University of Delhi, New Delhi, India
e-mail: chandrakantajnu@gmail.com

V. R. Sharma and Chandrakanta (eds.), *Making India Disaster Resilient*,
https://doi.org/10.1007/978-3-031-50113-5_19

systems (GIS). These steps can help to make early preparation and prediction about any potential flood occurrences. A second high priority is awarded to the integrated approach that needs to inculcate watershed management in the plans. In order to achieve a maximum of disaster management plans, specific address to curation on a regular basis of safety criteria for rivers, dams, and canals should be maintained. Another important implementation is the accumulation of the Sendai Framework for Disaster Risk Reduction. Central and state governments need to come up with long-term plans with a major focus on integrated disaster management.

Earthquake in India is one of the most unpredictable and destructive geo-hazards. The incident of Bhuj Earthquake 2001 (Gujarat-India) and the very recent Nepal earthquake have inflicted great loss of life and material among the affected communities. Thus, an extensive earthquake preparedness plan is obligatory. There is an urgent need to create an earthquake-related geological hazards management department locally in high and high-risk seismic zones of India. The population should be provided with incentives to build earthquake-resilient buildings. Need for local participation, training, and capacity building plans through the incorporation of Panchayati Raj Institutions in rural India and a municipal corporation in urban India.

Cyclone is also one of the major climatic-hydrological disasters of India that every year affects the coastal population. To protect the coastal population from the adversity of cyclones, there is a need to develop a comprehensive strategy for the affected communities. India has strengthened the climate preparedness and mitigation mechanism which can be witnessed from the cyclones disaster management which occurred recently in 2020 (during COVID-19 pandemic) on the eastern coast of India. The western coast is flattering vulnerable and is experiencing cyclones. Hence, it becomes crucial to focus on early warning, preparedness, and mitigation that are required more comprehensively.

Urban fire is another emerging challenge in Indian urban cities. Micro-level studies conducted in Delhi NCR have highlighted the need for retrofitting and awareness. Bio-hazards are the potential hazards that are occurring and, in the future, may cause serious challenges as cities as mostly congested and densely populated. Smog in urban and peri-urban affects critical patients, children, and especially the pregnant women. During cropping seasons especially through paddy harvesting, the problem intensifies due to stubble burning in north India.

In order to understand the disaster, it becomes crucial to understand gender. Gender and climate change are inter-linked. Women belong to the vulnerable group due to their economic, social, and cultural settings which affect their awareness level and coping capacity in a disaster situation. The other perspective of connecting gender to disasters is that though women are worst affected by disasters but at the same time, they work at capacity building during and post-disaster. Women work as the main sources of arranging necessities for the dependent families anywhere from cooking to managing numerous issues. International and regional cooperation during the advent of disasters have been highlighted in the last chapter.

The requirement to exchange knowledge and technology is necessary to generate funds for preparedness and capacity building.

Maintaining peace and justice is another significant step in this direction. An appropriate disaster-resilient model is the need of time. A paradigm shift has already taken place where the focus is more on pre-disaster rather than post-disaster management. Multi-stakeholder and inclusive/community participation are required to reduce the risk of disaster. In this regard, India's strategy to manage disasters is multi-layered where national, state, district, and local-level administration hold hands together with the participation of various relevant ministries and government departments. There is a requirement to make disaster management an integral part of the whole development process which should visualize and plan by keeping in mind the disasters and future challenges.

This book is beneficial for researchers, planners, academicians, and social scientists engaged in research related to various urban issues.

Printed by Printforce, the Netherlands